UNDERSTANDING DIABETES

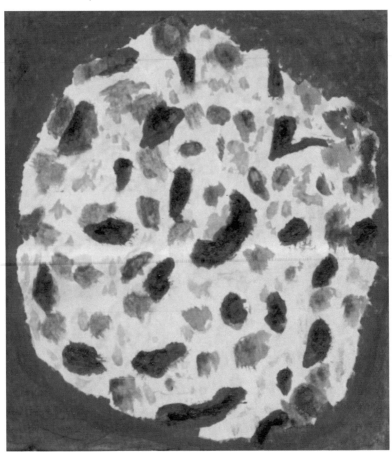

This image depicts a watercolor representation (by the author) of a single pancreatic islet of Langerhans as seen through a light microscope. The nuclei are visualized through staining as blue; the glucagon produced by the α-cells is stained as red-orange and the insulin produced by the β-cells as green.

Published by John Wiley & Sons, Inc., Hoboken, New Jersey.
Published simultaneously in Canada.

For general information on our other products and services or for technical support, please contact our Customer Care Department within the United States at (800) 762-2974, outside the United States at (317) 572-3993 or fax (317) 572-4002.

Wiley also publishes its books in a variety of electronic formats. Some content that appears in print may not be available in electronic formats. For more information about Wiley products, visit our web site at www.wiley.com.

Library of Congress Cataloging-in-Publication Data:

Dods, Richard F., 1938-
    Understanding diabetes : a biochemical perspective / by Richard F. Dods.
        p. ; cm.
    Includes bibliographical references and index.
    ISBN 978-1-118-35009-6 (cloth)
    I. Title.
    [DNLM: 1. Diabetes Mellitus. 2. Glucose–metabolism. WK 810]
    616.4'62–dc23

                                                                    2012040252

Printed in the United States of America

10 9 8 7 6 5 4 3 2

CONTENTS

7 DIAGNOSIS OF DIABETES MELLITUS   201

8 COMPLICATIONS OF DIABETES MELLITUS AND THEIR PATHOPHYSIOLOGY

249

PREFACE

In this book we commence a fascinating journey that will describe attempts to conquer a disease, really a scourge, which began as far back in history as earliest man, and continues today. Unfortunately we do not presently have a cure, but I feel we are on the brink of one.

This book is written for the individual who wants to learn about the underlying biochemistry/physiology of diabetes mellitus, its history, its detection, its complications, and its treatment.

Alan Alda in an editorial published in *Science*,[1] states "Scientists urgently need to speak with clarity to funders, policy-makers, students, the general public, and even other scientists. . . . clarity in communicating science is at the very heart of science itself". This book was written with this statement in mind.

Although not absolutely required, a basic knowledge of biochemistry/physiology will help the reader understand certain concepts presented in this book. However, in order to make technical language, objectives, and concepts more easily understood this book includes for each chapter a *Glossary*, *Summary*, and *Problems*. The *Glossary* defines key terms used in the chapter. The *Summary* highlights the essential ideas presented in each section of the chapter. *Problems* for each chapter located in Appendix B present the principal goals of the chapter in the form of questions for the reader. Technical terminology is presented in simple, easily understandable terms with the aid of *Footnotes*.

---

[1] Alan Alda, actor, writer, and founding board member of the Center for Communicating Science, State University of New York at Stony Brook from The flame challenge. Editorial. Science 2012;335:1019.

MY GOALS IN WRITING THIS BOOK

My goals in writing this book were

- To give the reader a view of diabetes mellitus, a disease that is in epidemic proportions worldwide, by identifying its biochemical basis, classification types, causes, diagnosis and monitoring, complications, and present and future treatment modalities.
- To provide students enrolled in a university-level biochemistry course the materials to understand glucose metabolism and what occurs when the metabolism goes astray.
- To provide incentive for further research on this disease by presenting what we presently know about diabetes.
- To provide a fundamental understanding of the tests used for the diagnosis and monitoring of diabetes—assays that Medical Technologists and Clinical Chemists perform every day.
- To give diabetes counselors and educators a text and reference book that they can use with confidence.
- To provide to medical students and physicians an understanding of the underlying basis for the disease that they treat.
- To provide policy-makers with an understanding and appreciation of the disease that promotes the support of public funds for the research, treatment, and eventual cure of the disease.
- For persons afflicted with diabetes an appreciation of how science is researching this disease and the many breakthroughs that have recently occurred in comprehending the causes, complications, diagnosis, and treatment of this disease.

After reading this book I hope you will agree that these goals have been achieved.

CONTENT

This book is intended to acquaint the reader with diabetes mellitus, a disease that is becoming pandemic.

In Chapter 1, "Diabetes Mellitus: A Pandemic in the Making", diabetes mellitus is introduced as a disease that is attaining epidemic proportions in the United States and across the world. Parallel to the outbreak of diabetes there is another developing pandemic: overweight and obesity. The connection between these two developing pandemics is discussed in this chapter and will be further elaborated on in Chapter 6.

In Chapter 2, "A Historical View of Diabetes Mellitus", the history of the disease is presented from the caveman to its recognition in ancient Greek medicine and the early days of the Roman Empire to Banting and Best's Nobel Prize winning discovery of insulin to Sanger's determination of the structure of insulin to Cuatrecasas's purification of the receptor site for insulin.

Chapter 3, "A Primer: Glucose Metabolism", contains the pathways for the metabolism of glucose. It includes the principal pathways by which glucose is metabolized: glycogenesis and glycogenolysis, glycolysis, the tricarboxylic acid pathway, electron transport system and oxidative phosphorylation, phosphogluconate oxidative cycle, uronic acid cycle, hexosamine biosynthesis pathway, and gluconeogenesis are described. Included in this chapter are "beautiful concepts" as seen through the elegance of many of these metabolic pathways. The notion of "beautiful pathways" is elaborated on in the Prolog to Chapter 3.

Chapter 4, "Regulation of Glucose Metabolism" relates the mechanisms that permit glucose to enter the cell from the blood. Included in this chapter are descriptions of insulin action, its manufacture in β-cells, the insulin signaling pathway, incretins, and other hormones that regulate insulin production, and the actions of AMP-activated protein kinase.

In Chapter 5, "Glucose Metabolism Gone Wrong", the altered metabolism of glucose in diabetics is presented.

Chapter 6, "Classification System for Diabetes Mellitus", deals with the classification scheme for diabetes that has been developed over the years. Described in this chapter are type 1, type 2, impaired glucose tolerance, impaired fasting glucose, gestational diabetes, statistical risk class, potential abnormality of glucose tolerance, and secondary causes of diabetes mellitus.

Chapter 7, "Diagnosis of Diabetes Mellitus", is divided into three parts—Part 1 deals with the approach to establishing the normal range; Part 2 the modern laboratory tests for glucose; and Part 3 symptoms, diagnostic tests, and criteria used to identify diabetes.

Chapter 8, "Complications of Diabetes Mellitus and Their Pathophysiology", describes the complications of diabetes retinopathy, angiopathy, nephropathy, infection, hyperlipidemia, atherosclerosis, ketoacidosis, lactic acidosis, hyperglycemic hyperosmolar nonketotic coma, and hypoglycemia. Their pathophysiology and prevalence will also be discussed.

In Chapter 9, "Hereditary Transmission of Diabetes Mellitus", the hereditary factors that are involved in the susceptibility and resistance to diabetes are discussed. The histocompatibility antigens (HLA) and their association with diabetes are described.

Chapter 10, "Treatment", goes into a discussion of advancements in the treatment of the disease. Some of what is discussed in this chapter represents ongoing research into the disease. Also treated in this chapter are measures to delay and prevent the occurrence of diabetes.

Postscript "The Future".

LEARNING AIDS

- Throughout the chapters, Problems, Summary, and Key Terms are listed. These aids are to guide readers as they navigate through the chapters. They permit the reader a shortcut that may be used to scan chapters that are not totally relevant to the reader's interest.
- The goals of each chapter are contained in the Problems located in Appendix B. In addition, summaries are included throughout each chapter.
- Key terms (including medical terms) are explained at each point as they are introduced in the chapter. A glossary of key terms is also included at the end of each chapter.
- Each chapter has a preamble as to its importance in understanding diabetes mellitus. In addition each chapter has a summary.
- Each chapter has numbers within parenthesis relating to references, which are listed in the reference section at the end of the chapter. Included are the URLs for many of the citations. Also the Digital Object Identifier (DOI$^{©}$) for many articles is included. DOI has been around since 2000. DOIs identify electronic objects such as journal articles, books, and scientific data sets in a particular location on the Internet. The system is managed by the International DOI Foundation (IDF), a consortium of commercial and non-commercial partners. A DOI name consists of a prefix and a suffix, for example, 10.1089/jwh.2010.2029; the prefix is 10.1089 and the suffix is jwh.2010.2029. One way to use this system is to go to the URL of the IDF, which is http://www.doi.org/ and insert the DOI you are in search of in the place provided and violá the document pops up. The other approach is to use the URL, http://dx.doi.org/ followed by the DOI name; for example: http://dx.doi.org/10.1089/jwh.2010.2029.

TECHNICAL TERMS

Technical terms are translated into simple language in this book. When I read an article or book, I find myself spending a considerable amount of time trying to learn the meaning of technical terms with which I am not familiar. I often turn to reference books and textbooks to learn the meaning of the term. In this book I think I have remedied this by having footnotes and a glossary defining any technical terms that you may come across in the text.

MY BACKGROUND AND INTEREST IN DIABETES MELLITUS

As you can see from the image shown below I had an interest in diabetes mellitus early during my education. The image is from a science notebook while I was in high school, Lafayette High School in Brooklyn, New York, to be specific.

Although Lafayette no longer exists it still remains alive in my heart as to where I started my career in science.

The pancreas is both a duct and ductless gland. The duct gland gives off pancreatic juice to the small intestine and the ductless gland gives off insulin which controls the amount of sugar in the blood stream. Shortage of insulin causes diabetes.

Pancreas

← duct   ductless produce insulin (Islands of Langerhans)

The hormone insulin was discovered by Banting in 1922. Earlier the Langerhans discovered the Islands of Langerhans. Banting discovered insulin when he saw flies gathering near the urine of dogs who had their pancreas removed. Banting remembered that sugar causes flies to gather and that sugar in urine indicates diabetes. Thus, he pinpointed the sickness to the pancreas. Diabetes is cured by injections of insulin which must be kept up throughout the life of the victim.

I earned a B.S. in science at Brooklyn College, an M.S. in organic chemistry at New York University and a PhD in biochemistry at the University of Connecticut. I was a post-doctoral fellow in cancer research at Sloan Kettering Institute for Cancer Research before joining New York University Medical School as a research associate. It was at the medical center that I first became acquainted with diabetes mellitus and published papers on the biochemistry of the beta cell. My paper[2] was one of the early publications characterizing beta cell protein kinase and protein phosphatase.

I studied clinical biochemistry as an NIH fellow under the esteemed Dr Samuel Natelson at Michael Reese Medical Center in Chicago. As Director of Clinical Chemistry at Louis A. Weiss Memorial Hospital in Chicago (a position which included an Adjunct Assistant Professorship with the University of Illinois Medical School), I published papers on the use of HbAc$_1$ as a test for monitoring diabetes mellitus. This article[3] was one of the earliest suggesting HbAc$_1$ as a tool for the diagnosis of diabetes. While at Weiss, I earned a Diplomate in Clinical Biochemistry from the American Board of Clinical Chemistry.

I wrote the chapter on Diabetes Mellitus for four of the five editions (the exception being the first edition) of *Clinical Chemistry: theory, analysis, and correlation*, edited by Lawrence Kaplan and Amadeo J. Pesce. I have also authored two audiocassette courses for the American Chemical Society entitled *"Clinical Chemistry"* and *"Pathophysiology for Chemistry"*. I established a company, Clinical Laboratory Consultants, which advised hospital and commercial laboratories in the implementation and interpretation of assays and the use of instruments for the diagnosis and monitoring of disease. Lastly I taught organic chemistry and biochemistry for 17 years at the Illinois Mathematics and Science Academy (IMSA), a world renowned secondary school funded by the Board of Higher Education of the State of Illinois. While at IMSA I published several papers on problem-based learning and its utilization in content-rich courses.[4,5]

DEDICATIONS

This book is dedicated to

My wife, Linda who supported and encouraged me throughout the writing of this book and helped me when I grappled with sentences that were so convoluted that they made little or no sense.

My cousin, Stanley Menson, who helped initiate my interest in science with his turtle tank. He was a biology teacher for the deaf and succumbed from the complications of type 2 diabetes too soon.

My grandchildren, Rachel and Shannon, who I hope will follow in my footsteps into the wonderful world of science.

My son, Steven, who has already followed me into science as an electrical engineer.

[2]Dods RF, Burdowski A. Adenosine 3′5′-cyclic monophosphate dependent protein kinase and phosphoprotein phosphatase activities in rat islets of Langerhans. Biochem Biophys Res Commun 1973;51:421.
[3]Dods RF, Bolmey C. Glycosylated hemoglobin assay and oral glucose tolerance test compared for detection of diabetes mellitus. Clin Chem 1979;25:764.
[4]Dods RF. A problem-based learning design for teaching biochemistry. J Chem Educ 1996;73:225.
[5]Dods RF. An action research study of the effectiveness of problem-based learning in promoting the acquisition and retention of knowledge. J Educ Gifted 1997;20:423.

ACKNOWLEDGMENTS

I sincerely appreciate the contributions of those who initially reviewed portions of the text and found them worthy enough of being incorporated into a book. They were Edward Hobart, M.D., Lawrence Kaplan, Ph.D., and Amadeo Pesce, Ph.D. I thank Professor Anne Cooke who contributed the remarkable cover micrograph of T-cells attacking beta islets. I greatly appreciate the information supplied to me by Thomas F. Mich, Ph.D, retired Vice-President of Chemical Development World-Wide, Warner Lambert Pharmaceutical Company regarding medicinals prescribed for diabetics. He passed away on October 22, 2012 due to a complication of diabetes mellitus, type 2. He will be missed.

No acknowledgement is complete without mentioning the team of experts who turned the text, figures, and other supplementary materials into a book. Anita Lekhwani, Senior Acquisitions Editor, who oversaw the entire operation and provided me with very wise suggestions, her assistant Cecilia Tsai, Editorial Assistant, Kellsee Chu, Senior Production Editor who coordinated the production phase, Haseen Khan, Project Manager of Laserwords who did the editing, and Dean Gonzalez, Illustration Manager who took my crude line drawings and put them into publishable form. To all of the above my genuine thanks for without you there would be no book.

CHAPTER 1

DIABETES MELLITUS: A PANDEMIC IN THE MAKING

It is a capital mistake to theorize before one has data. Insensibly one begins to twist facts to suit theories, instead of theories to suit facts.

Sir Arthur Conan Doyle, British mystery author & physician (1859–1930)

On December 20, 2006, the General Assembly of the United Nations passed resolution 61/225, the United Nations World Diabetes Day Resolution, designating November 14 as World Diabetes Day. On October 29, 2010, the President of the United States, Barack Obama, declared November 2010 as National Diabetes Month in the United States.[1]

Diabetes mellitus[2] is an array of diseases that have a common symptom—abnormally high blood glucose levels. Diabetes mellitus is a noncommunicable disease. It is not transmitted from person to person by viruses or bacteria as is HIV or cholera. Diabetes mellitus is a chronic, costly, and often debilitating disease. This will be our working definition of diabetes mellitus until later in the book where we shall learn more specifics about the disease. The President, in his Proclamation, uses the terms type 1 and type 2 diabetes. By the end of Chapter 6,

---

[1] See Appendix A for texts of the UN Resolution and President Obama's Proclamation, respectively.
[2] Diabetes mellitus should not be confused with diabetes insipidus, which is caused by vasopressin deficiency. When the term *diabetes* alone is used in this book it will always refer to diabetes mellitus.

Understanding Diabetes: A Biochemical Perspective, First Edition. Richard F. Dods.
© 2013 John Wiley & Sons, Inc. Published 2013 by John Wiley & Sons, Inc.

you will fully understand both these terms.[3] In this chapter we will learn about the extent of the diabetes problem both in the United States and globally. Later in this chapter we will learn of a related pandemic in the making—obesity and overweight. Finally, we will describe the connections between the diabetes and obesity/overweight pandemics.

DIABETES PREVALENCE[4] AND COST IN THE UNITED STATES

A Dire Prediction Based on Alarming Data

The Centers for Disease Control and Prevention (CDC) estimated that as of 2008 there were as many as 18.1 million Americans who had been diagnosed with diabetes (8 in 100 Americans). This number is presumed low because it is estimated that 6.0 million cases remain undetected. Thus, 10 in 100 adult Americans actually had the disease in 2008 (1). Between 1980 and 2008, the number of diagnosed diabetic Americans has nearly tripled.

An analysis of this data according to age, gender, and race is revealing. All of the following data derived from the CDC is for civilian, noninstitutionalized individuals with diagnosed diabetes.

Figure 1.1 shows the age-adjusted percentage of diagnosed cases of diabetes by sex. You may notice that percentages were similar for males and females until 1999, at which time the percentage for males with diabetes began to increase at a greater rate than for females.

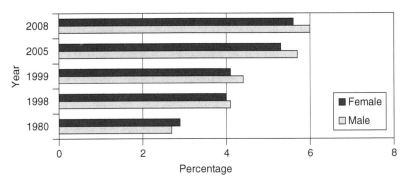

Figure 1.1   Age-adjusted percentage of civilian, noninstitutionalized persons with diagnosed diabetes by sex for selected years. (*See insert for color representation of the figure.*)

---

[3]Type 1 diabetes refers to hyperglycemia (high blood glucose levels) due to insufficient insulin secretion caused by destruction of pancreatic β-islet cells by an autoimmune response and type 2 refers to hyperglycemia due to insulin resistance (relative insulin deficiency). Type 1 diabetics exhibit blood insulin levels that are low or nonexistent and type 2 diabetics have levels that are high.

[4]Prevalence relates to the number of individuals who have diabetes at the time of the study. Incidence refers to the frequency of occurrence and is associated with a defined period of time.

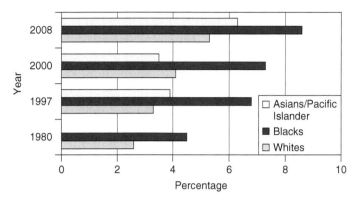

Figure 1.2   Age-adjusted percentage of civilian, noninstitutionalized persons with diagnosed diabetes by race: whites, blacks, and Asians/Pacific Islanders for selected years. (*See insert for color representation of the figure.*)

As you can see from Figure 1.2, the number of white diabetics increased 104% in the period 1980–2008; that of blacks increased 91% and of Asians/Pacific Islanders 62%. Blacks were diagnosed with diabetes at consistently higher percentages than whites and Asians. All races increased in percentage from 1980 to 2008. For Hispanics (Fig. 1.3), the largest increase in percentage was for Mexican/Mexican-Americans, 42.2%. All Hispanic groups, Puerto Ricans, Mexican/Mexican-Americans, and Cubans had percentages that significantly increased from 1997 to 2008.

Most alarming are the statistics presented in Figure 1.4. The CDC estimates that by the middle of this century the total number of diagnosed cases of diabetes will increase to between 1 in 3 and 1 in 5 Americans. These ratios correspond to 61–102 million Americans in 2050 assuming a total US population of 306.3 million adult persons (2). This estimate is based on an aging population—with increased age, there is a greater likelihood of developing diabetes; increases in

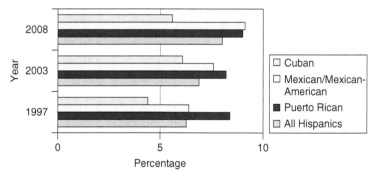

Figure 1.3   Age-adjusted percentage of civilian, noninstitutionalized persons with diagnosed diabetes among Hispanics: Puerto Ricans, Mexicans/Mexican-Americans, and Cubans for selected years. (*See insert for color representation of the figure.*)

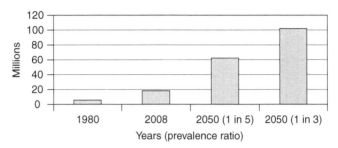

Figure 1.4 Number (in millions) of projected cases of diabetes for 2050 assuming prevalence of one in five and one in three compared to those reported in 1980 and 2008. (*See insert for color representation of the figure.*)

minority groups—minority groups have a higher prevalence of diabetes; longer life spans of people diagnosed with diabetes mellitus; and the exclusion in most studies of people younger than 18 years—an age bracket in which there have been significant increases in diabetes cases. This study also assumes 4.5–5.2% of the total population of Americans as having undiagnosed diabetes, which itself maybe an underestimated statistic. A poorer diet, overeating, and a sedentary lifestyle add credence to the prediction that by 2050 the number of cases of diabetes will at least triple.

Additional support for this prediction derives from the estimate (3) that in 2010 there were 67 million Americans (90% undiagnosed) who had prediabetes. Prediabetes (defined in Chapter 6) is the precursor to full blown diabetes.

Summary Box 1.1

- Diabetes mellitus is a noncommunicable disease that causes abnormally high blood glucose levels.
- Diabetes has a high prevalence in the United States across all racial groups.
- Projections of the increase in diabetes by 2050 are alarming.

The Increase of Diabetes in Youths

The statistics shown in Figure 1.5 with respect to children and adolescents are quite scary. The increase in diabetes cases is occurring in greater prevalence in younger persons. In 1980, the percentage of diagnosed diabetics under the age of 45 years was 0.6%. The increase began in 1986, and gradually has increased since 1986 to 1.4% in 2008. Also shown in this figure is that the greatest increases over time have occurred in the 65–74 age bracket.

Data shows (4, 5) that hospitalizations for diabetes increased 102% for young adults, 30–39 years between the 14-year period, 1993–2006.

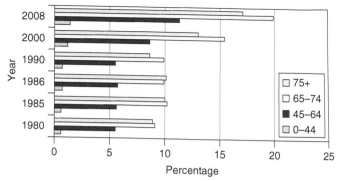

Figure 1.5 Percentage of civilian, noninstitutionalized persons with diagnosed diabetes by age (0–44, 45–64, 65–74, 75+) for selected years. (*See insert for color representation of the figure.*)

This alarming data suggests that diabetes is occurring at a younger age. Over the same period, charges for hospitalizations for diabetes increased 220%.

Additional data from Search for Diabetes in Youth (6) shows the same trends among American youth. The study, the first extensive one focused on diabetes, is specifically aimed at persons younger than 20 years. Funded by the CDC and the National Institute of Diabetes and Digestive and Kidney Diseases (NIDDK), it is located in six centers—Kaiser Permanente, Southern California; the University of Colorado Health Sciences Center, Denver, Colorado; the Pacific Health Research Institute, Honolulu, Hawaii; Children's Hospital Medical Center, Cincinnati, Ohio; University of South Carolina School of Public Health, Columbia, South Carolina; and Children's Hospital and Regional Medical Center, Seattle, Washington. The study's goals (7) are to determine diabetes prevalence in youths under 20 years, to classify the types of diabetes and their individual prevalence, to identify the types of complications, to determine the current care and treatment given to children and adolescents with diabetes, and to determine the quality of life for youths diagnosed with diabetes.

Figure 1.6 shows the prevalence of diabetes in non-Hispanic whites, African Americans, Hispanics, and Asian/Pacific Islanders younger than 20 years (8–11). This data combines males and females, age, and types of diabetes. In Chapter 6, we analyze this data according to gender, age, and type. As may be seen from the figure, non-Hispanic white youth had the highest prevalence, second were Hispanics, third African Americans, and fourth were Asian and Pacific Islanders. On the basis of this data compiled in 2002–2003, a total of 18,700 children and adolescents were diagnosed with diabetes. In total, approximately 150,000 children and adolescents have diabetes—1 in 400–500 American youths (12, 13). Combined with the data for individuals with diabetes who are older than 20 years, these statistics are ominous.

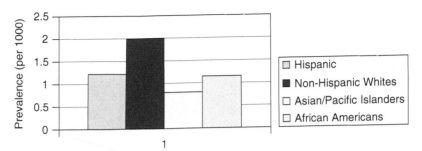

Figure 1.6 Diabetes prevalence in African Americans, Asian and Pacific Islanders, non-Hispanic whites, and Hispanics under the age of 20 years. The data for gender, age intervals, and type of diabetes were combined. This data will be expanded upon in Chapter 5. (*See insert for color representation of the figure.*)

The Cost

Using 2007 figures, diabetes was the seventh leading cause of death among Americans[5] and the total medical costs were estimated at $174 billion annually. Direct medical costs were estimated at $116 billion and indirect costs, which include disability compensation, work loss, and premature mortality, were estimated at $58 billion. By 2020, the cost is estimated (3) to grow to $500 billion. The disease is quite costly. A Consumer Reports Health survey (14) reported in 2009 that the cost of routine care for a diabetic (pharmaceuticals, testing, supplies, and doctor's visits) is in the vicinity of $6000 per year. The cost escalates for those who have any of the serious complications (see Chapter 8) associated with diabetes. This is part of the reason why only one-quarter of diabetics in the United States are obtaining optimal care.

Summary Box 1.2

- Since 1986 there has been a steady increase in diabetes in American children and adolescents.
- For young adults the number of hospitalizations for diabetes increased significantly between 1993 and 2006.
- A study on American youth called SEARCH demonstrates that diabetes is becoming more prevalent in youths under 20 years irrespective of race.
- The cost of diabetes care currently estimated at $174 billion annually is expected to increase to $500 billion annually by 2020.

---

[5]Heart disease, cancer, stroke, chronic lower respiratory disease, accidents, and Alzheimer's disease have greater numbers as the cause of death in the United States, according to the CDC.

DIABETES PREVALENCE AND COST WORLDWIDE

A Worldwide Epidemic

These increases in incidence, mortality, and cost are not only forecast for the United States but also are predicted worldwide. The International Diabetes Federation (IDF)[6] estimated that 285 million people worldwide were afflicted by diabetes in 2010 and by 2030 this figure will have increased to 438 million (15). Seventy percent of individuals diagnosed with diabetes live in low income to middle income areas of the world. These are alarming numbers, and if you add to this the fact that half of these go undiagnosed until complications have developed, you further understand why diabetes is considered a very serious public health problem worldwide.

Numbers of Cases of Diabetes

Until 2010, India was considered to have the largest diabetic population (50.8 million) with China second (43.2 million). However, a new study in China (16), which uses better methods to detect diabetes, indicates that there are 92.4 million Chinese adults with the disease. Those younger than 20 years of age were not included in this study.

Figure 1.7 shows the worldwide distribution of the number of cases of diabetes estimated in 2009 contrasted with the predicted number in 2030 (17). The number of diabetes cases is 1.54-fold greater in 2030 than in 2009. For 2009, the number of cases of males and females was approximately equal. The greatest number of cases was in the 40- to 59-year-old bracket. For the 2030 estimate, there are slightly more females than males (1.4% difference) and the age bracket for the number of cases of diabetes has advanced to 60–79 years. Looking purely at the number of cases of diabetes since 2009, China had the most in 2010, followed by India, the United States, the Russian Federation, and Brazil. In 2030, the alignment is expected to be China first, followed by India, the United States, Pakistan, and Brazil.

The IDF reported that the highest diabetes prevalence in adults in 2010 was in Saudi Arabia, Bahrain, and the United Arab Emirates in the Middle East; the North African region; Mauritius in the South-East Asian region; and Nauru in the Western Pacific region. Some of these figures may be underestimated in countries where healthcare is more limited and diagnosis more unreliable.

Cost

The costs of medical care in the poorest countries is borne almost entirely by the family, making it less likely that medical help will be sought until the condition of the family member becomes serious. Thus, in low income and middle income

---

[6]The IDF represents over 200 diabetes associations located in more than 160 countries. The IDF is associated with the World Health Organization and the United Nations Department of Public Information.

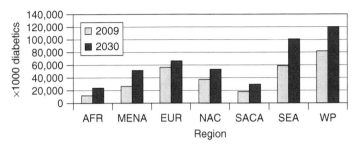

Figure 1.7 Estimated numbers of diabetics for 2030 contrasted with 2009 data for seven regions of the world; AFR, African Region; MENA, Middle East and North African Region; EUR, European Region; NAC, North America and Caribbean Region; SACA, South and Central American Region; SEA, South-East Asian Region; WP, Western Pacific Region. Data is compiled from the "IDF Diabetes Atlas," 4th ed., November 2009. The WP Region was recalculated to represent the data from the more recent estimates from Reference 14. (*See insert for color representation of the figure.*)

countries, the complications of diabetes result in a greater degree of disability and loss of life than in wealthier countries. In turn, diabetes exacts huge losses in productivity and economic growth for the countries that can least afford it. For example, in most countries of Latin America the family bears 40–60% of the cost of medical care. In fact, a recent study of seven countries—Colombia, England, the Islamic Republic of Iran, Mexico, Scotland, Thailand, and the United States—found that financial access to care was a strong predictor of diagnosis and treatment (18).

Using international dollars (ID),[7] the predicted net loss in income from diabetes and cardiovascular disease[8] during 2005–2015 is estimated for Brazil as 49.2 billion ID; China, 557.7 billion ID; the Russian Federation, 303.2 billion ID; India, 336.6 billion ID; and Tanzania, 2.5 billion ID (19).

Summary Box 1.3

- The number of diabetes cases worldwide was 285 million in 2010; it is projected to increase to 438 million by 2030.
- China has surpassed India in greatest number of diabetic cases.
- The greatest number of cases of diabetes is in China, India, the United States, Russian Federation, and Brazil.
- Diabetes is most prevalent in Saudi Arabia, Bahrain, the United Arab Emirates, Mauritius, and Nauru.
- In many nations the cost of diabetes is borne mostly by the family.

---

[7]ID is a unit of currency that has the same purchasing power as the United States dollar at a specific period in time.
[8]In these statistics from the World Diabetes Foundation (WDF), both diabetes and cardiovascular disease were combined.

OBESITY AND OVERWEIGHT; ANOTHER EPIDEMIC IN THE UNITED STATES

A Parallel Pandemic

In the previous section, we considered the global prevalence of diabetes. Next, we will investigate another seemingly unconnected pandemic, overweight and obesity. Later, we will investigate a strong connection between the diabetes pandemic and the overweight–obesity pandemic and describe the pathophysiology that links the two.

Definitions of Overweight and Obesity

Overweight and obesity are defined as more than the normal body fat accumulation (adiposity) relative to height. It is measured by the body mass index (BMI). BMI does not measure body fat accumulation directly. There is a significant difference in the correlations of BMI with adiposity between black and white persons. Nonetheless it is an easily calculated yardstick to identify overweight and obese individuals. A person's BMI is equal to w/h^2, where w equals the weight of the individual expressed in kilograms and h^2 the height squared expressed in meters squared. In the United States, where the metric system is not used except in scientific circles, the conversion from pounds (lb) to kilograms (kg) is obtained by multiplying the weight in pounds (lb) by 0.454 and the height by multiplying the height in inches (″) by 0.0254. For example, a person weighing 176 lb with a height of 68″ would have a BMI $= 79.9/1.73^2 = 79.9/2.98 = 26.8$.

The World Health Organization (WHO) (20) defines overweight as a BMI ≥ 25 and obesity ≥ 30. Although the WHO has developed BMI charts for infants and children younger than age 5, no corresponding charts have been developed for children in the 5–14 age bracket.

Overweight and Obesity among Adults in the United States

An increasing sedentary lifestyle coupled with poor eating habits and overeating has led to an increasingly overweight and obese American population. The prevalence of obesity according to ethnicity, gender, and region of the United States was surveyed by the Behavioral Risk Factor Surveillance System (BRFSS) in 2006–2008. BRFSS is a self-reported, random-dialed telephone survey of non-institutionalized Americans aged ≥ 18 years residing in the 50 states, Washington, DC, and the three territories. The CDC analyzed this data and published the results in the CDC Morbidity and Mortality Weekly Report (MMWR) (21, 22) in 2009.

As shown in Figure 1.8, there were nine states in 2009 that had prevalence of obesity $\geq 30\%$ of their adult populations. The nine states were mostly in the South and were Mississippi (34.4%), Louisiana (33.0%), Tennessee (32.3%), Kentucky

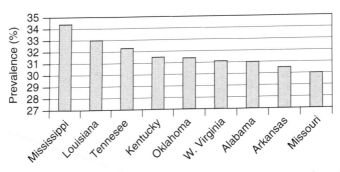

Figure 1.8   States with ≥30% obesity prevalence in 2009. (*See insert for color representation of the figure.*)

(31.5%), Oklahoma (31.4%), West Virginia (31.1%), Alabama (31.0%), Arkansas (30.5%), and Missouri (30.0%). The remainder of the states shown in Table 1.1 ranged in prevalence from 18.6% to 29.6%.

Table 1.2 shows the trends for the United States from 1994 to 2009. As you can see from this table, the number of states in the 10–14% prevalence range decreased from 33 in 1994 to nil in 2002, while those in the ≥30% range increased from 0 in 2004 to 9 in 2009. The state with the lowest prevalence was Colorado, although this figure increased significantly to 15–19% in the ensuing years. Colorado and Washington, DC had the lowest prevalence in 2009.

Figure 1.9 represents the racial/ethnic breakdown of this data. As you can see from the figure, non-Hispanic blacks had age-adjusted overall prevalence of obesity (35.7%) greater than Hispanics (28.7%) and non-Hispanic whites (23.7%). Non-Hispanic black females had a greater prevalence of obesity than their male counterparts. Hispanic females had a slightly greater percentage of obesity, while among non-Hispanic whites the opposite was true; the prevalence of obesity was slightly greater in males than in females.

For non-Hispanic blacks, the prevalence of obesity was greater in the South (36.9%) followed by the Midwest (33.1%), West (33.1%), and Northeast (31.7%). Among Hispanics, prevalence was highest in the Midwest (29.6%), South (29.2%) and West (29.0%), and lowest in the Northeast (26.6%). For non-Hispanic whites, the highest prevalence was found in the Midwest (25.4%), closely followed by the South (24.4%). Lowest prevalence was in the West (21.0%) and Northeast (22.6%).

Because of the uncertainty introduced into the survey by the manner in which the data was collected, that is, self-report by telephone, the number of persons successfully contacted, and the number of persons who gave complete interviews, the only principle conclusions from the study are the following:

TABLE 1.1   States with Prevalence of Obesity <30% in 2009

State	Prevalence, %	State	Prevalence, %
Colorado	18.6	Maine	25.8
Massachusetts	21.4	Nevada	25.8
Hawaii	22.3	Maryland	26.2
Vermont	22.8	Washington	26.4
Oregon	23	Illinois	26.5
Montana	23.2	Delaware	27
New Jersey	23.3	Georgia	27.2
Utah	23.5	Nebraska	27.2
New York	24.2	Pennsylvania	27.4
Idaho	24.5	Iowa	27.9
Minnesota	24.6	North Dakota	27.9
Rhode Island	24.6	Kansas	28.1
Wyoming	24.6	Texas	28.7
Alaska	24.8	Wisconsin	28.7
California	24.8	Ohio	28.8
Virginia	25	North Carolina	29.3
New Mexico	25.1	South Carolina	29.4
Florida	25.2	Indiana	29.5
Arizona	25.5	Michigan	29.6
New Hampshire	25.7	South Dakota	29.6

TABLE 1.2   Prevalence of Obesity (in Percentage) for the Years 1994–2009

	10–14%	15–19%	20–24%	25–29%	≥30%
1994*	33	16	0	0	0
1995	23	27	0	0	0
1996	20	30	0	0	0
1997	15	32	3	0	0
1998	10	33	7	0	0
1999	6	26	18	0	0
2000	1	27	22	0	0
2001	1	20	28	1	0
2002	0	18	29	3	0
2003	0	15	31	4	0
2004	0	7	34	9	0
2005	0	4	29	14	3
2006**	0	4	25	20	2
2007	0	1	20	27	3
2008	0	1	18	26	6
2009	0	2	16	28	9

*No data from Rhode Island.
**First year that data includes Washington, DC.
Source: Behavorial Risk Factor Surveillance System, CDC.

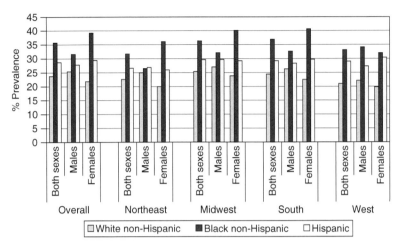

Figure 1.9 Obesity prevalence according to ethnicity, gender, and region. Data analyzed by the CDC from the Behavioral Risk Factor Surveillance System. Data collected by a random-dialed telephone survey of the US civilian noninstitutionalized ≥18 years. Surveys conducted in states, Washington, DC, and three territories. Pregnant women and those ≥500 lb or a height ≥7 ft were excluded. Surveys were conducted in 2006–2009. The data was age-adjusted to the US 2000 standard population. The prevalence relative standard error was less than 30%. (*See insert for color representation of the figure.*)

- The order of prevalence of obesity from greatest to least is non-Hispanic blacks, Hispanics, and non-Hispanic whites.
- While there are probably no differences in prevalence of obesity between genders for Hispanics and non-Hispanic whites, non-Hispanic black women have a slightly higher prevalence of obesity than non-Hispanic black men.
- Non-Hispanic blacks have the highest prevalence of obesity in the South.
- Non-Hispanic whites have the highest prevalence of obesity in the Midwest and South, with lowest prevalence in the West and Northeast.
- Hispanics had prevalence of obesity equally distributed throughout all regions.

Obesity and Overweight among Children and Adolescents in the United States

The BMI used earlier for adults is also used for children and adolescents. The definitions of overweight and obese are based on the 2000 CDC BMI age-adjusted growth charts. Children and adolescents aged 2–19 ≥ the 95th percentile for age are considered obese, and those who fall between the 85th and 95th percentiles are considered overweight (23). The CDC has published (24) a convenient BMI calculator for children and adolescents on their web site. All one has to do to calculate the BMI for age percentile is enter the birth date, date of measurement, sex, height, and weight.

The present generation of American adults who are overweight and obese will soon be joined by the next generation of Americans. Analysis of data from the National Health and Nutrition Examination Survey (NHANES), conducted by the National Center for Health Statistics (NCHS) of the CDC, estimates that in 2007–2008 about 16.9% of children and adolescents 2–19 years old were obese and 31.7% were overweight. In addition, the same report states that 9.5% of infants and toddlers are obese (25). This is an alarming statistic as obese children and adolescents are likely to remain obese into adulthood (26), thus augmenting the prevalence of obesity in adulthood.

Figure 1.10 was produced from the data presented in Reference 25. As you may see from the figure,

- Obesity is more prevalent among Mexican-Americans and Hispanics.
- Among non-Hispanic black males, obesity increases as age increases for all age brackets.
- Among non-Hispanic whites, obesity in Mexican-Americans and Hispanics is more prevalent in the 6–11 age bracket (elementary school age) than in the 2–5 age bracket (preschool age). Prevalence then decreases significantly in the 12–19 age bracket.
- For females, obesity is more prevalent as age increases from 2–5 to 6–11 and decreases in the 12- to 19-year range with one exception, non-Hispanic blacks.
- Mexican-American and Hispanic males had a greater prevalence for obesity in all age brackets than for females of the same race.

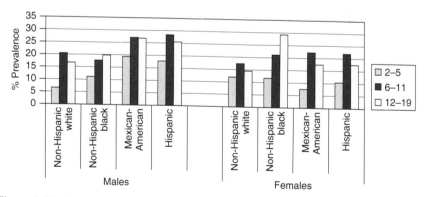

Figure 1.10 Obesity prevalence in US youths aged 2–19 by gender, age, and ethnicity in 2007–2008. Obesity defined as BMI at the 95th percentile or higher. Data are from Reference 22. (*See insert for color representation of the figure.*)

Summary Box 1.4

- An increasing sedentary life style with poor eating habits and overeating is creating a population of Americans who are overweight and obese.
- Obesity is most prevalent in the South. Southern states such as Mississippi, Louisiana, Alabama, and Arkansas have an obesity prevalence of $\geq 30\%$.
- Since 1994 obesity has been on the increase.
- During 2004–2009, BRFSS surveyed obesity in the United States, the District of Columbia, and three territories and found that among Americans ≥ 18 years non-Hispanic Blacks had the greatest prevalence of obesity followed by Hispanics and non-Hispanic Whites.

OVERWEIGHT AND OBESITY WORLDWIDE

Overweight and Obesity Globally in Adults

That obesity is not just an American problem is suggested by statistics from the WHO (27). The worldwide data on obesity in the WHO report is derived from multiple sources and for many countries the data is fraught with a great deal of uncertainty. In many instances, there is missing data. There is a great deal of variation in the manner that the data was collected, the definitions used, coverage, and statistical methods and modeling used. There is fewer data available for overweight and obese children. However, there are some valid conclusions that can be drawn from the statistics that were compiled.

The author took the liberty of including in Table 1.3 those countries that had $\geq 10\%$ prevalence of persons (male or female or both) ≥ 15 years old who were overweight in 2000–2009, and $\geq 20\%$ prevalence of persons (male or female or both) ≥ 15 years old who were obese in 2000–2009. In contrast, countries that had less than 5% prevalence were also listed in the table. With respect to obesity, the prevalence was greater in females than in males. Of the 193 countries that participated in the survey, 30 countries placed in the overweight category, 38 in the ≥ 20 obese category, and 20 in the less than 5% obese category. Those of you who know your geography may recognize from the table the lowest prevalence of obesity countries were for the most part in Asia and Africa (an average BMI of 22) and the highest were in North America, Europe, Latin America, North Africa, and the Pacific Island (BMI of 26).

In summary, obesity has tripled since 1980 in many countries throughout most regions of the world. The increase has been occurring in developing countries as well as industrialized countries. The WHO reported (28) that there were over 1 billion overweight adults (>15 years), with 300 million of them obese.

TABLE 1.3   Countries with Overweight Prevalence ≥10% and Obesity Prevalence <5% and ≥20%

Overweight (≥10%)	Obesity	
	≥20%	<5%
Albania	Australia	Burkina Faso
Algeria	Bosnia and Herzegovina	Cambodia
Armenia	Brazil	Cameroon
Azerbaijan	Canada	Chad
Belize	Chile	Democratic Republic of the Congo
Benin	Cook Islands	Eritrea
Bosnia and Herzegovina	Croatia	Ethiopia
Botswana	Czech Republic	Guinea
Bulgaria	El Salvador	India
Central African Republic	Fiji	Indonesia
Comoros	Germany	Japan
Egypt	Greece	Lao People's Democratic
Georgia	Guyana	Republic
Indonesia	Iraq	Mozambique
Iraq	Isreal	Nepal
Kazakhstan	Jordan	Niger
Kyrgyzstan	Kiribati	Rwanda
Lebanon	Kuwait	Uganda
Libyan Arab Jamahiriya	Lithuania	United Republic of Tanzania
Malawi	Malta	Viet Nam
Mongolia	Mexico	
Montenegro	Nauru	
Morocco	New Zealand	
Nigeria	Nicaragua	
Serbia	Oman	
Sierra Leone	Panama	
Swaziland	Samoa	
The former Yugoslav Republic of Macedonia	Seychelles	
	South Africa	
Ukraine	Syrian Arab Republic	
Uzbekistan	Tonga	
	Turkey	
	Tuvalu	
	United Arab Emirates	
	United Kingdom	
	United States of America	
	Uruguay	
	Vanuatu	

Overweight and Obesity in Children

The seriousness of the overweight and obesity problem worldwide is emphasized by the data compiled on children. Although published reports on childhood weight worldwide are fewer for children than adults, the most reliable published data (29) indicates that in 2010, the number of overweight and obese children was 43 million. Approximately 35 million (81%) children were from developing nations. Indeed, the prevalence of overweight and obesity in children increased from 4.2% in 1990 to 6.7% in 2010. If this trend were to continue unabated, the prevalence would be 9.1% in 2020. This would correspond to 60 million children.

The data provided in this chapter[9] supports the conclusions by epidemiologists that diabetes mellitus and obesity are pandemics in the making. Next, we will describe the connections between these two pandemics.

Summary Box 1.5

- Overweight and obesity is a worldwide problem.
- A total of 38 of 193 member countries of WHO have populations with ≥20% that are classified as obese.
- There more than 1 billion adults who are overweight and 300 million obese.
- If the trend of overweight and obesity continues there will be by 2020 60 million overweight and obese children.

THE RELATIONSHIP BETWEEN OBESITY AND DIABETES

Now we shall connect the dots. The link between obesity and diabetes was remarked upon since the late 1800s, as shown in Figure 1.11. The Books Ngram Viewer developed by Googlelabs© (30, 31) has a database of 500 billion words compiled from 5.2 million books published since 1500. By putting in the phrase "obesity and diabetes," it returns the graph shown in Figure 1.11.

An earlier observation of the link between obesity and diabetes was reported by Lyman (32), stating in a chapter of a textbook published in 1895 that "The coexistence of obesity and diabetes is a matter of frequent observation." In 1906, Ebstein (33) reported "... fact must be expressly pointed out that gout, obesity and diabetes mellitus are intimately related, which is evident from the circumstance that two of these diseases, or not infrequently all three, simultaneous occur in the same person." Horsford reported in an article (34) published in 1920 that "Obesity and diabetes, as signified by overfatness and glycosuria, are frequently associated conditions."

However, most of these early reports were anecdotal—the observations of medical practitioners. An early report that utilized measurements to substantiate

---

[9]The statistics cited in this chapter may be dated by the time this text is published. Those who want more current data may go to the URL web sites cited in this book for updates.

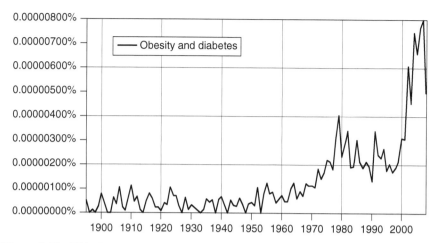

Figure 1.11 This figure shows the Googlelabs Books Ngram Viewer return for the phrase "obesity and diabetes" from the year 1895 to 2008. (*See insert for color representation of the figure.*)

conclusions measured the distribution of subcutaneous fat tissue in males and females. Using calipers to measure the degree of fatness in certain regions of the body, Vague (35) reported that a person with an abundance of upper body fat was more apt to suffer from atherosclerosis and diabetes. He stated "… it is also the usual cause of diabetes in 80 to 90 per cent of the cases."

More modern studies (since 1985) demonstrate the association of obesity and diabetes. Data collected from the NHANES indicated (36) that the relative risk of developing diabetes was 2.9 times greater for obese persons 20–75 years of age than for normal weight persons. A later publication (1996) reported (37) that the prevalence of morbidity and mortality from diabetes decreases with weight loss. In this report, it was recommended that a BMI of 20–25 should be maintained during a person's lifetime. A 12-year study (38) of 6917 men with no previous history of diabetes found an increased occurrence of diabetes in those who had a substantial weight gain as compared to those who had a stable weight. Furthermore, those who had a substantial weight loss were associated with a decreased prevalence of diabetes.

Convinced? As indicated by Figure 1.11, starting in about 1997, there has been a multitude of citations referring to the association between obesity and diabetes.

Summary Box 1.6

- The earliest reports that associated obesity with diabetes was anecdotal stories by medical practitioners.
- Individuals with excess upper body fat were found to be more likely to develop diabetes than those who had less fat.

This chapter is followed by a historical view of diabetes from caveman to modern times. In Chapter 3, we will look at the basics of normal glucose metabolism. In Chapter 5, we will look at glucose metabolism in the diabetic.

PROJECTS AND QUESTIONS

1.1 Describe the various approaches that are used for determining the prevalence of diagnosed diabetes using the original articles listed for this chapter.

1.2 Obtain data from an epidemiological diabetes study and determine if there are a significant number of persons with diabetes who are over the age of 40 years and of normal weight.

1.3 Choose a person (relative or friend) who has been diagnosed with diabetes. Learn what you can in regard to the following:

(a) When was the disease diagnosed?

(b) How was it diagnosed?

(c) What were the symptoms (if any)?

(d) The age, weight, and sex of the person.

(e) The classification of the diabetes (to be explained in later sections).

1.4 In order to have a BMI of 28, a person 75″ tall would weigh _____ kg.

1.5 Place in the Googlelabs Books Ngram Viewer the phrase "obesity causes diabetes" and analyze the returned graph. Try other words, and phrases involving overweight, obesity, and diabetes.

GLOSSARY

Adiposity   Body fat. Measured indirectly by BMI.

BMI   Body mass index. BMI equals weight in kilograms divided by height in meters squared.

BRFSS   Behavioral Risk Factor Surveillance System is a self-reported, random-dialed telephone survey of noninstitutionalized Americans aged ≥ 18 reporting the prevalence of obesity in 50 states, Washington, DC, and three territories during 2006–2008.

CDC   Centers for Disease Control and Prevention is a US federal agency under the Department of Health and Human Services directed to protect public health and safety by providing information to enhance health decisions.

Diabetes insipidus   A disease caused by vasopressin deficiency. This disease should not be confused for diabetes mellitus.

Diabetes mellitus   An array of noncommunicable diseases that have abnormally high blood glucose levels.

Epidemic   A disease or condition that spreads rapidly among many people in a community. Word used in this textbook for a disease or a condition occurring in a region or nation.

ID   International dollar equates foreign dollars to the same purchasing power as the US dollar at the same specific period of time.

IDF   International Diabetes Federation represents over 200 diabetes associations located in more than 160 countries.

Incidence   The frequency of occurrence of the disease during a defined period of time.

Morbidity   Number of cases of a specific disease per unit of time, per unit of population.

Mortality   Number of cases of death from a specific disease per unit of time, per unit of population.

MMWR   The Morbidity and Mortality Weekly Report published by the CDC analyzes and publishes the morbidity and mortality due to diseases and conditions in the United States.

NHANES   National Health and Nutrition Examination Survey conducted by the National Center for Health Statistics of the CDC. NHANES combines interviews and physical examinations designed to assess the health and nutritional status of adults and children in the United States.

Obesity   More than the normal fat accumulation relative to height according to WHO, a BMI ≥ 30.

Overweight   More than the normal fat accumulation relative to height according to WHO, a BMI ≥ 25.

Pandemic   Epidemic over a large region of the world. Used in this textbook to represent a disease or condition spreading across a large international region.

Pathophysiology   Deals with the biochemical structural and functional changes caused by disease.

Prevalence   The number of individuals who have the disease at the time of the study.

SEARCH   SEARCH for Diabetes in Youth provides the first extensive diabetes study focused on persons younger than 20 years in the United States.

REFERENCES

1. Centers for Disease Control and Prevention: Prevalence of diagnosed diabetes. Data and trends, United States, 1980–2008. Available at www.cdc.gov. Accessed 2010 Oct.

2. Boyle JP, Thompson TJ, Gregg EW, et al. Projection of the year 2050 burden of diabetes in the US adult population: dynamic modeling of incidence, mortality, and prediabetes prevalence. Popul Health Metr 2010;8:29. DOI: 10.1186/1478-7954-8-29.

3. United Health: Center for Health Reform & Modernization: The United States of Diabetes: Challenges and opportunities in the decade ahead. Working Paper 5; November 2010. Available at www.unitedhealthgroup.com/hrm/UNH_WorkingPaper5.pdf. Accessed 2010 Nov.

4. Lee JM. Why young adults hold the key to assessing the obesity epidemic in children. Arch Pediatr Adolesc Med 2008;162:682.

5. Lee JM, Davis MM, Gebremariam A, et al. Age and sex differences in hospitalizations associated with diabetes. J Womens Health 2010;19. DOI: 10.1089/jwh.2010.2029.

6. Centers for Disease Control and Prevention and National Institute of Diabetes and Digestive and Kidney Diseases: SEARCH for Diabetes in Youth Study. Available at www.searchfordiabetes.org/. Accessed 2010 Oct.

7. Mayer-Davis EJ, Bell RA, Dabelea D, et al. The Many faces of diabetes in American youth: type 1 and type 2 diabetes in five race and ethnic populations: the SEARCH for diabetes in youth study. Diabetes Care 2009;32(Suppl 2):599. DOI: 10.2337/dc09-S201.

8. Bell RA, Mayer-Davis EJ, Beyer JW, et al. Diabetes in non-Hispanic white youth: prevalence, incidence, and clinical characteristics: the SEARCH for diabetes in youth study. Diabetes Care 2009;32(Suppl 2):S102. DOI: 10.2337/dc09-S202.

9. Mayer-Davis EJ, Beyer J, Bell RA, et al. Diabetes in African American youth: prevalence, incidence, and clinical characteristics: the SEARCH for diabetes in youth study. Diabetes Care 2009;32(Suppl 2):S112. DOI: 10.2337/dc09-S203.

10. Lawrence JM, Meyer-Davis EJ, Reynolds K, et al. Diabetes in Hispanic American youth. Diabetes Care 2009;32(Suppl 2):S123.

11. Liu LL, Yi JP, Beyer J, et al. Type 1 and type 2 diabetes in Asian and Pacific Islander U.S. youth. Diabetes Care 2009;32(Suppl 2). DOI: 10.2337/dc09-S205.

12. Dabela D, Bell RA, D'Agostino RB Jr, et al. Incidence of diabetes in youth in the United States. JAMA 2007;297:2716.

13. Department of Health and Human Services, Center for Disease Control and Prevention. National diabetes fact sheet, 2007. General information. Available at www.cdc.gov/diabetes/pubs/pdf/ndfs_2007.pdf. Accessed 2010 Oct.

14. I could do better at...controlling costs. Consumer Reports Health.Org. Available at www.consumerreports.org/health/conditions-and-treatments/type-2-diabetes/controlling-costs/diabetes-costs.htm. Accessed 2010 Nov.

15. International Diabetes Federation. Diabetes facts. Available at www.idf.org. Accessed 2010 Nov.

16. Yang W, Lu JL, Weng J, et al. Prevalence of diabetes among men and women in China. N Engl J Med 2010;362:1090.

17. *IDF Diabetes Atlas*. 4th ed. International Diabetes Federation; 2009. Available at www.diabetesatlas.org. Accessed 2010 Nov.

18. Gakidou E, Mallinger L, Abbott-Klafter J. Management of diabetes and associated cardiovascular risk factors in seven countries: a comparison of data from National Health Examination Surveys. Bull World Health Organ 2011;89:172. DOI: 10.2471/BLT.10.080820.

19. World Diabetes Foundation: Diabetes facts; diabetes costs-a burden for families and society. Available at www.worlddiabetesfoundation.org/composite-35.htm. Accessed 2010 Oct.

20. World Health Organization. Obesity and overweight. Fact Sheet No 311; 2006. Available at www.who.int/mediacentre/factsheets/fs311/en/print.html. Accessed 2010 Nov.

21. U.S. obesity trends. Centers for Disease Control and Prevention, Available at www.cdc.gov/data/trends.html. Accessed 2010 Nov.

22. Centers for Disease Control and Prevention (CDC). Differences in prevalence of obesity among black, white and Hispanic adults; United States, 2006–2008. MMWR Morb Mortal Wkly Rep 2009;58:740 Available at www.cdc.gov/mmwr/preview /mmwrhtml/mm5827a2.htm. Accessed 2010 Nov.

23. Ogden CI, Carroll MD, Curtin LR, et al. Prevalence of high body mass index in US children and adolescents, 2007–2008. JAMA 2010;303:242.

24. BMI percentile calculator for child and teen. CDC. Available at http://apps.nccd.cdc. gov/dnpabmi/Calculator.aspx. Accessed 2010 Dec.

25. Ogden C, Carroll M. Prevalence of obesity among children and adolescents: United States, trends 1963–1965 through 2007–2008. 2010. Available at www.cdc.gov/nchs /data/hestat/obesity_child_o7_08.pdf. Accessed 2010 Dec.

26. Whitaker RC, Wright JA, Pepe MS. Predicting obesity in young adulthood from childhood and parental obesity. N Eng J Med 1997;337:869.

27. World Health Statistics 2010. World Health Organization. Available at www.who .int/whosis/whostat/2010/en/print/html. Accessed 2010 Dec.

28. World Health Organization, Obesity and overweight, 2010. Available at www.who.int /dietphysicalactivity/publications/facts/obesity/en/. Accessed 2010 Dec.

29. De Onis M, Blossner M, Borghi E. Global prevalence and trends of overweight and obesity among preschool children. Am J Clin Nutr 2010;92:1257. DOI: 10.3945 /ajcn.2010.29786.

30. Books Ngram Viewer. Googlelabs©; 2010. Available at http://ngrams.googlelabs.com. Accessed 2010 Dec.

31. Michel J-B, Shen YK, Aiden AP, et al. Quantitative analysis of culture using millions of digitized books. Science 2010. DOI: 10.1126/science.1199644. Available at www.sciencexpress.org. Accessed 2010 Dec.

32. Lyman HM. Saccharine diabetes. In: Pepper W, editor. Volume 2. *A Textbook of the Theory and Practice of Medicine*. Philadelphia: W.B. Saunders; 1895. p 107.

33. Ebstein W. Gout. In: Cabot RC, Salinger JL, editors. *Modern Clinical Medicine: Diseases of Metabolism and of Blood Animal Parasites Toxicology*. New York: D. Appleton & Co; 1906. p 144.

34. Horsford FC. Obesity. J Med Soc NJ 1920;17:375.

35. Vague J. The degree of masculine differentiation of obesities: a facor determining predisposition to diabetes, atherosclerosis, gout and uric calculous disease. Am J Clin Nutr 1956;4:20.

36. Van Itallie T. Health implications of overweight and obesity in the United States. Ann Intern Med 1985;103:983.

37. Pi-Sunyer FX. Weight and non-insulin dependent diabetes mellitus. Am J Clin Nutr 1996;63(suppl):4265.

38. Wannamethee SG, Shaper AG. Weight change and duration of overweight and obesity in the incidence of type 2 diabetes. Diabetes Care 1999;22:1266.

CHAPTER 2

AN EARLY HISTORY OF DIABETES MELLITUS

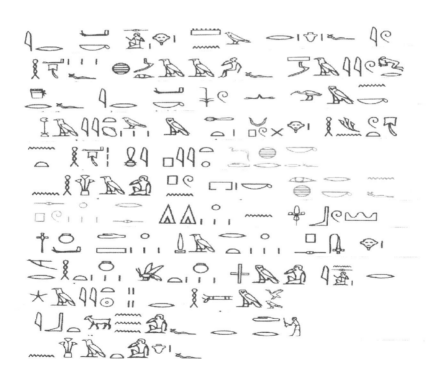

Understanding Diabetes: A Biochemical Perspective, First Edition. Richard F. Dods.
© 2013 John Wiley & Sons, Inc. Published 2013 by John Wiley & Sons, Inc.

TRANSLATION

If you examine someone sick (in) the center of his being (and) is his body shrunken with disease at its limit; if you examine him not (and) you do find disease in (his) body except for the surface of the ribs of which the members (are) like a pill you should then recite (a spell) (against) disease this in your house; you should (also) then prepare for him ingredients for (treating) it: blood stone of Elephantine, ground; red grain; carob; cook in oil (and) honey; (it) should be eaten by him over mornings four for the suppression of his thirst (and) for curing his mortal illness.

MORE SIMPLY STATED

If you examine someone mortally ill, with his body shrunken with serious or extreme illness and his body does not show disease except for the surface of his ribs which protrudes like pills, you should recite a spell against the disease. You should prepare ingredients for treating the disease composed of ground blood stone of Elephantine [possibly related to the red ochre (red dye) found on the Isle of Elephantine; an island found in the Nile River near Aswan], red grain (type of rice) and carob (chocolate tasting fruit) cooked in oil and honey. It should be eaten for four mornings in order to suppress thirst and for curing his illness.

All of the above was adapted from Reference (1) with permission from Dr. John Ferguson, Bard College.

THE EBERS PAPYRUS

I have opened this chapter with Rubric No. 197 (Column 39) from the Ebers Papyrus written in hieratic, a cursive form of hieroglyphics. Hieratic is the earliest known example of Egyptian writing and dates back to 3400 BCE. In 1872, Georg Ebers came across a papyrus, wrapped in mummy cloths in a state of superb preservation. The text found on the papyrus was written principally in black ink, with shorter sections in red ink. A calendar was included dating to 1536 BCE. The author was probably a physician as the writings include sections on what we, in modern times, would call ophthalmology, dermatology, obstetrics and gynecology, dentistry, surgery, psychiatry, and diseases involving parasites such as pinworm, hookworm, tapeworm, or roundworm (helminth infections) (2).

The Ebers Papyrus is believed by some scholars to contain the first known reference to diabetes mellitus. If the interpretation of the writing into English (see earlier text) is correct, the papyrus refers to an emaciated patient; he is "shrunken" and in the last sentence it is revealed that the afflicted patient needs to have his "thirst" suppressed—hallmarks of diabetes. As these symptoms could be ascribed to other conditions, we must take our diagnosis of diabetes with a grain of salt. There is a second reference that may be construed as implicating diabetes as the disease affecting an individual. Rubric No. 264 (Column 49) is translated, in part, "Another for correcting urine that (is in) excess: cyperus grass, one; seeds grass, one; root of shrub, one; beat into consistency a uniform, steep in beer sweet; drunk is along with the dregs a bowl of this." Polyuria is a major symptom of diabetes. I do not know but strongly doubt the prescribed drinks actually worked.

NEANDERTALS

If we are being a little overzealous in our interpretation of the Ebers Papyrus, we must nevertheless assume that diabetes has been present with us humans from the dawn of time. And, in fact, there is some evidence that this is true. Investigators have produced a DNA sequence (4 billion nucleotides, about two-thirds of the total genome) pieced together from three female Neandertals who lived in Croatia more than 38,000 years ago. By comparing the sequence to five humans from different regions of the world, they concluded that 1–4% of European and Asian DNA were shared in common with Neandertals. Africans do not share any DNA in common. This data can be construed as indicating that humans interbred with Neandertals after they left Africa but before populating Europe and Asia (3).

What is significant in respect to diabetes is that a region on Neandertal DNA chromosome 2 contains a gene corresponding to the modern human gene called *thyroid-adenoma-associated gene* (THADA) (OMIM Entry*611800[1]). Single-nucleotide polymorphisms (SNPs[2]) in the vicinity of THADA have been associated with type 2 diabetes (see Footnote 3 in Chapter 1 for definition of type 2 diabetes). SNPs near THADA may have caused diabetes in Neandertals (4). This is not to suggest that THADA is the only gene for diabetes or even the principal gene. We will find in Chapter 9 that there are a significant number of genes that make us susceptible to diabetes. What it does mean is that diabetes may have existed since our earliest presence on earth.

HIPPOCRATES, ARETAEUS, AND DEMETRIUS

Ancient Egypt is the first civilization to have a recognizable study of medicine. After the Ebers papyrus, the next mention of diabetes occurs in the writings of Hippocrates of Cos (about 450–380 BCE). Hippocrates' writings describe an ailment that involved wasting of the body concurrent with excessive urination. He does not name the condition. But Aretaeus of Cappadocia (81–138 AD) gave the first accurate account of the disease in about 150 AD and named it *diabetes*

---

[1]Online Mendelian Inheritance in Man (OMIM) is a compendium of human genes and genetic phenotypes. It contains information on all known mendelian disorders and focuses on phenotype and genotype. Its website is http://www.ncbi.nlm.nih.gov/omim.
[2]A SNP refers to a genetic variation consisting of a single base (or nucleotide change); for example, thymine substituted for adenine.

(diabetis, which means to siphon through). Aretaeus described in great detail (in Ionic Greek) the symptoms of diabetis. He states (translated):

> *Diabetes is a remarkable disorder and not one very common to man. It consists of a moist and cold wasting of the flesh and limbs into urine, from a cause similar to that of dropsy; the secretion passes in the usual way, by the kidneys and the bladder. The patients never cease making water, but the discharge is as incessant as a sluice let off. This disease is chronic in character, and is slowly engendered, though the patient does not survive long when it is completely established for the marasmus produced is rapid and death is speedy.* (5, 6)

Historians differ as to the individual who first characterized the disease and named it diabetes. The major contenders are Aretaeus and Demetrius of Apamea. There is not much first-hand information on Demetrius of Apamea as all of his writings were lost. One has to rely on second- and third-hand information.

During the time of Demetrius, physicians defined dropsy as having two forms. In the first type, water was retained. In the second, water was excreted in the urine almost as soon as it was ingested. Caelius of Aurelianus (a Roman physician and author who lived during the fifth century[3]) translated from Greek to Latin a manuscript by Soranus of Ephesus, "*On Acute and Chronic Diseases.*" In the translation, Caelius quotes Demetrius as naming the second form of dropsy, diabetes (7, 8).

GALEN

Galen of Pergamum (130–200 AD) lived about the time of Aretaeus (and a little after). He spent most of his life as physician to Emperor Marcus Aurelius. Galen studied the kidneys and spinal cord and was one of the first experimental physiologists. Galen used the term *diabetes* after Aretaeus but did not give Aretaeus credit.

> *I am of the opinion that the kidneys too are affected in the rare disease which some people call chamber-pot dropsy, other again diabetes or violent thirst. For my part I have seen the disease till now only twice, when the patients suffered from an inextinguishable thirst, which forced them to drink enormous quantities; the fluid was urinated swiftly with urine resembling the drink.*

Note that Galen writes that he only saw these symptoms twice. The rarity of these symptoms places some doubt in my mind as to whether he described the condition that we call diabetes mellitus or the more rare disease called *diabetes insipidus*. Diabetes insipidus has some of the symptoms of diabetes mellitus such as extreme thirst, wasting, and polyuria but is caused by a deficiency in the secretion of a hormone from the pituitary gland called *vasopressin* (formerly

---

[3]Some historians place him two to three centuries earlier.

called antidiuretic hormone). We will never know which illness, diabetes mellitus, or diabetes insipidus was described by Aretaeus and Galen. However, the advent of the urine tasting test in cases exhibiting polyuria makes it clear which condition was being dealt with.

Summary Box 2.1

- The Ebers Papyrus dates back to 3400 BCE. It makes the earliest known reference to diabetes.
- Scientists have produced a sequence of 4 billion nucleotides from three female Neandertals. A region on the Neandertal DNA contains a gene that shows polymorphisms that are associated with type 2 diabetes.
- Hippocrates in about 450–380 BCE described diabetes.
- Aretaeus and Demetrius were among the first to name diabetes.
- Galen wrote about the symptoms of diabetes but whether he saw diabetes insipidus or diabetes mellitus is uncertain.

SUSHRUTA

The sweet taste of the urine produced by those suffering from polyuria was first noticed in India and China as early as the fifth-century AD. Excessive urination and its sweet taste became the primary symptoms of diabetes for the ancients. Sushruta was an Indian surgeon who taught and practiced medicine in the fifth-century AD. He is considered the father of Indian surgery and is famous for classification of eye diseases, cataract surgery, and dental surgery. In addition, he is also noted for his description of diabetes. He was the author of a systematic and extensive compendium of his work called *Sushruta Samhita*. It contained 184 chapters, described 1120 illnesses, and had recipes for 121 remedies for the maladies described. The treatise was translated from the original Sanskrit to English in 1911 (9). Sushruta uses the word "Prameha"[4] for urinary tract diseases (probably polyuria) and specifically the word Madhu-Meha for diabetes. He describes the excretions of a Madhu-Meha patient as "sweet is the urine, the sweat and the phlegm." He also describes a patient with Madhu-Meha as one who has "acquire(d) a sweet taste and smell like that of honey."[5] He was well in advance of his times as shown in the next quote:[6]

This disease may be ascribed to two causes, such as the congenital (Sahaja) and that attributable to the use of injudicious diet. The first type (Sahaja) is due to

---

[4]Prameha translates as excess of urine, both in frequency and volume.
[5]Could he be describing ketoacidosis in which acetone is produced and thus a sweet smell emanates from the patient's breath?"
[6]Bold font for emphasis (was added by the author).

a defect in the seeds of one's parents and the second is originated from the use of unwholesome food. The symptoms, which mark the first of these two types, are emaciation and a dryness (of the body), diminished capacity of eating, too much thirst and restlessness; while the symptoms, which usually attend the latter type of the disease, are **obesity, voracity, gloss of the body, increased soporific tendency and inclination for lounging in bed or on cushions.** *A case of emaciation, etc., (viz., the first kind of Prameha) should be remedied with nutritious food and drink, etc.,* **whereas Apatarpana, etc. (fasting, physical exercise, depletory measures etc.), should be adopted in cases of obesity viz., the second kind of (Prameha).**

The type of diabetes (the Sahaja) could be interpreted as type 1 diabetes mellitus and the second (Apatarpana) could be type 2. The quotation was a prophetic declaration that there were two types of diabetes.

IBN SINA (AVICENNA)

Ibn Sina or Avicenna (Latinized name) (960–1037 AD) was born in Bukhara, then a city in the Persian Empire. It is located in what was southern Russia and is currently called Uzbekistan. He was brilliant and excelled in every intellectual pursuit he undertook. In his early years (10–18 years) he became a gifted philosopher and poet and by his late teens to early twenties an accomplished physician-scientist. I use the term *physician-scientist* because he not only treated patients but also developed new methods of treatment. His most noteworthy achievement was the 14-volume '*The Canon of Medicine.*' This work is a compilation of all that was known at the time about Greek and Islamic medicine and was written when Ibn Sina was in his early 30s. Ibn Sina classifies the diseases and makes assumptions on their causes as well as presents evidence-based experimental medicine. He divided his manuscript into five books and divided each into 'funun' (chapter). In Book 1 (4 funun) he presents the general principles of medicine; Book 2, drugs and their properties and uses, Book 3 (21 funun), treatment; Book 4 (7 funun), diseases that affected different parts of the body; and Book 5, recipes for his remedies. '*The Canon of Medicine*' remained the standard medical textbook in Islam and Europe until the 1600s.

In this work Ibn Sina details the characteristics of diabetes. He describes the abnormal appetite, the reduced desire for sexual intercourse, and remarks about the sweet taste of the urine. He, as did Aretaeus, recognized two forms of the disease. In Book 1, Ibn Sina states that "when the urine of diabetics is left to strand in ambient air, it leaves a residue that is particularly sticky and tastes like honey." (8). In Book 3, Part 19, Treatise 1, Chapters 7–15 (10) he describes bladder conditions including infravesical obstruction, infection, ulcers, urethritis,[7]

---

[7]Urinary urgency, discharges, blood in urine, and burning sensation constitute urethritis. Urethritis is a condition caused by irritation, possibly by bacterial infection.

intravesical hematoma,[8] and neurogenic bladder.[9] Urinary tract infection and obstruction is a common characteristic of diabetes and often leads to neurogenic bladder. Also, diabetes can cause neurological damage to the bladder, another condition causing neurogenic bladder.

THE YELLOW EMPEROR

The earliest known Chinese medical book is the *HuangTi*[10] *Nei Ching Su Wen* (The Yellow Emperor's Classic of Internal Medicine) (11). Huang Ti lived in 2697–2597 BCE. The work is a dialog between Huang Ti, the Emperor, and his physician Ch'I Po.[11] It consists of 24 books and 81 chapters. The first 34 chapters of the Nei Ching were translated by Ilze Veith (12), who notes in the Introduction that "the chief means of diagnosis employed in the Nei Ching is the examination of the pulse." That diabetes mellitus is described in this book is stretching the limits of the translation. Veith states that "the idea of a disease entity as it is known to Western medicine did not exist in ancient Chinese medical thought. The analysis of the diseases that are mentioned in the Nei Ching is extremely difficult." However, I have found in the translation some passages that could be construed as pertaining to diabetes. For example, in Book 1 (p. 109) Ch'I Po states, "If sweetness exceeds the other flavors, the breath of the heart will be [asthmatic and] full and the appearance will be black and the force of the kidneys will be unbalanced," a reference to polyuria? In Book 7, Ch'l Po states (p. 206), "When the bladder does not function efficiently, it causes retention of urine; when it functions without restraint, it causes copious urination"; again, a description of polyuria? In Book 9, the comment (p. 244) "...there is suffering of thirst [in spite of] frequent drinking, and the body is hot" can be construed as pertaining to polyuria. In summary, the Nei Ching is the earliest known Chinese medical textbook but it is difficult to ascribe illnesses and disease to it in Western medical terms.

JAPANESE MEDICINE

Chinese medicine was transported into Japan in the late fourth to the early fifth century. By the sixth century, Buddhism was fully incorporated into the Japanese culture and accompanying it was Chinese medicine. Previous to this, Japanese medicine consisted mostly of folklore and herbal remedies. In fact, beginning in the early 600s, Japanese physicians were sent to China for the study of Chinese medicine. The first Japanese medical text was a 30-scroll (equivalent to

[8]The hematoma is commonly caused by bladder infection. It is also caused by trauma to the bladder.
[9]Increased frequency of urination, urinary urgency, and leakage constitute a neurogenic bladder.
[10]In some sources, the emperor is known as Huangdi.
[11]In some references, the term Pai is used. In one reference (11), Huang Ti dialogs with his minister who is given the name, Qibo.

volumes) called the *Ishinhô*[12] [Methods from (or at) the Heart of Medicine] written by Yusuyori[13] Tamba in 984 AD. It describes internal medicine, drugs, and prevention of disease based on Chinese texts that were previously written (13, 14). No doubt the Chinese acquaintance with diabetes is included in the Ishinhô.

Summary Box 2.2

- Ibn Sina characterized diabetes in his 14-volume work called *"The Canon of Medicine."*
- Huang Ti wrote about diabetes in "The Yellow Emperor's Classic of Internal Medicine."
- Japan incorporated Chinese medicine by the sixth century when Buddhism was fully assimilated into the culture.

PARACELSUS (PHILIPPUS AUREOLUS THEOPHRASTUS BOMBASTUS VON HOHENHEIM)

A very colorful character, Philippus Aureolus Theophrastus Bombastus von Hohenheim, born in Switzerland (1493–1541), owes his nickname, Paracelsus, to his claim that he was superior ("para") to a Hellenistic poet by the name of Celsus. Bombastus is not a description of his personality and has nothing to do with the word "bombastic." His paternal grandfather was of the Bombast family. Paracelsus or von Hohenheim had a confrontational personality and often his personality bordered on braggadocio. His prescriptions consisted primarily of heavy-metal salts that were effective against bacterial infections. Paracelsus is considered the father of toxicology. Many historians consider him an alchemist (an early chemist). He conducted distillations (or evaporation) and one of the liquids he distilled was urine. He found that when he distilled the urine of someone suffering from polyuria (and, therefore, probably a diabetic) a white powder was left behind. Paracelsus concluded that this substance caused irritation of the kidneys and thus promoted excessive urination (15).

Many of the early investigators were practitioners who observed diabetes in their patients. They were pioneers who tried different approaches to treat and alleviate the effects of diabetes in their patients. Many were interdisciplinary, with interests in distinctly different fields. They made their discoveries known by publishing them as papers or monographs. At this point in time, it was accepted that although many who were afflicted with diabetes lived almost normal life spans (which in those days were short to begin with in comparison to the life spans today) there were some for which the diagnosis of diabetes was a death warrant.

---

[12]Some scholars westernize the name as Ishimpō or Ishinpō.
[13]Some scholars westernize the name as Yusunori.

THOMAS WILLIS

Thomas Willis (1621–1675), as many scientists of his time, was competent in several specialties including neuroanatomy. In fact, he is considered one of the greatest neuroanatomists and the father of clinical neuroscience (16). But that is not all. Willis also comprehensively described the symptoms of malaria, typhoid fever, myasthenia gravis, and diabetes. It is the latter disease for which he is best known. Willis observed that flies were attracted to samples of urine from certain patients who displayed polyuria. The urine of these patients tasted like honey (17). Thus, he tagged on the name mellitus (from the Latin meaning sweet-like honey) to the name diabetes in order to distinguish it from chronic kidney disease and diabetes insipidus (known during his time as *water diabetes*), diseases that lacked the urine sweetness. However, this description of the symptoms of diabetes mellitus as we have already seen was not a new one. He writes, "But why it should be so wonderfully sweet, like sugar or honey, is a knot not easy to untie." In the next several decades this question would be pursued, making incremental steps progressing toward the solution.

JOHANN CONRAD BRUNNER

Johann Conrad Brunner (1623–1727), a Swiss anatomist, came very close to discovering the importance of the pancreas in diabetes. He describes the pancreas and comes close to suggesting its association with diabetes in *Experimenta Nova circa Pancreas: Accredit diatribe de lympha & genuine pancreatic usu* (17), published in 1709. In the article, he describes incomplete pancreatectomies performed on eight dogs and the subsequent excessive urination, and increased thirst in the animals. Nonetheless, he does not explicitly link the symptoms to diabetes mellitus. As we will see later in this chapter, the connection between the pancreas and diabetes mellitus was left to Mehring and Minkowski from their own experiments two centuries later (18).

MATTHEW DOBSON

Matthew Dobson (∼1735–1784) was born in Yorkshire. He was an experimental physiologist whose observations were very keen. Dobson wrote a manuscript on what was then called "fixed air" (carbon dioxide) and on the treatment of hydrocephalus.[14] He made observations on the properties of water from Matlock[15] and experimented on the effect of high temperatures on humans (himself and his colleagues). But he appears in this historical account because, he, as

---

[14]Hydrocephalus is an abnormal expansion of the ventricles within the brain that is derived from birth or acquired from meningitis, cysts, and other acquired conditions.

[15]The waters of Matlock refer to a spa whose waters were supposed to have healing qualities.

scientists before him, observed that urine from diabetics was sweet and when evaporated left a white residue (remember Paracelsus' experiments?) that tasted sweet. He also noted that the blood of diabetics was also sweet to the taste. He thus discovered, although he did not know it, hyperglycemia (excessive blood sugar).

Dobson contended that the sweetness first occurred in the blood and later appeared in the urine. This was a significant paradigm shift. Diabetes was not a kidney disease but a systemic condition. In *Medical Observations, Chapter 27, Experiments and Observations on the Urine in Diabetes* (17), Dobson describes a patient, Peter Dickonson, who is a diabetic. Dickonson excreted 28 pints of urine every 24 hours (13.2 1/24 h, normal is about 1.5 1/24 h (19)[16]). This statistic must be either an exaggeration or a miscalculation. In addition, Dobson states that Dickonson "drank large quantities of water, and made large quantities of urine." He had "a perpetual gnawing sense of hunger; the palms of his hands and the soles of his feet were frequently hot." "When he came to the hospital, he was emaciated, weak and dejected." Dobson evaporated a portion of his urine and recovered a "white cake" that "smelled sweet-like brown sugar, neither by taste be distinguished from sugar."

Dobson connected this urinary sugar to the sweet taste of blood from diabetics and concluded that the sugar was dietary related. This explained the wasting effect found in diabetics; as the nutrient sugar was being drawn off by the kidneys, to the neglect of providing nutrition to the body. Dobson's discoveries led to the conclusion that diabetes was dietary in nature. On the basis of Dobson's writing, many physicians, upon diagnosing diabetes in a patient, prescribed a diet that lacked sweet tasting foods and emphasized meats. This diet unfortunately excluded fruits and vegetables. Yet, this did not help ease the disease with the victims of the nutritional approach being undernourished in addition to being afflicted with the typical symptoms of the disease.

JOHN ROLLO AND WILLIAM CRUICKSHANE

Two of the physicians who adhered to this approach to the treatment of diabetes were John Rollo (?–1809) and William Cruickshane (?–1810 or 1811). Both were of Scottish origin and surgeons in the British Royal Artillery. They conducted a longitudinal study on two officers who were diagnosed with diabetes on the basis of polyuria and the finding by taste of sugar in the urine. In addition, during "their progress of the cure" they used the degree of effervescence caused by adding yeast to the urine to semiquantitate the amount of sugar present. By this time, chemists had described the action of yeast on sugar and sugar had been definitely determined as the residue left behind when a diabetic's urine had been evaporated to dryness. Using a high meat, low carbohydrate diet the weight recorded for one of the officers as originally 105 kg (232 lb) was significantly

---

[16]To be more exact, the urine volume for adults is 0.6–1.6 1/24 h and for older adults 0.25–2.4 1/24 h.

reduced. And lo and behold, his output of urine and sugar was reduced (20). Rollo published a book called *An Account of Two Cases of the Diabetes Mellitus: With Remarks As They Arose During the Progress of the Cure* (17). Included in the title are "*some observations on the nature of sugar*" by William Cruickshane. Unfortunately Rollo gave credit to his "cure" not to weight loss but to diet (which lacked what we currently call *carbohydrates*).

Summary Box 2.3

- Paracelsus, the father of toxicology, distilled the urine of diabetics and found a white powder.
- Willis observed that flies were attracted in greater number to the urine of diabetics than to the urine of normal persons. Urine from diabetics tasted like honey and therefore he tagged on the name mellitus to diabetes.
- Brunner described incomplete pancreatectomies and their subsequent polyuria and increased thirst but failed to see the connection between the pancreas and diabetes.
- Dobson determined that sweetness first occurred in the blood and then in the urine. He concluded that diabetes was a dietary condition.
- Rollo and Cruickshane treated their patients by using a high meat, low carbohydrate diet. Reduction in output of urine and sugar of a patient on this regimen was looked at as proof that the diet was working. What they did not recognize as significant was the weight loss of the patient.

THOMAS CAWLEY

Thomas Cawley (birth and death not reported) was another surgeon who did not fully comprehend the significance of his investigations. Only in retrospect do we give his investigations the correct interpretation. He published a manuscript entitled (17) "*A singular case of diabetes, consisting entirely in the Quality of the Urine: with an Inquiry into the different Theories of that Disease. By Thomas Cawley, M.D. late chief Surgeon to the Forces in Jamaica, published in the London Medical Journal volume 9 pages 286–308 in 1788.*" In this paper, he describes an overweight ("corpulent"), 38 years of age, Allen Holford, Esq. who is "seized with diabetes." His disease began in March 1787 and it took three months before it was diagnosed as diabetes. Although Allen did not manifest polyuria, his urine was "fermentable with yeast" and upon evaporation yielded a "black extract, exactly resembling, that preparation of melasses made by confectioners for children . . ." Treatment with the usual medications of the day was unable to prevent Mr. Holford's demise on the "18th of June." The medicines described were "decoction of bark[17] with vitriolic acid[18] and alum,[19] with astringents and aromatics with chalybeates,[20] with sace,[21] saturni[22] and opium and with cantharides,[23] together with cold bathing in salt water." Prescribing a "decoction of bark" suggests that Cawley diagnosed diabetes as an inflammatory disease. Upon autopsy, Cawley found that the pancreas of the deceased contained stones (calculus formation) and tissue damage. But he did not link the pancreas to diabetes.

MICHEL EUGENE CHEVREUL

Michel Eugene Chevreul (1786–1889), a French chemist, witnessed the French revolution during his childhood. He was the first to isolate and characterize "cholesterine" (cholesterol)[24] from the bile of humans and bovines. Chevreul studied the properties and structure of lipids such as butyric, caproic, oleic and stearic acids, and glycerol. He published these findings as "*Recherches chimiques sur les corps gras d'origine animale*" in 1823. However, in addition, he studied the structure of the sugar isolated from diabetic urine and came to the conclusion that it was identical to "grape sugar" (glucose) (21). Chevreul identified the sugar found in the urine of diabetics as glucose in 1815. By this time in history diabetes was defined by the presence of "glucose" in the urine. But the source of the glucose was still under investigation.

CLAUDE BERNARD

Claude Bernard (1813–1878) in his youth was devoted to writing plays for the theater. But when he was 21 years of age, with a play in hand he visited a theatrical critic in Paris who persuaded Bernard not to become a playwright but to become a physician. And this he did. By 1855 he was full professor of physiology at the Collége de France. Perhaps an interesting sidelight is that after Bernard was appointed the chair of physiology at the Sorbonne he had an interview with Louis Napoleon. He impressed Napoleon so greatly that Napoleon built a laboratory and created a professorship for Bernard at the Jardin des Plantes;

---

[17]Decoction is a method of extraction using boiling. Decoction of bark means boiling the bark to extract substances to be used in the treatment. The vitriolic acid and alum probably were present to extract the portion of bark that Cawley thought would act as an anti-inflammatory agent. I do not believe that they were active ingredients.
[18]Sulfuric acid.
[19]Hydrated potassium aluminum sulfate.
[20]Chalybeates means containing iron salts.
[21]Unknown.
[22]Saturni most likely refers to an extract of lead.
[23]Cantharides refers to an extract from the cantharide beetle, which is an inflammatory agent. It also is supposed to be an aphrodisiac (commonly known as Spanish fly).
[24]The name cholesterine originates from chole = bile and stereos = solid.

a position he accepted in 1868. Bernard discovered that the pancreas secreted a "juice" that promoted the process of digestion. But this discovery was followed by his most important investigation, that the liver secretes a substance that affects blood sugar levels. By the time of Bernard the analysis of sugar had gotten beyond the yeast fermentation experiments described above. Bernard was able to estimate the sugar content of extracts by titrating with a copper reagent.[25] He learned that the liver had an extractable substance that did not test for sugar but upon fermentation did test positive for sugar. He then isolated the substance and named it "la matiére glycogéne." In a paper communicated to the Société de Biologie in Paris on March 21, 1857, Barnard described this starch-like substance isolated from the liver (22). He believed that the breakdown of the glycogéné from the liver resulted in the glucose found in the blood and urine of diabetics.

Bernard made other important discoveries in addition to the above-mentioned. He studied the changes (blood flow and temperature) that occur in various parts of the body when certain nerves are severed and when nerves are stimulated by electrical currents. But possibly the most important conclusion that he enunciated was the process he called "milieu intérieur." Milieu intérieur refers to "La fixité du milieu intérieur est la condition d'une vie libre et indépendente," which translates as "the constancy of the internal environment is the condition for a free and independent life." This defines an important concept in physiology—homeostasis.

PAUL LANGERHANS (EDOUARD LAGUESSE AND EUGENE L. OPIE)

Paul Langerhans (1847–1888) was basically an anatomical microscopist. He was born in Berlin. He was afflicted with tuberculosis at the age of 27. He died at the early age of 41 from a kidney infection. While a student, he developed the use of gold chloride as a stain for nerve endings in the Malpighian layer[26] of the skin. But he is most noted for his observation of polygonal-shaped cells scattered throughout the pancreas using one of his unique staining techniques. Langerhans used his observations for his doctoral dissertation, which he named *Contributions to the microscopic anatomy of the pancreas*. Langerhans did not realize the significance of these cells nor did he name them after himself. In fact he thought that they were simply lymph nodes. Five years after the death of Langerhans, G. E. Laguesse (1861–1927) recognized the significance of these "polygonal-shaped cells" and named them Islands of Langerhans. And after Laguesse, in 1901 Eugene Opie (1873–1971) of Johns Hopkins University associated deterioration

---

[25]The reagent consisted of alkaline copper sulphate with potassium tartrate added to prevent decomposition during heating. The reagent was standardized against pure grape sugar of known strength. It was added drop-wise to the sample until the blue color disappeared.
[26]The skin is composed of several layers called *strata*. They are from the outermost layer, the corneum, lucidum, granulosa, spinosum and the basale. The Malpighian layer refers to the deepest layers, the spinosum and basale and occasionally just the stratum basale.

in the Islands of Langerhans with diabetes. Opie's paper *"Hyaline Degenera-tion of Islands of Langerhans"* in the Journal of Experimental Medicine,[27] shows "drawings" in low and high magnification of the hyaline destruction of the islands of Langerhans cells. He states in the article that "destruction of the pancreas in animals and man is accompanied by diabetes" (17). The photomicrograph presented on the cover of this book gives a clearer representation of β-islet cells (in this instance, being attacked by T-cells). The photomicrographs are gra-ciously presented from the laboratory of Professor Anne Cooke, University of Cambridge.

OSCAR MINKOWSKI AND JOSEF VON MERING

Often, scientists make discoveries that go unrecognized until another discov-ery suggests the importance of the initial discovery. In hindsight the original discovery is recognized. This was the case in Langerhans' discovery of pan-creatic cells later named by Laguesse as the islets of Langerhans. At the time, the importance of this finding went unnoticed. The link of diabetes to the pan-creas was finally established in 1889 by the eloquent experiments of Minkowski (1858–1931) and Mering (1849–1908). Interestingly, Minkowski and Mering set out to analyze the absorption of lipids in pancreatectomized dogs. They were not attempting to link pancreatic function and diabetes. However, the pan-createctomized dogs were observed to increase their volume and frequency of urination. Minkowski and Mering tested the urine and found it to have high levels of sugar (glucose) and ketones. They linked this symptom to diabetes. Further experiments were conducted in which the diabetic condition was reversed by sub-cutaneous implantation of fragments of the previously removed pancreas. Mering and Minkowski published their findings in a paper entitled *"Diabetes mellitus nach Pankreasextirpation"* in the journal, Zentralblatt für klinische Medizin, vol-ume 10, pages 393–394, Leipzig, 1889. Thus, the importance of diabetes and the pancreas was established. One of their most historic articles, *"Untersuchungen über des Diabetes Mellitus nach Exstirpation des Pankreas"* appeared in the jour-nal Archiv für experimentelle Patholgie und Pharmakologie, volume 26, page 37, 1890 (17).

Oskar Minkowski was born in the city of Kaunas in the Russian Empire, currently Lithuania. Minkowski, a pathologist, performed the surgery on the dogs. Minkowski had previously been involved in diabetic research. He was the first investigator to describe excessive quantities of β-hydroxybutyrate in diabetic coma (23). Joseph Freiha von Mering was born in Halle an der Saale, Germany. He was well-versed in pharmacology, laryngology, and internal medicine and learned biochemistry from Hoppe-Seyler. He teamed up with Minkowski at the University of Strasbourg (24).

---

[27]Also, "Diabetes mellitus associated with hyaline degeneration of the pancreas." Johns Hopkins Hospital Bull. 12: 263, 1901.

Summary Box 2.4

- Crawley used the medicines of the day to treat Allen Holford, Esq. Upon the death of Mr. Holford, Esq., Cawley found that the pancreas of the deceased contained stones and damaged tissue.
- Chevreul identified the sugar found in the urine of diabetics as glucose.
- Bernard discovered glycogen in the liver and believed the liver to be the source of glucose in the urine of diabetics.
- Langerhans discovered the cells in the pancreas now known as the islets of Langerhans. He did not associate these cells with diabetes.
- Mehring and Minkowski used pancreatectomized dogs to prove the link between the pancreas and diabetes.

ADVANCES IN SUGAR (GLUCOSE) DETERMINATIONS

Progress in understanding disease is always accompanied or preceded by major analytical advances. Advances in the understanding of diabetes were dependent on the ability of investigators to detect and quantify glucose. We shall deviate from our historical approach to diabetes for a moment to look at the technological advances that took place in glucose analysis.

The earliest glucose determinations were taste; did the urine and blood taste sweeter in a suspected diabetic as contrasted to a nondiabetic? The next developed procedure was fermentation. Wine and beer making has gone on for nearly 8000 years. In 1856, Louis Pasteur linked yeast with fermentation. We now know that in fermentation glucose is converted to ethanol and carbon dioxide. The formation of the carbon dioxide is observed as bubbles and can be used as a semiquantitative determination of the concentration of glucose present in the substance being tested. As the sugar content of the urine of nondiabetics is negligible, a positive fermentation test would clearly identify urine from a diabetic. This was one of the earliest assays that determined that the urine of diabetics contained large quantities of glucose.

Earliest Approaches—Taste and Fermentation

The procedure for tasting glucose in urine—well, I will leave it to your imagination! The procedure for fermentation as described in a publication in 1860 is as follows (25):

> Two test tubes, of the same form, and of equal size, are to be taken. One is nearly filled with water, and into the other a corresponding quantity of the urine is to be poured. An equal amount of yeast is now to be added to the liquids in the tubes, and after pouring in just sufficient fluid to fill the tubes, the thumb is to be carefully

placed over the opening, and the tube inverted in a small cup of mercury. The best plan, however, is to cut out a little india-rubber pad, slightly larger than the upper extremity of the tube. When the tubes have been filled up to the brim with a little water, the pad is allowed to float on the surface; next a little cup or beaker is inverted and carefully placed over the end of the tube. The india-rubber being pressed against the open end, the fluid is prevented from escaping. The whole may be inverted, and a little mercury having been poured into the beaker, the india-rubber may be removed, with forceps, without any escape of the liquid. The tubes maybe supported in position by a wire stand. Both tubes are then to be exposed for a few hours, to a temperature of from 80^0 to 90^0, and the comparative size of the bubble of gas in the upper part of each may then be noted. If an appreciable quantity of sugar be present, the bubble of gas in the tube containing the urine will be many times larger than that in the tube which contains the yeast and water. In the latter tube the bubble of gas merely arises from the small quantities of air previously mixed with the yeast, becoming disengaged, and floating to the surface. Fermentation, when carefully performed, is positive evidence of the presence of sugar, although it does not indicate the kind of sugar present.

This was the gold standard for sugar (glucose) in urine for many years (probably decades) because sugar is not present in appreciable amounts in healthy individuals. It was not a quantitative test but it served to diagnose diabetes in conjunction with symptoms such as polyuria. As fermentation tubes were not being manufactured in 1860, the instructions used test tubes, india-rubber pads, and mercury instead. This test for sugar in the urine was used by experimenters at least as early as the late 1700s.

Evaporation of Urine to Yield Sugar Crystals

Frequently, the observation of sugar crystals after the evaporation of urine samples was used to identify diabetic urine (25).

Moore's Test

The first chemical test for sugar (glucose) was probably Moore's test. As reported in 1860 (25), Moore's test (for "grape sugar") was conducted by adding to the urine "half its bulk liquor potassae."[28] The mixture is then boiled. If sugar is present the solution "becomes of a rich brown color." The brown color is due to the formation of mellassic acid (sacchulmic acid[29]). Glucic acid[30] is also produced. Presently, this test is described in various college chemistry handbooks using sodium hydroxide in place of liquor potassae.

---

[28]Liquor potassae is produced by dissolving potassium bicarbonate in water and boiling. Lime (calcium hydroxide) is added to water and heated and then added to the potassium hydroxide solution. The mixture is heated and then strained.

[29]Mellassic acid (sacchulmic acid) is an acid produced by the breakdown of the glucose.

[30]Glucic acid is now called *triose reductone*, CHO–C(OH)=CHOH.

Trommer's Test

It had been known by early physicians that a solution of verdigris (copper acetate)[31] and honey over time changed slowly from green to red. This concoction was used as an ointment to heal wounds. The red substance was a precipitate that was copper oxide. The precipitate did not form in the absence of honey. From this information, Trommer devised a test for sugars using this fundamental approach.

Trommer's test consisted of the following:

> To this saccharine urine in a test tube I add a drop or two of a solution of sulfate of copper, and to this an excess of liquor potassae. The oxide of copper, which is first thrown down is redissolved, and forms a clear liquid. Now, on applying heat the oxide of copper is reduced to a suboxide, which forms a dense red or yellow deposit. An excess of copper, not being dissolved, may cause confusion, and the dark brown color, from the action of potash on sugar,[32] may interfere with the result (26).

Moore's and Trommer's tests and fermentation were clearly qualitative tests. But they were also very insensitive tests. Only urine could be tested for sugar using these methods. Moore's test is not reliable for testing for small quantities of sugar because of the many other substances found in urine that would give a false-positive result. Prolonged heating in Trommer's test could result in a false positive due to other substances in the urine that can reduce the suboxide. In addition, as indicated in the last sentence in the above-mentioned quote, excess copper and a brown color due to potash reacting with sugar can interfere with the test.

The Moore's and Trommer's tests went unnoticed in textbooks until the middle of the nineteenth century. Through the 1850s, the presence of sugar in urine was commonly determined by the sweet taste of the urine or the presence of sugar crystals left behind after evaporation of urine (27). However, the reduction of copper by sugar as a method to determine the presence of sugar continued to be pursued by chemists.

Barreswil and Fehling's Solutions

In 1841, Charles-Louis Barreswil (1812–1870) added cream of tartar and carbonate of soda to the Trommer solution. His solution and procedure consisted of the following (25):

Cream of tartar (potassium tartrate, $KHC_4H_4O_4$)	96 grains (6.22 g)
Crystallized carbonate of soda (sodium carbonate)	96 grains (6.22 g)
Sulfate of copper (copper sulfate)	32 grains (2.07 g)
Caustic potash (potassium hydroxide)	64 grains (4.15 g)
Water	2 fl oz (59.14 ml)

[31]Verdigris (copper acetate) was made from copper sulphate added to vinegar (acetic acid).

[32]The observation of a brown color caused by the reaction of potash and sugar is actually the basis for Moore's test.

Procedure: Add an equal volume of the solution to the urine in a test tube. Boil. If sugar is present, the precipitate of suboxide occurs immediately.

Hermann von Fehling (1812–1885) in 1849 modified Barreswil's test by substituting sodium hydroxide (caustic soda) for potassium hydroxide (caustic potash) and potassium bitartrate for potassium tartrate. His recipe consisted of

Sulfate of copper (copper sulfate)	90.5 grains (5.86 g)
Neutral tartrate of potash (potassium bitartrate, $K_2C_4H_4O_6$	364 grains (23.59 g)
Caustic soda specific gravity 1.12 (sodium hydroxide)	4 fl oz (118.3 ml)
Cold water to make up a total solution	6 fl oz (177.4 ml)

Procedure: A column of about three-quarters of an inch is poured into a test tube and heated until it boils, and then a drop or two of the urine to be tested is added. In a few seconds, if the urine contains much sugar, the liquid becomes of an opaque yellow color, and a copious red or yellow precipitate falls (26).

The Fehling solution was unstable, making it necessary to make it up before a test. But analytical chemists soon learned to divide the reagent into two separate solutions, the Fehling A reagent consisting of copper sulfate and the Fehling B reagent composed of tartrate and sodium hydroxide.

Frederick Pavy

Frederick Pavy (1829–1911) had a thriving practice in London consisting primarily of diabetics. He was one of the few prominent physicians of his time to use glucose assays to make diagnosis and to monitor his patients. His publication was entitled *"Researches on the Nature and Treatment of Diabetes. Part I. On the Detection of Sugar: Qualitative and Quantitative Analysis"* (17). Pavy describes his modification of the Fehling's solution in order to make it into a quantitative assay. Pavy reasoned that as urine containing glucose was added to the Fehling reagent, the blue color of the cupric ion (from copper sulfate) would disappear as it was converted to cuprous ion (cupric oxide). The more glucose in the urine the less urine has to be added to a standard volume of Fehling's reagent before all the cupric ion is converted to cuprous ion, the blue color disappearing. But the red cupric oxide precipitate that forms hides observation of the blue color. Thus, Pavy added an ammoniacal salt to the Fehling reagent to prevent precipitation of cuprous oxide. Without the precipitation of the red colored cuprous oxide, it is easier to determine the disappearance of the blue color of the cupric ion (28). Pavy used his modified Fehling's test to analyze glucose in blood and urine as well as the livers of deceased diabetics.

Benedict's Solution

In 1907, Stanley Rossiter Benedict (1889–1936) modified Fehling's solution and Pavy's modification of it by using sodium citrate for tartrate and sodium carbonate for sodium and potassium hydroxide. The following recipe was formulated by Benedict (29–31):

Copper sulfate	17.3 g
Sodium citrate	173.0 g
Sodium carbonate (anhydrous)	100.0 g
Distilled water	1000 ml

With the aid of heat, dissolve the sodium citrate and carbonate in about 600 ml of water. Pour (through a folded filter if necessary) into a graduate and make up to 850 ml. Dissolve the copper sulfate in about 100 ml of water. Pour the carbonate-citrate solution into a large beaker or casserole and add the copper sulfate solution slowly, with constant stirring. The mixture is ready for use.

Procedure: To about 5 ml of the reagent in a test tube are added 8 (not more) drops of the urine to be examined. The fluid is then heated to boiling, kept at this temperature for from 1 to 2 min, and allowed to cool spontaneously. In the presence of dextrose, the entire body of the solution will be filled with a precipitate, which may be red, yellow, or green. If the amount of dextrose is small, the precipitate forms only upon cooling. If no dextrose is present, the solution either remains absolutely clear or a very faint turbidity, due to precipitated urates, may be apparent. Even small quantities of dextrose in urine (0.1%) yield precipitates of surprising bulk with this reagent and the positive reaction consists in the filling of the entire body of the solution with a precipitate so that the solution becomes opaque.

Benedict's substitution of citrate for tartrate holds the cupric hydroxide in solution as did the ammonium salts used by Pavy. The advantage over Pavy's modification of Fehling's solution is that sodium citrate does not evaporate because of boiling as do ammonium salts. His substitution of carbonate for potassium or sodium hydroxide promotes a greater sensitivity to the reagent.

As analytical chemists improved the sensitivity of the glucose assays, attention turned from urine to blood. Urine had some substances that interfered with the current assays including "albumen" (albumin). Albumin was mostly removed in the aforementioned assays by the boiling step. But the assay of blood glucose remained a challenge because of higher concentrations of proteins which interfere in these assays, other blood constituents such as uric acid that interfere in the assay, and the low total amount of glucose in the blood relative to the quantity of blood that could be sampled. But, as we will learn in a later chapter, glycosuria is an insensitive marker for diabetes. Otto Folin and Hsien Wu in 1919 made a significant breakthrough in the analysis for glucose in the blood.

Folin–Wu Determination of Blood Glucose

Otto Folin (1867–1934) and Hsien Wu (1893–1959) published (32, 33) in 1919–1920 a method that included a protein precipitation step. The protein precipitation method appears as follows:

Transfer a measured amount of blood into a flask having a capacity of fifteen to twenty times that of the volume taken. Dilute the blood with 7 volumes of water and mix. With an appropriate pipette add 1 volume of 10 percent solution of sodium tungstate ($Na_2WO_4 \cdot 2H_2O$) and mix. With another suitable pipette add to the contents in the flask (with shaking) 1 volume of 2/3 normal sulfuric acid. Close the mouth of the flask with a rubber stopper and give a few vigorous shakes.

Next, the precipitate is separated from the solution by filtration. The filtered solution is assayed for glucose. However, "The precipitate is very fine, yet does not go through good filter paper and does not stop up the pores. The filtration is slow ... If the precipitated mixture is heated in a water bath for 2 or 3 min, the precipitate settles spontaneously."

Russell L. Haden (1888–1952) in 1923 (34) modified the precipitation step in the Folin–Wu procedure to reduce the filtration time.

We have found that the procedure may be simplified by diluting 1 volume of blood with 8 volumes of N/12 sulfuric acid and then adding 1 volume of 10 percent sodium tungstate. The protein is precipitated in a very much more granular form than when the reagents are added in the manner directed by Folin and Wu, filtration is more rapid, a larger amount of filtrate is obtained, and the filtrate is more nearly neutral.

Shortly after this, other means of precipitating proteins from blood were advanced. For example, in the Nelson approach (35), the "blood proteins are precipitated by means of a mixture of zinc sulfate and barium hydroxide, which produces zinc hydroxide. The two reagents are so balanced that no excess of either remains in solution. The barium sulfate and zinc hydroxide are both insoluble, thus allowing a complete removal of proteins with no interference by excess foreign reagents or change in pH. The glucose is determined by heating with alkaline copper reagent, and the cuprous oxide formed is treated with an arsenomolybdate reagent, which is reduced to a colored compound whose intensity is proportional to the cuprous ion and thus to the glucose present (36)."

By 1860, the method of testing urine for glucose was sufficient to make it the principal procedure for the diagnosis of diabetes. By 1890, urinary tests for albumin and glucose were well under way at New York Hospital, one of New York City's oldest and progressive hospitals (27). By 1923, the assay for glucose in the urine and blood was an established approach to diagnosing diabetes. As far as the cause of diabetes is concerned, we have advanced from the kidneys to the liver to the pancreas as the site of the disturbance that leads to the disease. Mering and Minkowski defined the pancreas as the culprit organ. It was further pinned down to the site in the pancreas known as the islets of Langerhans. At this time in biochemical history, hormones were being characterized and understood. Thus, in 1916, Sharpley–Schafer suggested that the islets of Langerhans produced a glucose-regulating hormone. And although that hormone had as yet not been isolated or characterized, he named it insulin. (17, 37)

Summary Box 2.5

- Understanding disease is accompanied by major analytical advances.
- The earliest glucose determinations were taste, fermentation tests, and evaporation of the urine to yield sugar crystals.
- The earliest chemical test was Moore's test, which produced a brown color if sugar was present in the sample. It utilized a very basic solution.
- The red color produced by cupric oxide from cuprous salts was the basis of Trommer's test for glucose and all other tests that followed. The changes that followed were to make the tests for glucose more sensitive and applicable to blood samples.
- Barreswil added potassium tartrate and sodium carbonate to essentially Trommer's solution to analyze for glucose.
- Fehling added sodium hydroxide and potassium bitartrate to the solution in place of potassium hydroxide and potassium tartrate.
- Pavy added an ammoniacal salt to the Fehling's test to keep the red colored cupric oxide in solution so that it was easier to see the blue color of the cuprous ion. Pavy's test is a titration assay that looks for the disappearance of the blue color.
- Benedict substituted citrate for tartrate to keep the cupric ion in solution.
- Folin–Wu made blood sugar feasible by a precipitation step using sodium tungstate.
- Haden modified the Folin–Wu precipitation step to reduce the filtration time. Nelson followed with a modification that increased the quantity of proteins precipitated.

BANTING, BEST, AND MACLEOD

Sir Frederick Grant Banting (1891–1941) and John James Rickard MacLeod (1876–1935) received the 1923 Nobel Prize in Physiology or Medicine for their isolation of insulin. The person forgotten by the awards committee was a second-year medical student, Charles Harold Best (1899–1978). However, the isolation of insulin is often described in the literature as belonging to Banting and Best. Banting was so bothered by the exclusion of Best by the awards committee that he voluntarily gave him one-half of his monetary award. The research was conducted at the University of Toronto, Canada. Banting, a Canadian physician, was born in Ontario, Canada. MacLeod was a physician from Clunie, Perthshire, Scotland, and Best was born in West Pembroke, Maine (17, 38). Numerous investigators had previously failed in the preparation of pancreatic extracts. Their principal error was that digestive enzymes from the "acinous" cells[33] (cells

---

[33]The acinous cells are presently called *exocrine cells*.

that produce proteolytic enzymes) as contrasted to "insular" cells[34] (the islets of Langerhans that produce insulin) destroyed the insulin when the extract was produced. Banting, Best, and MacLeod made two principal modifications to previous attempts:

- Ten weeks after ligation of the pancreatic ducts the pancreas is removed. The acinous cells degenerate before the insular cells during the 10-week interval.
- The pancreas is maintained at low temperatures after removal and during the production of the extract.

The procedure was as follows:

- Under general anesthesia the pancreas is ligated[35] and then allowed to degenerate for 10 weeks.
- The dog was given a lethal dose of chloroform. The degenerated pancreas was swiftly removed and sliced into chilled mortar containing Ringer's solution. The mortar was placed in freezing mixture and the contents partially frozen. The half frozen gland was then completely macerated. The solution was filtered through paper and the filtrate, having been raised to body temperature, was injected intravenously.

In the course of our experiments we have administered over seventy-five doses of extract from degenerated pancreatic tissue to ten different diabetic animals. Since the extract has always produced a reduction of the percentage sugar of the blood and of the sugar excreted in the urine, we feel justified in stating that this extract contains the internal secretion of the pancreas (39).

Leonard Thompson

Leonard Thompson was a diabetic who, at the age of 14 years, was very close to entering a coma when he received an injection of Banting's and Best's crude extract. His blood glucose levels decreased but he showed an allergic reaction. Twelve days later, he received an injection of purified insulin. His health improved and he lived 13 additional years until he died from pneumonia, a complication of diabetes. Thompson was the first diabetic to receive injections of insulin. The medical community looked upon this as a great advance in the treatment of diabetes. Until then the diagnosis of diabetes was quite dire, with death ensuing within a few months.

---

[34]The insular cells are presently called *endocrine cells*.
[35]Ligature involves using a thread or wire to bind or tie a vessel such as an artery, vein, ducts, intestine, etc.

John Jacob Abel

John Jacob Abel (1857–1938) was a biochemist born in Cleveland, Ohio. He was trained at the University of Michigan and Johns Hopkins University and received his MD from the University of Strasbourg. He was Professor of Pharmacology and Director of the Endocrine Laboratory at Johns Hopkins. He had already isolated epinephrine when he turned his attention to insulin. He was the first to isolate crystalline insulin, although his work was not accepted by a skeptical scientific community until the middle 1930s. A description of his procedure follows (40):

> One gram, approximately, of the so-called Fraction IV[36], is dissolved in a little more than the required volume of N/6 acetic acid, enough water is added to bring the volume up to 60 cc. or thereabout, and the contaminating substances (together with some insulin) are then precipitated by the addition of an acidulated solution of brucine[37] containing 6 g of brucine in 95 cc. of N/6 acetic acid. The resulting clear supernatant fluid which contains nearly pure insulin is separated from the precipitate by centrifugalization[38]. Insulin remaining in the precipitate may be removed by dissolving in N/6 acetic acid and precipitating with the brucine solution as before. ... The clear colorless centrifugalate containing the insulin is then precipitated with N/6 pyridine and the precipitate and fluid immediately centrifuged. The precipitate which settles out is largely crystalline in character, the sides of the tube are found to be coated with glistening highly refractive crystals and the topmost layer of the precipitate consists of similar crystals. It is not a difficult matter to remove this topmost layer of crystals by means of a rubber-mounted pipette and to free them from adherent pyridine and acetic acid by frequently washing them with distilled water at room temperature in which medium pure crystalline insulin is quite insoluble.

The purification and crystallization of insulin made it possible to study the molecule in detail. Pure, crystalline insulin led to Frederick Sanger's determination of the amino acid sequence for insulin.

Frederick Sanger

Frederick Sanger (1918–), born in Rendcomb, England, was an English biochemist who is the only scientist to win two Nobel Prizes in chemistry. The first was in 1958 for his work on the amino acid sequence of insulin and the second in 1980 for the determination of base sequences in nucleic acids. He received his PhD from Cambridge University. Previous investigators had determined the amino acid composition of insulin. Acid hydrolysis of a protein yields a hydrolysate that is analyzed both qualitatively and quantitatively by ion-exchange chromatography. The amino acid composition of human insulin is given here. It may be useful to refer to Figure 4.1 for the following information.

---

[36] In an earlier paper, Abel and his associates had described the preparation of a rabbit insulin fraction called *Fraction IV*.

[37] Brucine is an alkaloid with the formula, $C_{23}H_{26}N_2O_4$.

[38] Centrifugalization is synonymous with centrifugation.

The amino acid composition of human insulin

Amino acids	Residues	Amino acids	Residues
Alanine	1	Leucine	6
Arginine	1	Lysine	1
Asparagine	3	Phenylalanine	3
Cystine	6	Proline	1
Glutamate	4	Serine	3
Glutamine	3	Threonine	3
Glycine	4	Tyrosine	4
Histidine	2	Valine	4
Isoleucine	2		
		Total	51

Fragmentation involves the cleavage of the protein at several points along the polypeptide chain in order to produce low molecular weight polypeptide fragments. Sanger used di-, tri- and tetrapeptides as they are more easily characterized than the higher molecular weight polypeptide fragments.

Fragmentation of insulin by Sanger was accomplished by the use of enzymes and chemical reagents that hydrolyze the polypeptide chain at specific peptide bonds. Cleavage of peptide bonds is shown here:

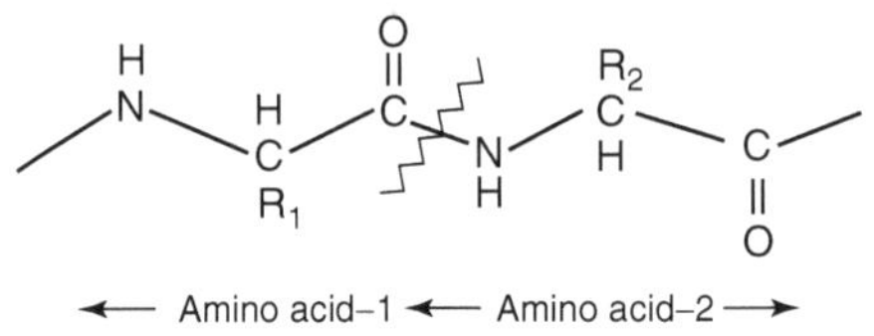

← Amino acid–1 ← Amino acid–2 →

The enzymes and chemicals shown hydrolyze the peptide bond at the carboxyl end of the amino acids listed.

Reagent	Amino acid-1
Pepsin	Phe, Tyr, Trp, Leu
Chymotrypsin	Try, Phe, Trp, Leu
Trypsin	Arg, Lys
Cyanogen bromide	Met

Treatment of the polypeptide by one of the above-mentioned reagents is followed by column chromatography, electrophoresis, etc. to separate the fragments produced. Large fragments are hydrolyzed even further.

Identification of the N-terminal amino acid in each fragment is determined by reacting the fragment with 1,2,4-fluorodinitrobenzene as shown. The dinitrophenyl (DNP) group attaches at the free amino groups of the peptide. The DNP-peptide is hydrolyzed with acid, thus freeing the N-terminal amino acid as the DNP derivative. DNP-amino acids are colored yellow and can be separated from the amino acids that are underivatized using ether and subsequently can be separated by paper chromatography.

Fluorodinitrobenzene (DNFB)

Protein---aa-NH$_2$ $+$
(N-terminal amino acid)

F / NO$_2$ ... NO$_2$ → Protein---aa–NH- ⟨O⟩- NO$_2$ —Acid→
NO$_2$

NO$_2$

aa-NH –⟨O⟩–NO$_2$ + aa$_1$, aa$_2$, aa$_3$ ....etc.

Sanger's strategy for the determination of the amino acid sequence of bovine insulin is shown using the above-mentioned techniques (41).

1. Sanger and his associates separated the A and B subunits by treatment with performic acid. This converts cystine disulfide bonds to
   cysteine sulfate.
2. Complete hydrolysis of the A-subunit followed by paper and column
   chromatography of the hydrolysate yielded the composition of the sub-unit.
3. Treatment with DNFB followed by hydrolysis identified the N-terminal
   residue as glycine. Carboxypeptidase treatment identified the C-terminal
   amino acid as aspartic acid.
4. Partial acid hydrolysis of subunit A gave the following fragments.
   The useful di-, tri-, and tetra peptides are arranged under the structure
   found for subunit A.

Subunit A –Beef Insulin

N-terminal   C-terminal
Gly. Ilu. Val. Glu. Glu. Cys. Cys. Ala. Ser. Val. Cys. Ser. Leu. Tyr. Glu. Leu. Glu. Asp. Tyr. Cys. Asp

Peptides isolated
from the acid hydrolysis

Tripeptide I	Glu.Glu.Cys
Tripeptide II	Glu.Cys.Cys
Tripeptide III	Cys.Cys.Ala
Tripeptide IV	Ser.Val.Cys
Tripeptide V	Ser.Leu.Tyr
Tripeptide VI	Leu.Tyr.Glu
Tripeptide VII	Glu.Leu.Glu
Tripeptide VIII	Glu.Asp.Tyr
Dipeptide I	Cys.Asp
Dipeptide II	Tyr.Cys

Gly.Ilu.Val.Glu.Glu   Pentapeptide I

5. Hydrolysis of subunit A with pepsin results in cleavage at the peptide
   bonds indicated by arrows.
   A polypeptide composed of 13 amino acids was isolated from the pepsin
   digestion of subunit A. In retrospect this polypeptide was probably
   Gly. Ilu. Val. Glu. Glu. Cys. Cys. Ala. Ser. Val. Cys. Ser. Leu. Acid hydrolysis of
   this polypeptide yielded a tripeptide which was sequenced as Ser. Val. Cys.
   Also isolated from the hydrolysate was the dipeptide Ser. Leu.
   This last information allowed Sanger to complete the amino acid
   Sequence of subunit A. Subunit B was analyzed in a similar manner.

Pedro Cuatrecasas

Pedro Cuatrecasas (1936–) was born in Madrid, Spain, and earned his doctorate and medical degrees from Washington University in St. Louis. Cuatrecasas was Vice-President of Research and Development at Burroughs Wellcome, Director of Glaxo, President of the Research Division of Parke-Davis and Vice-President for Warner Lambert. In 2004, he was appointed to the Advisory Board of Aethlon Medical Scientific. While at these pharmaceutical companies, Cuatrecasas was involved in the discovery of 40 drugs including AZT for treatment of AIDS and atorvastatin (Lipitor). Cuatrecasas has so far authored over 400 publications. He is known for his discovery of affinity chromatography and using this novel technique for the isolation and characterization of the insulin receptors of fat and liver cells (42, 43).

Cuatrecasas proved the hypothesis that insulin conducted its physiological effects by binding to receptors on the surfaces of certain types of cells. A cell that bound a specific hormone was called the *target cell* for that hormone. Insulin was not an exception. Structures capable of binding insulin were first identified in adipose cells. Cuatrecasas and associates radiolabeled insulin with I-125, incubated the labeled insulin with intact fat cells and found that the I-125 insulin bound to the fat cells (42).

But the ultimate proof that insulin binds to receptor sites was proved a year later by the use of affinity chromatography. In order to fully understand the achievement of Cuatrecasas, we need to know what affinity chromatography entails. Affinity chromatography is a powerful procedure for purifying proteins. The technique takes advantage of the binding of proteins to specific chemicals such as coenzymes and allosteric modulators. These molecules are called *ligands*. In order to separate a specific protein, its ligand is covalently attached to a large molecular weight compound such as agarose (ligand-agarose is called the *affinity adsorbent*). A mixture of proteins is passed through a column containing the affinity adsorbent. The specific protein that is isolated binds to the ligand of the adsorbent while the rest of the proteins pass through the column. This protein binds to the ligand by ionic forces and hydrogen binding and is easily displaced by an appropriate buffer solution.

In Cuatrecasas' procedure for the isolation of the liver insulin receptor, the affinity adsorbent was insulin attached by its N-terminal amino acid to an agarose backbone. A liver homogenate was centrifuged and the pellet resuspended for further centrifugation. The final pellet was proved to contain insulin-binding activity by incubating I-125 insulin with a suspension of the pellet. The supernatants from the centrifugations were shown to contain no significant insulin-binding activity. A suspension of the final pellet was passed through a column containing the affinity adsorbent. In Cuatrecasas' experiments,

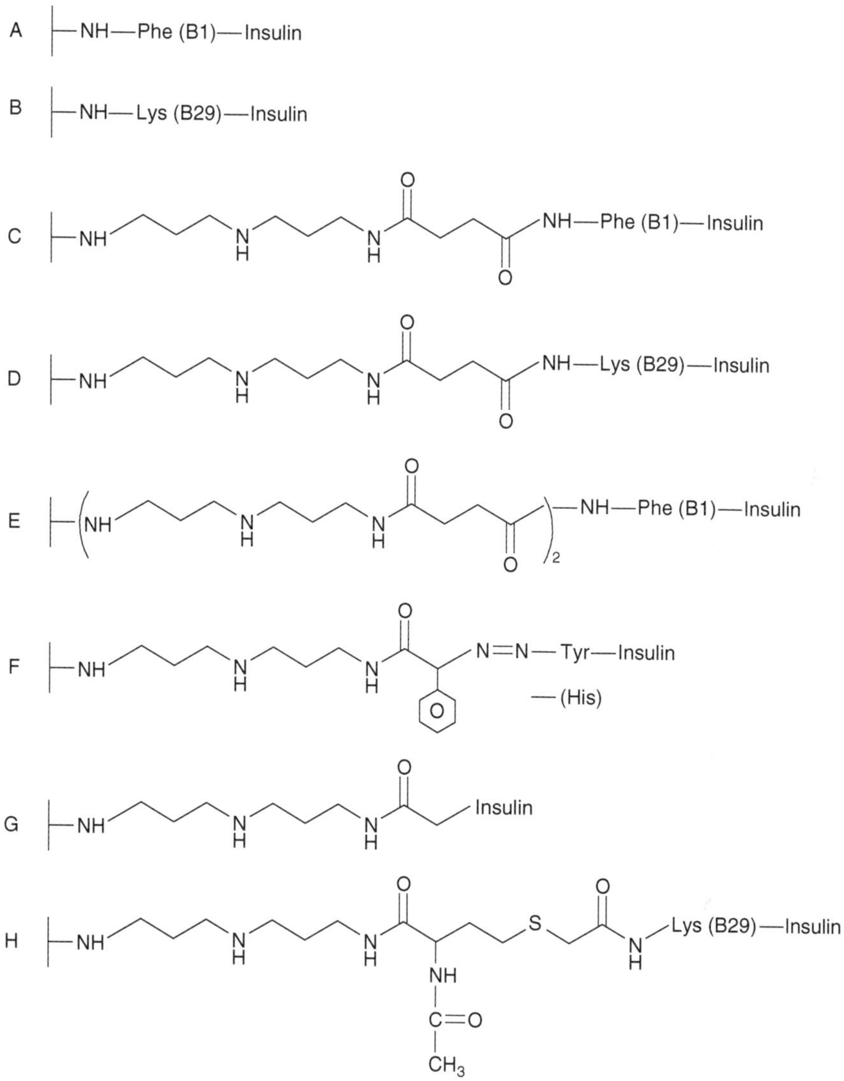

Figure 2.1   This figure shows the structures of the eight linkers (spacers) used as adsorbents linked to the agarose backbone in Cuatrecasas' isolation of the insulin receptor.

the affinity adsorbent consisted of one of eight adsorbents as shown in Figure 2.1.

As you can see from the figure, there is a "spacer" or "linker arm" that separates the insulin molecule from the agarose backbone. This is necessary for the efficient isolation of the insulin receptor; C, D, E and H being the most efficient. The insulin-binding receptor is eluted from the column by buffers at low pH and containing urea. The insulin receptor was purified 250,000-fold with respect to the liver homogenates. The importance of Cuatrecasas' isolation of the liver of the insulin receptor will be seen in Chapter 4 when we discuss the insulin signaling pathway.

In this chapter I have presented the history of diabetes from ancient times to the early 1970s. In subsequent chapters I will discuss the discoveries and inventions that take us to our understanding of diabetes as it stands in the twenty-first century. We have come a long way from the initial recognition of diabetes. In ensuing chapters the advances in the understanding of this disease will be discussed. How glucose is metabolized and regulated, how insulin effects the regulation of glucose levels, the pathophysiology of diabetes, its complications, how diabetes is diagnosed, types of diabetes, and modern-day methods of treating diabetes will be elaborated on in later chapters.

Summary Box 2.6

- Sharpley–Schafer named the hormone produced by the pancreas that regulates glucose, insulin even though it had not as yet been isolated.
- Banting, Best, and MacLeod used a procedure that involved the degeneration of the acinous cells. They administered the extract to diabetic dogs and reported a reduction in blood and urine glucose.
- Leonard Thompson was the first diabetic to be injected with the crude insulin.
- Abel was the first to crystal insulin.
- Sanger was the first to determine the structure of insulin.
- Cuatrecasas isolated by affinity chromatography pure samples of the insulin receptor sites from adipose and liver tissues.

QUESTIONS AND CROSSWORD PUZZLE

2.1 Suppose you are a physician living in 1800. You diagnose a patient as suffering from diabetes. Using the best knowledge of your time how would you treat her? What is the likely prognosis?

History of Diabetes

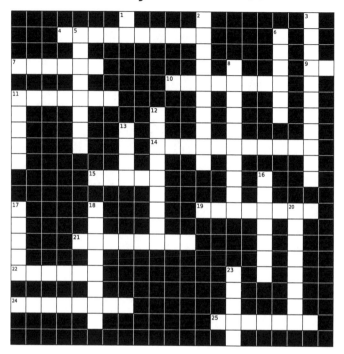

Across

4 Indian word for diabetes.
7 Added the name mellitus to diabetes.
9 Wrote the earlist known Chinese medical book.
10 Ancient Egyptian who named diabetes.
11 Shared the 1923 Nobel Prize for the isolation of insulin.
14 Pre-historic individuals that have a gene corresponding that is associated with diabetes.
15 Advised a high meat, low carbohydrate diet for diabetics.
19 Was a Indian scientist who recognized the presence of two different types of diabetes.
21 Added citrate and substituted sodium carbonate for sodium and potassium hydroxide in Fehring's solution.
22 Determined the amino acid sequence of insulin.
24 An Indian surgeon who taught and practiced medicine in the 5th century.
25 Modified Barreswil's test for glucose.

Down

1 Added a precipitation step to glucose determination.
2 Stated that destruction of the pancreas results in diabetes.
3 Proved the hypothesis that insulin binds to a receptor site in target cells.
5 Latinized name for Ibn Sina
6 Defined diabetes as a systemic condition.
8 The father of toxicolgy.
11 The first chemical test for glucose in the urine.
12 Established beyond that the pancreas was involved in diabetes.
13 First to crystallize insulin.
16 Believed that the breakdown of glycogen in the liver resulted in diabetes.
17 First known reference to diabetes.
18 Identified the sugar of diabetic urine as grape sugar.
20 First diabetic to be treated with insulin.
23 Served as the physician to Marcus Aurelius.

REFERENCES

1. Carpenter S, Riguad M, Barile M, et al. An interlinear transliteration and English translation of portions of the ebers papyrus: possibly having to do with diabetes mellitus. Bard College, Annendale-on-Hudson (NY); 2006. Based on the hieratic to hieroglyphic transcription by Walter Wreszinski. Leipzig; 1913. Available at biology.bard.edu/ferguson/course/bio407/Carpenter_et_al_(1998).pdf.

2. Loriax DL. Diabetes and the Ebers papyrus: 1552 BC. Endocrinologist 2006; 16:55.

3. Gibbons A. Close encounters of the prehistoric kind. Science 2010;328:680.

4. Green RE, Krause J, Briggs AW, et al. A draft sequence of the Neandertal genome. Science 2010;328:710.

5. Krall LP, Levine R, Barnett D. The history of diabetes. In: Weir GC, Kahn CR, editors. *Joslin's Diabetes Mellitus. s.l.* Vol. 1. Philadelphia: Lea & Febiger; 1994. p. 1.

6. Henschen F. On the term diabetes in the works of Aretaeus and Galen. Med Hist 1969;13:190.

7. Gemmill CL. The Greek concept of diabetes. Bull N Y Acad Med 1972;48:1033.

8. Eknoyan G, Nagy J. A history of diabetes mellitus or how a disease of the kidneys evolved into a kidney disease. Adv Chronic Kidney Dis 2005;12:223.

9. Bhishagratna KK An English translation of the Sushruta Samhita with A full and conprehensive introduction, additional texts, different readings, notes, comparative views, index, glossary and plstes. Calcutta: The Author 1911. 3. www.archive.org/details/englishtranslati00susruoft/englishtranslat00susruoft_djvu.txt.

10. Madineh SMA. Avicenna's Canon of Medicine and modern urology Part III: Other bladder diseases. Urol J 2009;6:138.

11. Lo V. Healing and Medicine. In: Shaughnessy EL, editor. *China; Empire and Civilization*. Vol. 11. New York: Oxford University Press; 2000. p. 148.

12. Veith I. *The Yellow emperor's Classic of Internal Medicine*. Los Angeles: University of California Press; 1949.

13. Selin H. *Encyclopaedia of the History of Science, Technology, and Medicine in Non-Western Culture*. Dordrecht: Kluwer Academic Publishers; 1997. p. 698–699.

14. Hsia E, Veith I, Geertsma R, editors. *The essentials of medicine in ancient china and japan*. Leiden: E.J. Brill; 1986. Yasuyori Tamba's ISHIMPO [trans.].

15. Salzberg HW. *From Caveman to Chemist: Circumstances and Achievements*. American Chemical Society: Washington; 1991.

16. Molnár Z. Thomas Willis (1621–1675), the founder of clinical neuroscience. Nat Rev Neurosci 2004;5:329. DOI: 10.1038/nrn1369.

17. J Dallas. Diabetes, Doctors and Dogs; An exhibition on diabetes and endocrinology by the college library for the 43rd St.Andrew's day festival symposium. Rare Books, The Sibbald Library. s.l.: Royal College of Physicians of Edinburgh; 2011. http://www.rcpe.ac.uk/library/exhibitions/diabetes.php.

18. Keck FS, Duntas LH. Brunners missing aha experience delayed progress in diabetes research for 200 years. Hormones 2011;6:251.

19. Kaplan JM, First MR. Renal Function. In: Pesce AJ, Kaplan LA, editors. *Clinical Chemistry: Theory, Analysis, and Correlation*. Vol. 5. St. Louis: Mosby; 2010. p. 571.

20. Witter LE, Luciano M, Williams C et al. Diabetes detectives s.l. Dartmouth Med 2008;33:36–41. dartmed.dartmouth.edu

21. Chevreul ME. Note sur le sucre de diabetes. Ann Chim 1815;95:319–320.

22. Young FG. Claude Bernard and the discovery of glycogen. Br Med J 1957;1:1431.

23. Jorgens V. Oskar Minkowski (1885–1931). An outstanding master of diabetes research. Hormones 2006;5:310.

24. Joseph von Mering (1849–1949). N Engl J Med 1949;240:699 Editorial.

25. Beale L. A course of lectures: Urine, urinary deposits and calculi. Delivered at the pathological laboratory, during the session 1857–58. Br Med J 1860:501–503.

26. Johnson G. Clinical lecture on the various modes of testing for sugar in the urine. Br Med J 1884;1:1–4.

27. Bolduan C. Public Health Problems. Bull N Y Acad Med 1933;9:523–531 Discussion.

28. Pavy FW. *Researches on the Nature and Treatment of Diabetes. Part 1 On the Detection of Sugar: Qualitative and Quantitive Analysis*. London: John Churchill; 1862.

29. Benedict SR. The detection and estimation of reducing sugars. J Biol Chem 1907;3:101.

30. Benedict SR. A note on the reduction of alkaline copper solutions by sugars. Biochem J 1907;2:408.

31. Benedict SR. A reagent for the detection of reducing sugars. J Biol Chem 1909;5:485.

32. Folin O, Wu H. A system of blood analysis. J Biol Chem 1919;38:81.

33. Folin O, Wu H. A system of blood analysis. Supplemental I. A simplified and improved method for determination of sugar. J Biol Chem 1920;41:367.

34. Haden RL. A modification of the Folin–Wu method for making protein-free blood filtrates. J Biol Chem 1923;56:469.

35. Nelson N. A photometric adaptation of the Somogyi method for the determination of glucose. J Biol Chem 1944;153:375.

36. Harrow B, Borek E, Mazur A, Stone GCH, Wagreich H. *Laboratory Manual of Biochemistry*. 4th ed. New York: WB Saunders; 1955.

37. Sharpey-Schafer EA. *The Endocrine Organs: An Introduction to the Study of Internal Secretion*. London: Longmans, Green and Co; 1916.

38. Banting FG, Best CH. The internal secretion of the pancreas. Ind J Med Res 1922;7:251.

39. Gabriel ML, Fogel S, editors. *Great Experiments in Biology*. Englewood Cliffs: Prentice-Hall; 1955. p. 64.

40. Abel JJ. Crystalline insulin. Proc Natl Acad Sci 1926;12:132.

41. Sanger F. Chemistry of insulin. Science 1959;129:1340.

42. Cuatrecasas P. Insulin–receptor interactions in adipose tissue cells: direct measurement and properties. Proc Natl Acad Sci 1971;68:1264.

43. Cuatrecasas P. Properties of the insulin receptor isolated from liver and fat cell membranes. Proc Natl Acad Sci USA 1972;247:1980.

CHAPTER 3

A PRIMER: GLUCOSE METABOLISM

PROLOG

Beauty is truth,
truth beauty—that is all
Ye know on earth,
And all ye need to know.

—John Keats (1795–1821)

From time to time during the discussion in this chapter, the author will identify "beautiful concepts" encompassed in the metabolic pathways that are presented. Jules Henri Poincare, France's premier mathematician-physicist, wrote[1]: "The scientist does not study nature because it is useful to do so. He studies it because he takes pleasure in it, and he takes pleasure in it because it is beautiful. If nature were not beautiful it would not be worth knowing, and life would not be worth living. I am not speaking, of course, of the beauty which strikes the senses, of the beauty of qualities and appearances. I am far from despising this, but it has nothing to do with science. What I mean is that more intimate beauty which comes from the harmonious order of its parts, and which a pure intelligence can grasp."

---

[1] Science et method, Part 1. Ch.1: The Selection of Facts. 1908. p. 22.

Understanding Diabetes: A Biochemical Perspective, First Edition. Richard F. Dods.
© 2013 John Wiley & Sons, Inc. Published 2013 by John Wiley & Sons, Inc.

The metabolic pathways that we are going to study are beautiful because they are as elaborate as any edifice built by humans; they are complex, and they fit elegantly the purpose that they are supposed to accomplish. An excellent example of a "beautiful concept" is pyruvate dehydrogenase. This multicomponent system is an important enzyme system in the tricarboxylic acid cycle. It is composed of 60 protein units that are organized into three enzymes. Structurally, the complex is arranged in 12 pentagonal sides, creating a dodecahedron. The 60 protein complexes are arranged exactly in the same way in each pyruvate dehydrogenase. Amazing! Further, this complex is extremely efficient; each enzyme component resides close to the other two so that the product of one enzyme can be transferred as the substrate for the next enzyme. You must admit there is a beauty in this.

Giovanni Vignale summarized it all with the words[2]: "A good scientific theory is like a symbolic tale, an allegory of reality."

It is the destiny of wine to be drunk, and it is the destiny of glucose to be oxidized. But it was not oxidized immediately: its drinker kept it in his liver for more than a week, well curled up and tranquil, as a reserve aliment for a sudden effort; an effort that he was forced to make the following Sunday, pursuing a bolting horse. Farewell to the hexagonal structure: in the space of a few instants the skein was unwound and became glucose again, and this was dragged by the bloodstream all the way to a minute muscle fiber in the thigh, and here brutally split into two molecules of lactic acid, the grim harbinger of fatigue: only later, some minutes after, the panting of the lungs was able to supply the oxygen necessary to quietly oxidize the latter. So a new molecule of carbon dioxide returned to the atmosphere, and a parcel of the energy that the sun had handed to the vine-shoot passed from the state of chemical energy to that of mechanical energy, and thereafter settled down in the slothful condition of heat, warming up imperceptibly the air moved by the running and the blood of the runner. 'Such is life,' although rarely is it described in this manner: an inserting itself, a drawing off to its advantage, a parasitizing of the downward course of energy, from its noble solar form to the degraded one of low-temperature heat. In this downward course, which leads to equilibrium and thus death, life draws a bend and nests in it.

—Primo Levi, an Italian chemist, who survived Auschwitz from his book "The Periodic Table," Schrocken Books, NY, 192–193, 1975, translated by R. Rosenthal, 1984.

THE CARBOHYDRATES AND THEIR FUNCTION

Dietary carbohydrates and, under certain conditions, endogenous stores of carbohydrates (polymers of glucose called *glycogen*) provide the energy needed by living organisms to carry out the many complex mechanical, chemical, and electrical processes that constitute life. This chapter describes the importance of glucose as an energy source. Glucose enters the blood from the digestion

[2]The Beautiful Invisible: Creativity, Imagination, and Theoretical Physics. Oxford University Press; 2011.

of polysaccharides and disaccharides. Glucose is a monosaccharide. Poly-, di-, and monosaccharides are members of a class of biomolecules termed *carbohydrates*. Carbohydrates are chemically defined as polyhydroxyl aldehydes and ketones. Glucose is a six-carbon polyhydroxyl aldehyde. A common polysaccharide found in food is starch and a common disaccharide is sucrose, also known as *table sugar.*

Starch comprises more than half of the carbohydrate ingested by humans. There are two forms of starch, amylose and amylopectin. Amylose is an unbranched polymer of glucose, with each glucose molecule connected to the other by an $\alpha(1 \rightarrow 4)$ linkage. Amylopectin is a branched form of glucose molecules linked together by $\alpha(1 \rightarrow 4)$ and $\alpha(1 \rightarrow 6)$ links in the ratio of $30:1$. The $\alpha(1 \rightarrow 4)$ and $\alpha(1 \rightarrow 6)$ bonds are depicted here:

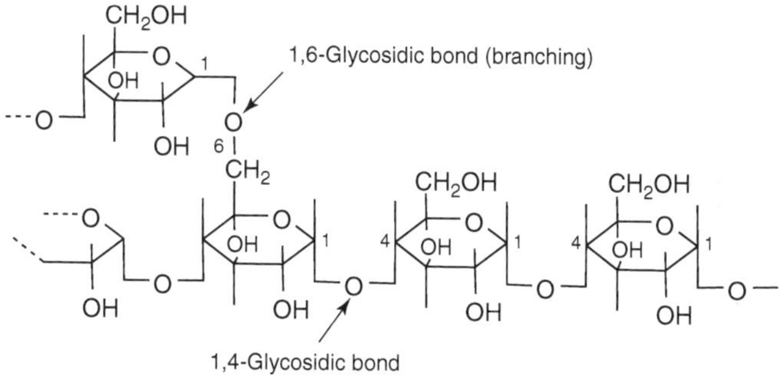

In addition, many macromolecules (gangliosides, cerebrosides, nucleic acids, and glycoproteins) contain a significant quantity of carbohydrate. The carbohydrate moiety in these macromolecules contributes to their special properties.

Starch and sucrose are digested to glucose and fructose. In fact, the digestion begins almost immediately in the mouth. Moreover, there rests the basis for the accurate parental admonition to chew your food thoroughly before swallowing.

DIGESTION AND ABSORPTION OF CARBOHYDRATES

Salivary and Pancreatic Amylase

First, let us look at the digestion and absorption of carbohydrates that were present in a starchy meal that you might have just eaten. You ingested, among other nutrients, a large complex array of polysaccharides. The mouth (or oral cavity) contains saliva. The saliva derives from three glands: submandibular (submaxillary), sublingual, and parotid. Saliva consists of 99% water; the other 1% is composed of electrolytes, mucus, and some enzymes such as salivary α-amylase, lysozyme, carbonic anhydrase, and peroxidase. Saliva has a pH equal to about 6.8. Salivary α-amylase is an important enzyme in the digestion of starch and other polysaccharides. It hydrolyzes the starch and other polysaccharides to short chains of glucose. The optimum pH for the action of α-amylase is 5.6–6.9 and, therefore, it is inactivated in the stomach by gastric juice, which has a pH as low as 1.0. Salivary α-amylase does not figure prominently in the digestion of carbohydrates.

Next, the gastric contents reach the small intestine. The alkaline content of the pancreatic and biliary secretions results in a pH equal to 7.8. The pancreatic secretion includes an amylase similar in properties and action to the salivary amylase previously discussed. Pancreatic α-amylase is the principal enzyme involved in the initial stages of the digestion of polysaccharides. It has an optimum pH of about 7.1. Thus, the small intestine α-amylase acts on polysaccharides such as starch. As shown subsequently, salivary and pancreatic α-amylases hydrolyze the $\alpha(1 \rightarrow 4)$ bond between the glucose units in polysaccharides to produce maltose, maltotriose, and α-dextrin.

Maltose is composed of two glucose molecules in $\alpha(1 \rightarrow 4)$ linkage, maltotriose consists of three glucose molecules in $\alpha(1 \rightarrow 4)$ linkage, and α-dextrin has both $\alpha(1 \rightarrow 6)$ and $\alpha(1 \rightarrow 4)$ links.

It should be emphasized that both salivary and pancreatic amylase do not simply cleave $\alpha(1 \rightarrow 4)$ bonds, an action that would produce glucose molecules. Amylase maintains an intact $\alpha(1 \rightarrow 4)$ bond adjacent to the bond that is hydrolyzed, thereby producing a disaccharide (two glucose residues linked together, maltose). As depicted earlier, the linkage cleaved is one glycoside bond from the end of the chain. The $\alpha(1 \rightarrow 6)$ glycoside bond is unaffected by amylase.

Humans lack enzymes that hydrolyze $\beta(1 \rightarrow 4)$ linkages. Thus, humans are incapable of digesting cellulose, a major constituent of vegetables that consists of glucose residues in $\beta(1 \rightarrow 4)$.

Disaccharidases

The disaccharidases are located in the brush border cells of the small intestine mucosa. They hydrolyze the disaccharides produced from the action of salivary and pancreatic α-amylase on polysaccharides. The three principal monosaccharides produced are glucose, fructose, and galactose. In addition, they act on disaccharides that were introduced directly by the diet.

Maltose and maltotriose are hydrolyzed to glucose by maltase. Isomaltase hydrolyzes the $\alpha(1 \rightarrow 6)$ bond to produce glucose. Sucrase hydrolyzes the $\alpha(1 \rightarrow 2)$ bond, resulting in glucose and fructose. Lactase hydrolyzes the $\beta(1 \rightarrow 4)$ bond to produce galactose and glucose.

Absorption

The monosaccharides glucose and galactose are absorbed across the intestinal epithelium by active transport, that is, transport into an area of higher concentration from a region of lower concentration. This process occurs by the binding of the monosaccharides with proteins present in the plasma membranes of the epithelial cells. These proteins change their conformation upon interaction with the monosaccharide and thus transfer it to the interior of the cell. The "binding" or "carrier" proteins are quite specific in their binding. A transport process that conducts molecules to an area of higher concentration necessitates an energy source. The energy necessary for transport is derived from the hydrolysis of adenosine 5′-triphosphate (ATP) by an epithelial-membrane-located adenosine 5′-triphosphatase (ATPase).

Fructose crosses the intestinal epithelium by a mechanism termed *facilitated transport*. Facilitated transport, unlike active transport, lacks an ATPase component. It conveys molecules from a high to a low concentration region in a manner similar to simple diffusion through a porous membrane. The rate of transport is greater than would be expected for a simple diffusion process. Transport proteins facilitate the movement of the sugar molecules across the membrane. Disaccharides are not transported across the intestinal epithelium. However, the observation of disacchariduria indicates that disaccharide absorption can occur in some instances. Fructose and galactose are converted to glucose in the liver.

In order to understand the diabetes disease state, we have studied the pathways in which glucose enters the blood from the food that we eat. Salivary and pancreatic amylase digests the starch present in food to smaller molecules, the disaccharides, whereupon the disaccharidases release glucose to the small intestine. The intestinal mucosa transports glucose via active transport to the blood. However, there is another source of glucose, an internal source called *glycogen*. Glycogen is a large, branched glucose polymer (molecular mass reported as high as 2.5×10^7 Da[3]) with $\alpha(1 \rightarrow 4)$ and $\alpha(1 \rightarrow 6)$ glycosidic bonds. Glycogen stored predominantly in muscle and liver cells undergoes a process termed *glycogenolysis* to send glucose into the bloodstream.

Summary Box 3.1

- Dietary carbohydrates and glycogen provide the energy for life processes.
- Glucose is derived from dietary polysaccharides and disaccharides and is the actual provider of energy.
- Glucose is a six-carbon polyhydroxyl aldehyde.
- Salivary α-amylase digests starch and polysaccharides.
- In the intestines, pancreatic α-amylase digests $\alpha(1 \rightarrow 4)$ linkages in di- and polysaccharides to maltose, maltotriose, and α-dextrin and disaccharidases

---

[3]Geddes R, Harvey JD, et al. The molecular size and shape of liver glycogen. Biochem J 1977;163:201.

hydrolyze disaccharides to monosaccharides (primarily glucose, fructose, and galactose).

- Active transport involves transport into higher concentration from lower concentration and derives its energy from ATP. Glucose and galactose cross the intestinal epithelium by active transport.
- Facilitated transport involves transport from high concentration to low concentration. Fructose crosses the intestinal epithelium by facilitated transport.

OVERVIEW OF GLUCOSE METABOLISM

Review the metabolism of glucose by examining Figure 3.1.

As stated earlier, the principal source of energy for human life processes is the monosaccharide, glucose. Glucose itself does not possess the high energy bonds necessary to drive chemical reactions. However, glucose metabolism results in the formation of a compound possessing high energy bonds. As indicated, the high energy molecule that serves as the immediate source of energy needed to drive endergonic reactions to completion is ATP. When there is an oxygen debt, that is, an inadequate oxygen supply, the metabolism of glucose to lactate results in the production of 2 mol of ATP per mole of glucose. This process is termed *anaerobic glycolysis*. In the presence of an adequate oxygen supply, aerobic glycolysis results in the formation of pyruvate, which next enters the tricarboxylic acid cycle.

Aerobic glycolysis produces 2 mol ATP/mol glucose; however, the metabolism of pyruvate to carbon dioxide and water by the tricarboxylic acid cycle and subsequent activation of the electron transport system results in the formation of an additional 34 or 36[4] mol ATP/mol glucose. The electron transport system is dependent on an oxygen source. The shunting of pyruvate to lactate provides a means by which the full energy potential of the glucose is conserved under conditions where the electron transport system is inoperable. The lactic acid cycle is responsible for the transport of lactate from oxygen-depleted tissues, for example, fatigued muscle, to the liver where it is converted back to glucose by gluconeogenesis. An alternative to this disposition of lactate is its reconversion to pyruvate when tissue oxygen levels have been restored. The pyruvate then enters the now reactivated tricarboxylic acid cycle and subsequently activates the electron transport system. It is obvious that the energy requirements of a living organism can be met only under conditions where there is an adequate supply of glucose.

In order to meet situations where there is a temporary depletion of circulating glucose, humans store glucose in a readily available form. This form is the glucose polymer, glycogen. Glycogenesis and glycogenolysis are the metabolic pathways that synthesize and catabolize glycogen, respectively. Another oxidative pathway for glucose is the phosphogluconate oxidative pathway.

---

[4] The total ATP formation (36 or 38 mol) will be discussed later in this chapter.

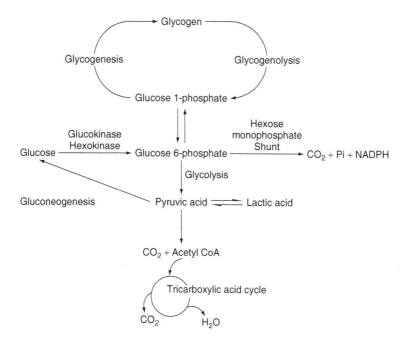

NADPH = Nicotinamide adenine
        dinucleotide phosphate (reduced form)
Pi = Inorganic phosphate

Figure 3.1   Pathways for glucose 6-phosphate.

Unlike the tricarboxylic acid pathway, the phosphogluconate oxidative pathway neither utilizes nor produces ATP. However, its importance lies in its reduction of nicotinamide adenine dinucleotide phosphate (NADP) to produce NADPH, a cofactor necessary for many biochemical transformations and the synthesis of several important monosaccharides.

As ATP and glucose 6-phosphate play prominent roles in these metabolic pathways, the chemical properties and functions of these molecules will be discussed before proceeding to a description of the individual pathways.

ADENOSINE 5′-TRIPHOSPHATE (ATP)

As you can see, ATP is composed of the purine, adenine, the carbohydrate, D (−) ribose and three phosphates covalently linked in a linear manner to the 5′ hydroxyl position of the sugar.[5]

---

[5]The prime notation (′) placed next to a numeral is used to distinguish a carbon position on the sugar moiety from a carbon on the purine ring, adenine.

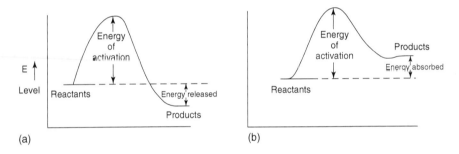

Adenosine 5'-Triphosphate

Shown subsequently are two chemical reactions, one that results in a product, which is at a lower energy level than the reactants (a) and another, which results in a product at a higher energy level (b). The former reaction releases energy and is termed *exergonic*, while the latter absorbs energy and is called *endergonic*. Each reaction has an energy barrier that must be surpassed in order for the reaction to proceed. This barrier is called the *energy of activation*.

Both reaction types, exergonic and endergonic, need a source of energy at least equal to, if not greater than, the energy of activation in order to proceed. However, once the exergonic reaction is initiated, it generates enough energy to perpetuate itself until the reactants are depleted. The endergonic reaction needs a constant source of energy in order to proceed. Many biological reactions are endergonic. They proceed by being coupled to an exergonic reaction. Most endergonic reactions in biological systems are coupled to the hydrolysis of ATP to ADP, an exergonic reaction that has an approximate standard free energy ($\Delta G°$) of -31.4 kJ/mol of ATP.

$$\text{ATP} \leftrightarrow \text{ADP} + \text{Pi} - 31.4 \text{ kJ/mol}$$

An example of a biological transformation, which is coupled to the hydrolysis of ATP, is the synthesis of the amino acid L-glutamine from L-glutaric acid in the presence of Mg^+ and glutamine synthetase.

$$\text{ATP} \xrightarrow{\text{Mg}^{+2}} \text{ADP}$$

coupled to

$$NH_3 + \text{L-Glutaric Acid} \rightarrow \text{L-Glutamine}$$

Summary Box 3.2

- Glucose metabolism results in the formation of ATP, the high energy compound that drives endergonic reactions to completion.
- When there is an inadequate oxygen supply (anaerobic), the metabolism of glucose to lactate produces 2 mol of ATP.
- When there is adequate oxygen (aerobic glycolysis), 36 or 38 mol of ATP are produced.
- Endergonic reactions are coupled to the hydrolysis of ATP to ADP. ATP hydrolysis provides 31.4 kJ/mol of ATP.

GLUCOSE METABOLISM

Glucose Transport into Cells

Owing to its polar nature, glucose cannot pass through the bilipid layer of the cell. Transport proteins are necessary to carry the glucose by facilitated diffusion through the cell membrane into the cytoplasm of the cell. Twelve monosaccharide transport proteins (GLUTs) have thus far been characterized, seven with high affinity for glucose and four with affinity for fructose. The high glucose affinity proteins are GLUT1, GLUT3, GLUT4, GLUT6, GLUT8, GLUT10, and GLUT12.

A somewhat lower affinity protein for glucose is GLUT2. The four that have high affinity for fructose and very low affinity for glucose are GLUT5, GLUT7, GLUT9, and GLUT11. Table 3.1 tabulates the transport proteins, their affinities, and locations. GLUT1, GLUT3, and GLUT4 have a Km between 2 and 5 mmol/l[6] and are below the normal range of blood glucose (about 7.5 mmol/l). These GLUTs function at rates close to maximum velocity. GLUT2 has a Km of about 25 mmol/l. Thus, GLUT2 transport of glucose is proportional to glucose concentrations in the blood. Hyperglycemia causes GLUT2 gene expression in beta cells and the liver and hyperinsulinemia decreases GLUT2 gene expression in the liver. GLUT4 is the most important glucose transport protein as it is found in highest concentrations in skeletal muscle, adipose cells, and cardiac muscle.

In addition, there are two sodium-glucose-linked transport (SGLT) proteins (SGLT1 and 2). The N-terminus of the protein transports sodium and the

---

[6]Low Km values indicate high GLUT affinities for glucose and high Km values indicate low GLUT affinities for glucose.

TABLE 3.1 **Monosaccharide Transport Proteins (GLUTs)**

GLUT Number	Monosaccharide(s) Transport	Location
1	Glucose	Erythrocytes, kidney, colon, glial cells
2	Glucose	Intestine, liver, kidney
3	Glucose	Macrophages, platelets, testis, neurons, placenta
4	Glucose	Skeletal muscle, adipose cells, cardiac muscle
5	Fructose	Small intestine
6	Glucose	Spleen, leukocytes, brain
7	Fructose	Small intestine, colon, testis, prostate
8	Glucose	Muscle, adipose cells, liver, testis
9	Fructose	Liver, kidney
10	Glucose	Placenta, liver, skeletal muscle, pancreas, lung, kidney, brain
11	Fructose	Cardiac muscle, skeletal muscle
12	Glucose	Skeletal muscle, cardiac muscle, adipose cells, prostate

TABLE 3.2 **Sodium-glucose transporters (SGLTs)**

SGLT Number	Monosaccharide Transported	Location
1	Glucose, galactose	Small intestine lumen
2	Glucose	Kidney

C-terminus carries the monosaccharide. As you can see in Table 3.2, both glucose and galactose are transported by SGLT1 and glucose by SGLT2.

Phosphorylation of Glucose

Once in the cell, glucose is phosphorylated at position 6 to produce glucose 6-phosphate (Fig. 3.1). This reaction is catalyzed by hexokinase. There are four known forms of hexokinase. Hexokinase I–III have low Kms ($10^{-6} - 10^{-1}$ mmol/l) for glucose. Hexokinase IV (also called *glucokinase*) has a Km 6–15 mmol/l. Table 3.3 describes the four hexokinases.

Glucose is positioned at the crossroads for glucose metabolism. Glucose can enter any of the five metabolic pathways: storage as glycogen by glycogenesis, conversion to pyruvic acid and lactic acid by glycolysis, oxidation to ribose 5-phosphate and carbon dioxide by the phosphogluconate pathway (also called the *pentose phosphate pathway* and the *hexose monophosphate shunt*), conversion to hexosamines and glycoproteins by the hexosamine biosynthesis pathway, and conversion to glucuronic acid by the uronic acid pathway.

TABLE 3.3 Phosphorylation of Glucose (Hexokinases)

Hexokinase Number	Glucose Affinity	Location
I	High	Skeletal muscle, beta cells, brain, kidney
II	High	Skeletal muscle, adipose cells
III	High	Spleen, lymphocytes
IV	Low	Liver, beta cells

Summary Box 3.3

- Glucose transport across the cell membrane is accomplished by nine transport proteins called *GLUTs*. Four additional GLUTs have a high affinity for fructose and a low affinity for glucose.
- There are two sodium–glucose transport proteins.
- Once in the cell, glucose is phosphorylated at position 6 by hexokinase.
- There are four forms of hexokinase.
- Glucose 6-phosphate can next undergo glycogenesis, glycolysis, oxidation by the phosphogluconate pathway, or be converted to glucuronic acid by the uronic acid pathway or to be converted to a hexosamine by the hexosamine biosynthesis pathway.

INTRODUCTION TO GLYCOGEN SYNTHESIS AND HYDROLYSIS

We will start our discussion of glucose metabolism by first looking at the synthesis of glycogen and its hydrolysis. In these two pathways, we will come upon an interesting biochemistry rule.

BEAUTIFUL CONCEPTS

Occasionally, while discussing the metabolism of glucose we will identify "beautiful concepts." "Beautiful concepts" are those concepts in biochemistry that are especially striking, complex, and infinitely tuned to maintain homeostasis. They are beautiful in how they function and respond to other signals from the organism.

An important maxim of metabolism and an example of a "beautiful concept" are those pathways that convert a substance (A) through intermediates to an end product (D), that is, $A \rightarrow B \rightarrow C \rightarrow D$, do not simply reverse for the conversion of D to A, that is, $D \rightarrow C \rightarrow B \rightarrow A$. Instead, another metabolic pathway that has unique intermediates and enzymes is utilized, that is, $D \rightarrow E \rightarrow F \rightarrow A$. The two separate and distinct pathways represented for substance D provide more regulatory sites for the metabolism of this substance than would a simple reversal of the pathway.

Glycogenesis and glycogenolysis

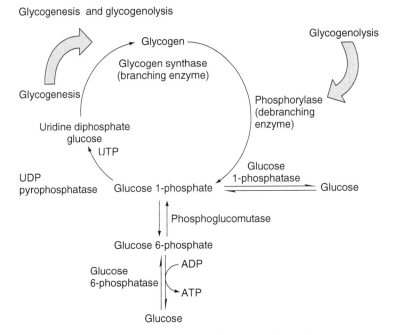

Figure 3.2   Glycogenesis and glycogenolysis pathways.

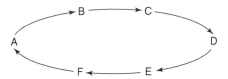

Glycogen synthesis and breakdown are good examples of this rule. A simplified schematic representation of glycogenesis and glycogenolysis is shown in Figure 3.2. Glycogenesis occurs primarily in resting muscle and the liver. Glycogenolysis occurs in contracted muscle, providing glucose for energy, and in the liver, where it provides glucose when blood glucose levels are low.

GLYCOGEN SYNTHESIS

A protein called *glycogenin* provides the initial step in glycogenesis. Glycogenin serves as a primer for glycogen synthesis by first attaching a glucose molecule to a tyrosine in the protein and next catalyzing the addition of up to seven glucose residues. Uridine diphosphate (UDP-glucose) serves as the donor molecule for these steps. There are two forms of glycogenin: glycogenin-1 found in skeletal muscle and glycogenin-2 located in the liver, cardiac muscle, and pancreas. Once sufficient glucosyl residues have been added, glycogen synthase catalyzes the transfer of the remaining glucose residues to the growing chain as described here:

- Condensation of glucose 1-phosphate to uridine 5′-triphosphate to form uridine 5′-diphosphate-glucose (UDP-glucose)
- Transfer of the glucose moiety of UDP-glucose to the glycogen molecule in a reaction catalyzed by glycogen synthase

$$\text{UDP-glucose} + (\text{glucose})_n \leftrightarrow (\text{glucose})_{n+1} + \text{UDP}$$

where $(\text{glucose})_n$ represents glycogen and $(\text{glucose})_{n+1}$ represents glycogen to which 1 mol glucose has been bonded in $\alpha(1 \rightarrow 4)$ linkage

- Branching of the glycogen carried out by the enzyme, $\alpha(1 \rightarrow 4)$-glucan 6-glycosyl transferase (branching enzyme).

Uridine Diphosphate Glucose (UDP-Glucose)

The substrate for glycogen synthase is UDP-glucose. It is derived from the reaction of α-D-glucose 1-phosphate with uridine triphosphate. This reaction is catalyzed by UDP-glucose pyrophosphorylase.

Uridine diphosphate glucose (UDP-glucose)

The α-D-glucose 1-phosphate is derived from α-D-glucose 6-phosphate in a reaction mediated by phosphoglucomutase.

Glycogen Synthase

Glycogen synthase catalyzes the formation of the $\alpha(1 \rightarrow 4)$ linkage between the glucose carried by UDP and the growing glycogen molecule. It is the rate-limiting enzyme in glycogenesis. Glycogen synthase exists in two forms, glycogen synthase a and b. The a and b represent the glucose 6-phosphate independent and dependent forms of the enzyme, respectively. The a form is the most active form of the enzyme. The b form is unreactive unless it is associated with glucose 6-phosphate.

Glycogen synthase b is a phosphorylated form of a. It contains six covalently bonded phosphate molecules. These six phosphates are covalently bonded to glycogen synthase by an activated protein kinase. *Kinase* is the term used for an enzyme that facilitates the transfer of inorganic phosphate (usually from ATP) to organic molecules. In the case of protein kinases, the recipient of the phosphate is a protein. Protein kinases also exist in active and inactive forms. The activating agent in this case is the small organic molecule, adenosine 3′,5′-cyclic monophosphate [cyclic adenosine 5′-monophosphate (cAMP)].

Cyclic AMP

Cyclic AMP is produced from ATP by adenylyl cyclase, an enzyme present in cell membranes.

$$\text{ATP} \xrightarrow[\text{Cyclase}]{\text{Adenylyl}} \text{cyclic AMP} + \text{PPi}$$

where PPi is pyrophosphate.

Thus, the concentration of glycogen synthase b is regulated largely by cyclic AMP. In summary, glycogen synthase a is converted to the b form in three steps (Fig. 3.3a); the activation of adenylyl cyclase resulting in the synthesis of cyclic AMP, activation of cyclic-AMP-dependent protein kinases A and C by cyclic AMP, and phosphorylation of glycogen synthase a to produce glycogen synthase b. In the absence of glucose 6-phosphate, the b form is inactive. Even in the presence of glucose 6-phosphate, glycogen synthase b is not as active an enzyme as the a compound. There have been several other kinases implicated in this phosphorylation: calmodulin-dependent kinases, glycogen synthase kinase 3 (GSK-3), and others.

As shown in Figure 3.3b, conversion of b to a is accomplished by a dephosphorylation mediated by a specific enzyme, phosphoprotein phosphatase. Cyclic AMP is hydrolyzed to AMP by cyclic AMP phosphodiesterase. The activated protein kinase is converted to the inactive form by a protein kinase phosphodiesterase. A phosphoprotein phosphatase converts glycogen synthase b to a. This phosphatase is dependent on the presence of insulin for its activation.

The Conversion of Glycogen Synthase a to b (Inactivation)

1. ATP $\xrightarrow{\text{Adenylyl Cyclase}}$ cyclic AMP + PPi

2. Protein kinase $\xrightarrow{\text{Cyclic AMP}}$ cyclic AMP-protein kinase
   (Inactive)   (active)

3. Glycogen synthase a + 6ATP $\xrightarrow{\text{Cyclic AMP-protein kinase}}$ Glycogen synthase b-6P
                                                                              + 6ADP

The Conversion of Glycogen Synthase b to a (Activation)

1. Cyclic AMP $\xrightarrow{\text{Cyclic AMP Phosphodiesterase}}$ AMP

2. Cyclic ATP-protein kinase $\xrightarrow{\text{Protein kinase Phosphodiesterase}}$ AMP + protein kinase
   (active)   (inactive)

3. Glycogen synthase b-P $\xrightarrow{\text{Phosphoprotein Phosphatase}}$ Glycogen synthase a

Figure 3.3   (a) Inactivation of glycogen synthase. (b) Activation of glycogen synthase.

Both the activation of protein kinase by cyclic AMP and glycogen synthase by glucose 6-phosphate are examples of allosteric activation. Allosteric effectors are usually small organic molecules that combine through hydrogen binding and ionic interactions at specific regions (allosteric-binding sites) of enzymes. The effect may activate the enzyme (positive effector) as in the above-mentioned examples or inhibit the enzyme (negative effector) as is the case of adenosine 5′-bisphosphate (ADP) and glycogen synthase a.

Branching Enzyme

As previously stated, glycogen is a highly branched molecule. Glycogen synthase catalyzes the formation of $\alpha(1 \rightarrow 4)$ bonds. This would result in a linear molecule. How then is branching achieved? The answer lies in the action of $\alpha(1 \rightarrow 4)$-glucan-6-glycosyl transferase (branching enzyme), an enzyme that hydrolyzes the $\alpha(1 \rightarrow 4)$ bond to produce segments six to seven glucose units in length and reattaches the fragments by $\alpha(1 \rightarrow 6)$ linkages to the glycogen chain (see Fig. 3.4). The branches are positioned approximately 10 units apart.

GLYCOGENOLYSIS

Glycogenolysis proceeds via four principal processes, as follows:

- Debranching of glycogen is accomplished by the debranching enzyme.
- Glucose 1-phosphate residues are removed stepwise by the enzyme, phosphorylase.

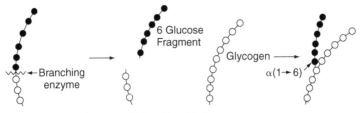

O—O Represents glucose units in α(1→4) linkage.

●—● Represents glucose units in α(1→4) linkage composing the fragment.

Figure 3.4   The action of branching enzyme.

- Phosphoglucomutase converts glucose 1-phosphate to glucose 6-phosphate.
- Glucose 6-phosphate is hydrolyzed to glucose by glucose 6-phosphatase.

Debranching Enzyme

Debranching of glycogen occurs through the action of a multifunctional enzyme. One function is an α(1 → 4) → α(1 › 4) glucan transferase activity that splits off trisaccharides from the branch, one glucose before the branch point, and transfers them to an end of the glycogen molecule, as shown in Figure 3.5. The enzyme is highly specific and acts only on four unit branches. Thus, a branch of more

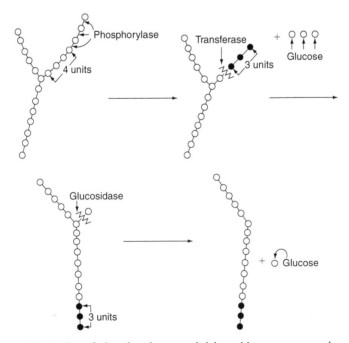

Figure 3.5   Action of phosphorylase a and debranching enzyme on glycogen.

than four units must be degraded before the action of this enzyme can occur. Phosphorylase carries out this function hydrolyzing $\alpha(1 \to 4)$ bonds stepwise until the branch is reduced to four units. Following the actions of phosphorylase and the transferase enzyme, the remaining glucose unit that is attached to the branch point by an $\alpha(1 \to 6)$ bond is released by the second function of the glycogen debranching enzyme. This function is an amylo-1,6-glucosidase activity. The debranching enzyme is a multifunctional enzyme known to consist of a single polypeptide chain. The catalytic centers are separated on the chain.

Glycogen Phosphorylase

Glycogen phosphorylase exists in two forms, phosphorylases a and b. Phosphorylase a is a tetramer. It is derived from phosphorylase b, a dimer, in two steps. First, phosphorylase b is phosphorylated by a protein kinase, one phosphate per unit, and then the phosphorylated dimer combines with another phosphorylated dimer to produce the tetramer, phosphorylase a. Phosphorylase a is active in the hydrolysis of glycogen; phosphorylase b is inactive unless activated in an allosteric manner by AMP. Note the parallels to the regulation of glycogen synthase. Pyridoxal phosphate is bound to each subunit of the phosphorylases. Enzyme activity is dependent on pyridoxal phosphate as its removal eliminates activity.

$$\underset{\text{Dimer}}{2 \text{ Phosphorylase b}} + 4\text{ATP} \xrightarrow{\underset{\text{kinase}}{\text{Protein}}} \underset{\text{Tetramer}}{\text{Phosphorylase a 4-ADP}}$$

$$\underset{\text{Tetramer}}{\text{Phosphorylase a}} + 4\text{H}_2\text{O} \xrightarrow{\text{phosphatase}} \underset{\text{Dimer}}{2 \text{ Phosphorylase b}} + 4\text{Pi}$$

The protein kinase involved in these phosphorylations has been given the more specific name phosphorylase b kinase. It is activated by a phosphorylation carried out by a cyclic-AMP-dependent protein kinase (Fig. 3.6).

$$\underset{\text{Inactive}}{\text{Phosphorylase kinase 4-ATP}} \xrightarrow{\underset{\text{protein kinase}}{\text{cAMP-dependent}}} \underset{\text{Active}}{\text{Phosphorylase kinase}} + \text{ADP}$$

Dephosphorylation of phosphorylase a is affected by phosphorylase a phosphatase, the same enzyme that dephosphorylated glycogen synthase b. These phosphatases also inactivate phosphorylase kinase. Phosphorylase a hydrolyzes the $\alpha(1 \to 4)$ glycogen linkage in a stepwise fashion releasing glucose 1-phosphate units. However, hydrolysis is limited to the fourth glucose from the $\alpha(1 \to 6)$ branch point.

Phosphoglucomutase

Phosphoglucomutase catalyzes the conversion of glucose 1-phosphate to glucose 6-phosphate. The reaction will not proceed in the absence of glucose 1,6-bisphosphate and Mg^{+2}.

1. ATP $\xrightarrow[\text{Cyclase}]{\text{Adenylyl}}$ cAMP + PPi

2. Phosphorylase kinase + ATP $\xrightarrow[\text{protein kinase}]{\text{cAMP dependent}}$ Phosphorylase kinase.P + ADP
    (Inactive)   (Active)

3. 2 Phosphorylase b + 4ATP $\xrightarrow{\text{Phosphorylase kinase (active)}}$ Phosphorylase a + 4ADP
    (dimer)   (tetramer)
    Inactive   Active

(a)

1. cAMP $\xrightarrow[\text{diesterase}]{\text{CAMP phospho}-}$ AMP

2. Phosphorylase kinase $\xrightarrow{\text{Phosphatase}}$ Phosphorylase kinase + Pi
    (Active)   (Inactive)

3. Phosphorylase a $\xrightarrow[\text{Phosphatase}]{\text{Phosphorylase a}}$ 2 Phosphorylase b + 4Pi
    (Active)

(b)

Figure 3.6   (a) Activation of phosphorylase b and (b) inactivation of phosphorylase a.

Glucose 6-Phosphatase

Glucose 6-phosphatase hydrolyzes glucose 6-phosphate to glucose.

$$\text{Glucose 6-phosphate} + H_2O \xrightarrow{Mg^{+2}} \text{Glucose} + Pi$$

Glucose 6-phosphate is located in the endoplasmic reticulum. Thus, glucose 6-phosphate must be transported to the endoplasmic reticulum for hydrolysis to glucose.

α(1 → 4)-Glucosidase

α(1 → 4)-Glucosidase hydrolyzes the α(1 → 4) glycoside bond, resulting in the formation of maltose and glucose. This enzyme, primarily found localized in the lysosomes of the liver, is present to some degree in most organs. The enzyme digests glycogen to a highly branched glucose polymer termed *macrodextrin*.

SYNCHRONIZATION OF GLYCOGENESIS AND GLYCOGENOLYSIS (A BEAUTIFUL PATHWAY)

Glycogenesis and glycogenolysis are examples of beautiful pathways. Glycogenesis and glycogenolysis afford an excellent example of the synchronization of two opposing metabolic pathways by a common regulator. Control of blood glucose and glycogen stores would be disorganized and energy release for life functions quite inefficient if glycogenesis and glycogenolysis were not synchronized in some manner. This bioregulation is accomplished by a reciprocal activation–deactivation of the principal enzymes involved in the pathways through a common regulatory molecule, cyclic AMP (calcium serves this role in muscle tissue). Cyclic AMP is a positive effector of cyclic-AMP-dependent protein kinases. Cyclic AMP is synthesized by the plasma-membrane-bound adenylyl cyclase, an enzyme that is activated by polypeptide hormones.

Protein kinase mediated phosphorylations of glycogen synthase a and phosphorylase b result in the formation of b and a forms, respectively. Thus, glycogenolysis is promoted and glycogenesis decreased. In organs that contain glucose 6-phosphatase activity, principally the liver, glucose accumulates and is transported into the blood. In organs that lack glucose 6-phosphatase, glucose 6-phosphate accumulates and enters the glycolytic pathway. Cyclic AMP is hydrolyzed to AMP by cyclic AMP phosphodiesterase, an enzyme that, paradoxically, is activated by association with cyclic AMP.

Dephosphorylation

Dephosphorylation of glycogen synthase b and phosphorylase a occurs through the action of phosphatases. Both these actions are catalyzed by phosphoprotein phosphatases. Phosphoprotein kinase is activated by insulin. Thus, insulin directs the activation of glycogen synthase (glycogenesis) and deactivates phosphorylase (glycogenolysis). The cyclic AMP dependent protein kinases are inactivated by cyclic AMP phosphodiesterases.

Effectors

A number of low molecular weight organic compounds serve as positive and negative effectors of glycogenolysis and glycogenesis. They are summarized in the following table:

Effector	Enzyme Involved	Effect
AMP	Phosphorylase b	+
	Glycogen synthase a	−
ADP	Glycogen synthase a	−
ATP	AMP-activated phosphorylated b	−
	Glycogen synthase	−
Cyclic AMP	Phosphorylase b kinase	+
	Glycogen synthase b protein kinase	+
Pi	Glycogen synthase a	−
Ca^{+2}	Phosphorylase b kinase	+
Glycogen	Phosphorylase b kinase	+
	Phosphoprotein phosphatase	−
Glucose 6-phosphate	AMP-activated phosphorylase b	−
	Glycogen synthase	+
Phospho-esterase inhibitor protein	Cyclic-AMP-dependent-phospho-diesterase	−
Glucose	Glucose phosphorylase b	−

+ denotes a positive effector; − denotes a negative effector.

Summary Box 3.4

- An important rule of metabolism is that a metabolic pathway simply does not reverse; rather, it uses a different pathway to reverse.
- A protein, named glycogenin, serves as the primer for glycogen synthesis.
- Glycogenesis occurs in three principal steps that include the intermediate UDP-glucose.
- UDP-glucose synthesis is catalyzed by uridine diphosphate glycogen pyrophosphorylase.
- Transfer of glucose from UDP-glucose to a growing glycogen chain is mediated by glycogen synthase.
- The glucose 6-phosphate dependent form of glycogen synthase (form b) is produced by phosphorylation from the glucose 6-phosphate independent form of glycogen synthase (form a). A cyclic AMP activated protein kinase phosphorylates the glycogen synthase.
- Reversal of the phosphorylation of glycogen synthase is mediated by phosphoprotein phosphatase.
- Glycogen synthase catalyzes the formation of $\alpha(1 \rightarrow 4)$ bonds. A transferase (branching enzyme) carries fragments of glycogen to reconnect them by $\alpha(1 \rightarrow 6)$ linkage.
- Glycogenolysis occurs in four steps.
- Debranching enzyme splits off trisaccharides from glycogen, while phosphorylase hydrolyzes branches off glycogen to four glucose units. The remaining glucose is removed by amylo-1,6-glucosidase.
- Phosphorylase is derived from phosphorylation by a protein kinase. It next dimerizes to form phosphorylase b. The protein kinase is activated by a cyclic AMP dependent protein kinase phosphorylation. Phosphorylase a is active and b is inactive unless activated by AMP. Dephosphorylation of phosphorylase a is mediated by phosphorylase a phosphatase.
- Phosphoglucomutase mediates the conversion of glucose 1-phosphate to glucose 6-phosphate and glucose 6-phosphatase hydrolyzes glucose 6-phosphate to glucose.
- $\alpha(1 \rightarrow 4)$ Glucosidase hydrolyzes the $\alpha(1 \rightarrow 4)$ glycoside bond.
- Glycogenesis and glycogenolysis are two pathways that are in opposition but are controlled by a common regulator.

The next principal pathway that glucose 6-phosphate can enter is glycolysis (glycolytic pathway), as shown in Figure 3.7.

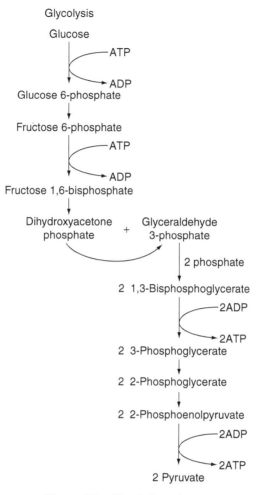

Figure 3.7 Glycolytic pathway.

GLYCOLYSIS (GLYCOLYTIC PATHWAY)

Phosphoglucose Isomerase

Glucose 6-phosphate enters the glycolytic pathway by its conversion to fructose 6-phosphate, a reaction catalyzed by phosphoglucoisomerase.

$$\text{Glucose 6-phosphate} \xrightleftharpoons[\text{Phosphoglucoisomerase}]{} \text{Fructose 6-phosphate}$$

Phosphofructokinase

The principal regulatory enzyme in the glycolytic metabolic pathway is phospho-fructokinase (PFK). Regulation is primarily by allosteric mechanisms. ATP, an important product of glycolysis, is a negative effector of PFK. It is also interesting to note that ATP is a substrate in the reaction. This rather common phenomenon is called *homotropic allosterism*. AMP and, to a lesser extent, ADP and Pi are positive allosteres of PFK. Citrate, an intermediate in the tricarboxylic acid cycle, also exerts a negative allosteric effect on PFK. Alanine and NADH are inhibitors of the enzyme. Liver and muscle isozymes of PFK have been reported. A similar, if not identical, enzyme is present in erythrocytes. This reaction in the glycolytic pathway proceeds with the utilization of 1 mol of ATP.

A fructose 1,6-bisphosphatase exists, which hydrolyzes fructose 1,6-bisphosphate to fructose 6-phosphate. The enzyme is inhibited by AMP. Thus, ATP depletion with a concomitant increase in AMP concentrations favors glycolysis by inhibiting fructose 1,6-bisphosphatase and activating PFK.

$$\text{Fructose 6-phosphate} \xrightarrow[\text{ATP, Mg}]{\text{phosphofructokinase}} \text{Fructose 1,6-bisphosphate}$$

$$\text{Fructose 1,6-bisphosphate} \xrightarrow{\text{Phosphatase}} \text{Fructose 6-phosphate} + \text{Pi}$$

Aldolase

Fructose 1,6-bisphosphate is cleaved by aldolase.

| Fructose 1, 6-bisphosphate | Dihydroxyacetone phosphate | D-glyceraldehyde 3-phosphate |

The products retain their phosphates, which are dihydroxyacetone phosphate and D-glyceraldehyde 3-phosphate. The reaction involves the intermediate binding of the ketone group of fructose 1,6-bisphosphate to the ε-amino group of lysine by a covalent bond.

Triose Phosphate Isomerase

Triose phosphate isomerase catalyzes the interconversion of dihydroxyacetone phosphate and D-glyceraldehyde 3-phosphate.

$$\text{Dihydroxyacetone phosphate} \xleftrightarrow{\text{Isomerase}} \text{D-Glyceraldehyde 3-phosphate}$$

Reduction of dihydroxyacetone phosphate to glycerol 3-phosphate by glycerol 3-phosphate dehydrogenase is the first step in lipid metabolism.

Glyceraldehyde 3-Phosphate Dehydrogenase

Glyceraldehyde 3-phosphate dehydrogenase both phosphorylates and oxidizes the aldehyde group of D-glyceraldehyde 3-phosphate, resulting in the formation of 1,3-bisphosphoglycerate. Nicotinamide adenine dinucleotide (NAD$^+$) is a cofactor for this reaction. A cofactor is an organic molecule that binds to an enzyme and is necessary for the catalytic activity of the enzyme. The cofactor is directly involved in the chemical change mediated by the enzyme and is chemically modified (unlike the enzyme) by the reaction, for example, NAD$^+$ → NADH.

$$
\begin{array}{l}
\text{HC} = \text{O} \\
| \\
\text{HCOH} + \text{NAD}^+ + \text{Pi} \\
| \\
\text{H}_2\text{COPO}_3\text{H}_2
\end{array}
\longrightarrow
\begin{array}{l}
\text{H}_2\text{COPO}_3\text{H}_2 + \text{NADH} + \text{H}^+ \\
| \\
\text{H COH} \\
| \\
\text{H}_2\text{COPO}_3\text{H}_2
\end{array}
$$

D-Glyceraldehyde
3-phosphate
Pi indicates phosphate ion

1,3-Bisphosphoglycerate

where Pi indicates the phosphate ion.

Erythrocyte Bisphosphoglyceromutase and Bisphosphoglycerate Phosphatase

Human erythrocytes contain the enzyme bisphosphoglyceromutase. It catalyzes the conversion of 1,3-bisphoglycerate (1,3-BPG) to 2,3-bisphosphoglycerate (2,3-BPG). The association of 2,3-bisphosphoglycerate with hemoglobin results in a complex possessing a greater affinity for oxygen than hemoglobin alone. The formation 2,3-bisphosphoglycerate constitutes a bypass of the glycolytic pathway and is called the *Rapoport–Luebering shunt*; it provides an interaction between erythrocyte glycolysis and hemoglobin oxygenation.

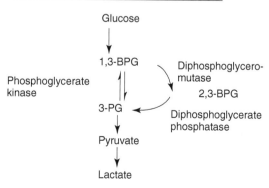

DPG formation and hydrolysis in erthrocytes

$$1,3\text{-Bisphosphoglycerate} \xleftrightarrow{\text{Mutase}} 2,3\text{-Bisphosphoglycerate}$$

Bisphosphoglycerate phosphatase hydrolyzes 2,3-bisphosphoglycerate to 3-phosphoglycerate.

3-Phosphoglycerate Kinase

The transformation of 1,3-bisphosphoglycerate to 3-phosphoglycerate is mediated by 3-phosphoglycerate kinase. One mole of ATP is produced by the reaction.

$$1,3\text{-Bisphosphoglycerate} + \text{ADP} \xleftrightarrow{\text{Mg}^{+2}} 3\text{-Phosphoglycerate} + \text{ATP}$$

If one assumes that all molecules of dihydroxyacetone phosphate are converted to D-glyceraldehyde 3-phosphate, then 1 mol of glucose yields 2 mol of D-glyceraldehyde, 2 mol of 1,3-bisphosphoglycerate, 2 mol of 2,3-bisphosphoglycerate, 2 mol of 3-phosphoglycerate, and 2 mol of ATP.

Phosphoglyceromutase

Phosphogyceromutase converts 3-phosphoglycerate to 2-phosphoglycerate through the intermediate 2,3-bisphosphoglycerate.

Enolase

Enolase converts 2-phosphoglycerate to phosphoenolpyruvate in a reaction that is dependent on the presence of Mg^{+2} or Mn^{+2}.

Pyruvate Kinase

The conversion of phosphoenolpyruvate to pyruvate is catalyzed by pyruvate kinase (PK). Two moles of ATP are produced from this reaction per mole of glucose 6-phosphate.

Compound	PFK	PK	Pathway Compound is derived from
Citrate	−	−	Tricarboxylic Acid Cycle
Succinyl CoA	NE	−	Tricarboxylic Acid Cycle
Alanine	NE	−	Alanine Synthesis
Fatty acids	−	−	Fatty Acid Synthesis
NADH	−	−	--------
ATP	−	−	--------
AMP	+	NE	--------
ADP	+	NE	--------

+ Activates enzyme
− Inhibits enzyme
NE No effect on enzyme activity

Figure 3.8 Effectors of PFK and PK.

Glycolysis presents two steps in metabolism that can also be considered "beautiful." They are the step mediated by PFK and the terminal step in the pathway mediated by PK. These two steps eloquently tune glycolysis to the needs of the body and also provide communication with other important biochemical pathways. Figure 3.8 summarizes the effects of several organic molecules on PFK and PK.

Pyruvate has four metabolic fates, namely, reconversion to lactate and thence via gluconeogenesis conversion to glucose, conversion to oxalacetate or malate that enters the tricarboxylic acid cycle, formation of alanine in a reaction mediated by alanine transaminase (ALT) and oxidation to CO_2 and acetyl coenzyme A (acetyl CoA). Acetyl CoA is utilized for the synthesis of long chain fatty acids.

At least one metabolite of each of these pyruvate metabolic pathways is represented in Figure 3.8 as an effector of PFK and/or PK. NAD^+ is converted to NADH in the oxidation of glyceraldehyde 3-phosphate to 3-phosphoglycerate; it is regenerated by lactate dehydrogenase (LDH). NADH is an inhibitor of glycolysis; thus, as the NADH concentration increases, its formation is reduced.

Lactate Dehydrogenase

In the presence of the cofactor, NADH, LDH reduces pyruvate to lactate.

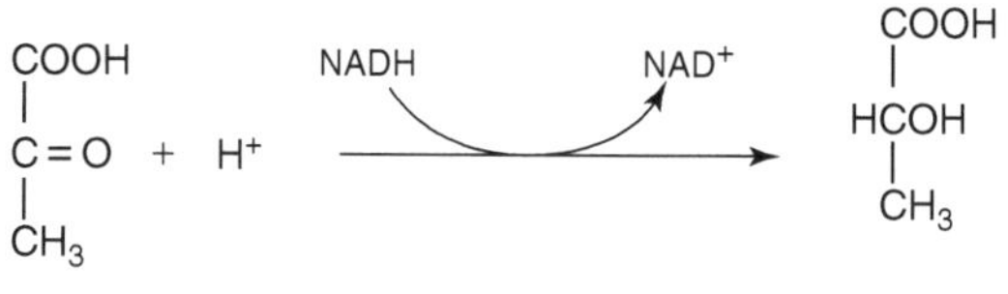

$$\begin{array}{ccccc}
\text{COOH} & & & & \text{COOH} \\
| & \text{NADH} & \text{NAD}^+ & & | \\
\text{C=O} + \text{H}^+ & \searrow \quad \nearrow & & & \text{HCOH} \\
| & \longrightarrow & & & | \\
\text{CH}_3 & & & & \text{CH}_3
\end{array}$$

LDH is a tetramer composed of two different subunits designated M and H. Many textbooks describe the reduction of pyruvate to lactate as a "metabolic dead-end." This is true with respect to the fact that other than the reversible conversion to pyruvate, lactate is not directly involved in any other metabolic pathways. However, lactate may be compared to glycogen, as a storage form of glucose as lactate generated in an oxygen-debt situation is carried by circulation to the liver, where gluconeogenesis converts it to glucose. As the erythrocyte lacks most of the tricarboxylic acid cycle enzymes, the product of glycolysis in erythrocytes can be only lactate.

Pyruvate next enters the tricarboxylic acid cycle.

Summary Box 3.5

- PFK is the principal regulatory enzyme in the glycolytic pathway.
- Aldolase cleaves the six-carbon molecule, fructose 1,6-bisphosphate to two three-carbon fragments dihydroxyacetone phosphate and D-glyceraldehyde 3-phosphate.
- Glyceraldehyde 3-phosphate dehydrogenases both phosphorylates and oxidizes D-glyceraldehyde 3-phosphate.
- Human erythrocytes contain an enzyme that catalyzes the formation of 2,3-bisphosphoglycerate (BPG) from 1,3-bisphosphoglycerate. BPG associates with hemoglobin to form a complex that has a greater affinity for oxygen.
- The last step in the glycolytic pathway converts phosphoenolpyruvate to pyruvate in a reaction catalyzed by PK.
- Pyruvate has four fates.
- One pyruvate fate is reduction to lactate in a reaction catalyzed by LDH.

TRICARBOXYLIC ACID CYCLE

The tricarboxylic acid cycle is the principal metabolic pathway that generates ATP. Figure 3.9 depicts the principal substances in the pathway and includes the major contributors from other metabolic pathways that fuel the cycle. As you can see from the figure, the principal metabolic pathways that contribute to the tricarboxylic acid cycle are the glycolytic, amino acid, and fatty acid metabolic pathways. Figure 3.10 shows the specific compounds that compose the pathway. As you can see from this figure, the tricarboxylic acid cycle does not directly produce ATP; it produces reducing units designated $[H:]^-$ in the figure. The reducing units are transferred to the cofactors, NAD^+ and flavin adenine dinucleotide (FAD).

The processes called *electron transport* and *oxidative phosphorylation* reoxidize the cofactors and simultaneously produce ATP from ADP and Pi.

Each cycle of the tricarboxylic acid pathway produces 2 mol of CO_2 if the cycle alone is considered and 3 mol if the decarboxylation of pyruvate to form acetyl CoA is included.

The reactions of the tricarboxylic acid cycle occur in an organelle of the cell called the *mitochondrion.* The mitochondrion is composed of two membrane structures, an outer membrane that is smooth in texture and gives the

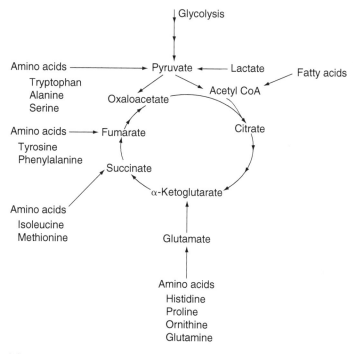

Figure 3.9 Substances that flow into the tricarboxylic acid cycle (citric acid cycle).

mitochondrion an oval shape, and an inner membrane space that is a highly convoluted structure resulting in folds termed *cristae*. Embedded in the cristae are the components associated with electron transport. The enzymes of the tricarboxylic acid cycle (with the exception of succinate dehydrogenase) are located in the fluid matrix of the interior of the mitochondrion.

The Coenzymes: Nicotinamide Adenine Dinucleotide (NAD⁺) and Flavin Adenine Dinucleotide (FADH)

NAD^+ and FAD are important components (coenzymes, or cofactors) of the tricarboxylic acid cycle. Thus, these compounds will be described before the individual steps of the tricarboxylic acid cycle. The structure of NAD^+ is presented in Figure 3.11.

As depicted, the oxidation–reduction reactions of NAD^+ occur at the niacinamide portion of the molecule.

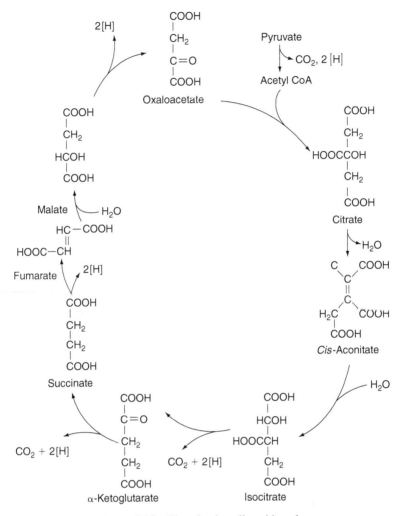

Figure 3.10 The tricarboxylic acid cycle.

In terms of electron movement, the following shifts occur:

For illustrative purposes, ethanol is used as a substrate. H:⁻ is known to organic chemists as a hydride ion and is composed of a proton and two electrons. H⁺ is a proton.

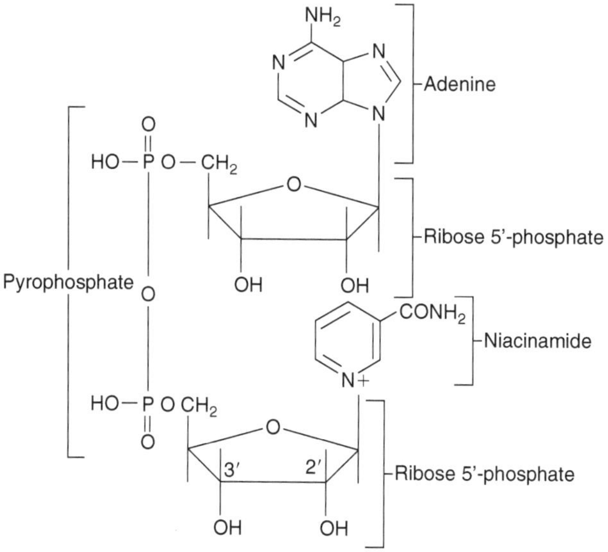

Figure 3.11   The structure of Nicotinamide adenine dinucleotide (NAD^+).

The structure of FAD is shown in Figure 3.12. When FAD is involved in redox (oxidation–reduction) reactions, the electron movements shown subsequently occur. Ethanol is the substrate used in the illustration. The FAD molecule combines with both the H:$^-$ hydride) and H^+ (proton).

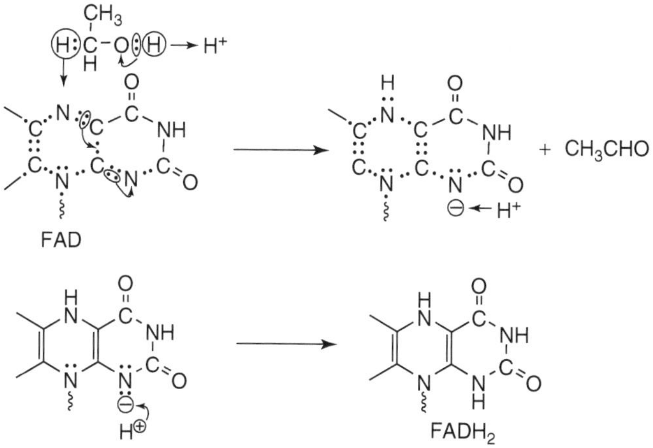

STEPS IN THE TRICARBOXYLIC ACID CYCLE

Pyruvate Dehydrogenase; Acetyl CoA

One of the most "beautiful concepts" in biochemistry is the large, multicomponent complex that converts pyruvate to acetyl CoA. Pyruvate dehydrogenase

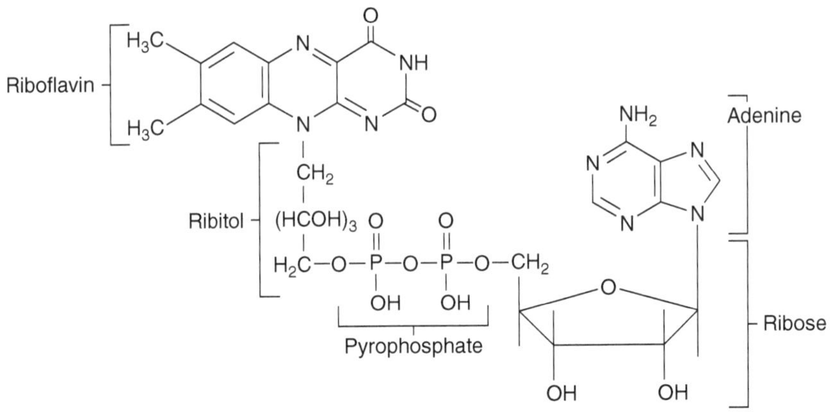

Figure 3.12 The structure of Flavin adenine dinucleotide (FAD).

is composed of 60 protein subunits organized into three enzymes, pyruvate decarboxylase, dihydrolipoyl transacetylase, and dihydrolipoyl dehydrogenase. It utilizes five small organic cofactors for its actions— thiamine pyrophosphate, lipoic acid, coenzyme A, FAD, and NAD^+. The complex is located in the matrix of the mitochondrion. The components are arrayed in 12 pentagonal sides, thus creating what would be called a *dodecahedron* by mathematicians. As seen here, the coordinated action of the components of the complex results in the oxidative decarboxylation of pyruvate and the production of acetyl CoA. A master of efficiency, the products of one enzyme are passed directly to the next.

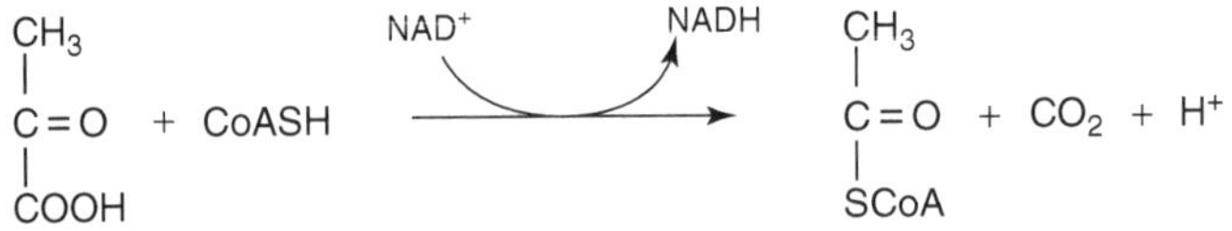

As shown in Figure 3.13, acetyl CoA is a molecule composed of adenosine 3-phosphate 5′-pyrophosphate, pantothenic acid, β-mercaptoethylamine, and an acetyl group. CoA lacks the acetyl group and therefore has a free sulfhydryl group ($-SH$) and may be represented as CoASH or CoA. The oxidative decarboxylation proceeds in five steps, utilizing three enzymes.

Pyruvate Decarboxylase

1. In a reaction mediated by the pyruvate decarboxylase shown, thiamine pyrophosphate reacts with pyruvate at the five-member thiazole ring, evolving CO_2.

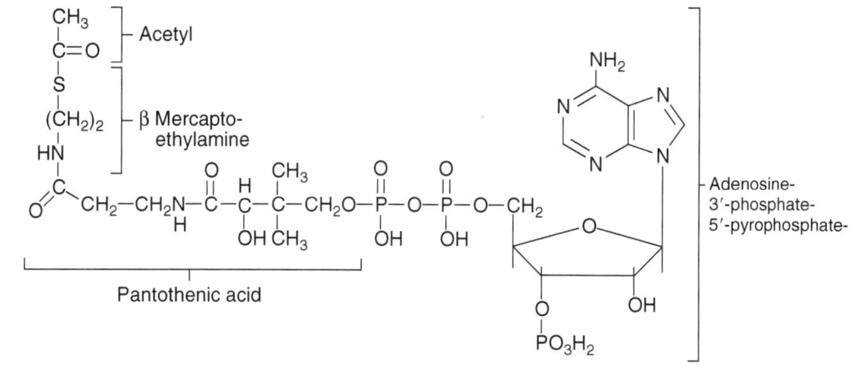

Figure 3.13 The structure of Acetyl CoA.

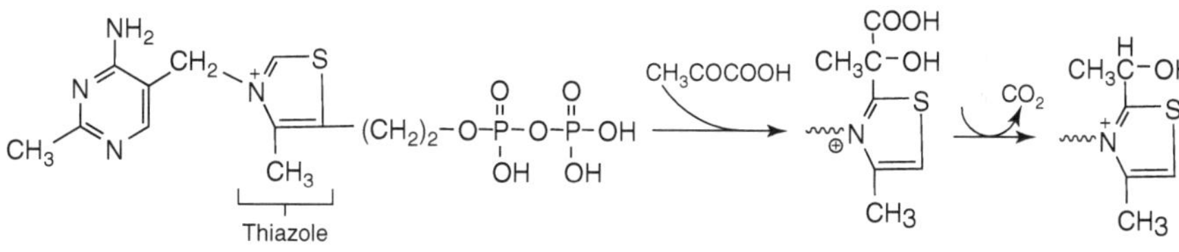

Dihydrolipoyl Transacetylase

2. As shown, the hydroxyethyl thiazole ring of thiamine diphosphate reacts with lipoate, transferring the acetyl group to the lipoate moiety. Note that the lipoate removes a hydride (bold-faced H) from the hydroxyethyl group, thereby oxidizing it to a carbonyl group. The thiazole ring picks up the proton released from the hydroxyl group.

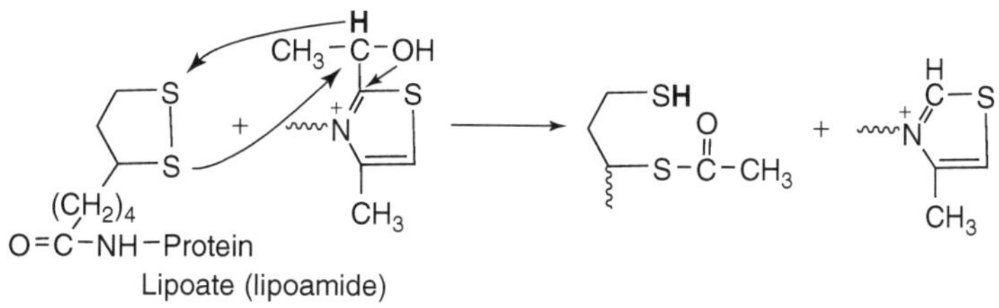

3. CoASH is next acetylated by acetyl lipoate. This part of the reaction is shown in this diagram.

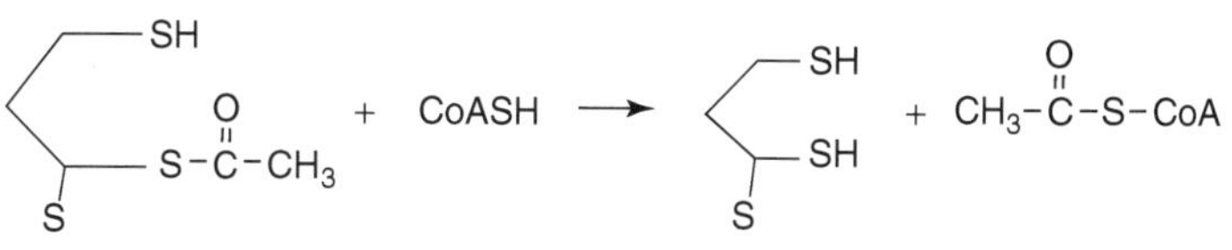

Dihydrolipoyl Dehydrogenase

4. In this step, catalyzed by dihydrolipoyl dehydrogenase, lipoate is reoxidized by reacting with FAD.

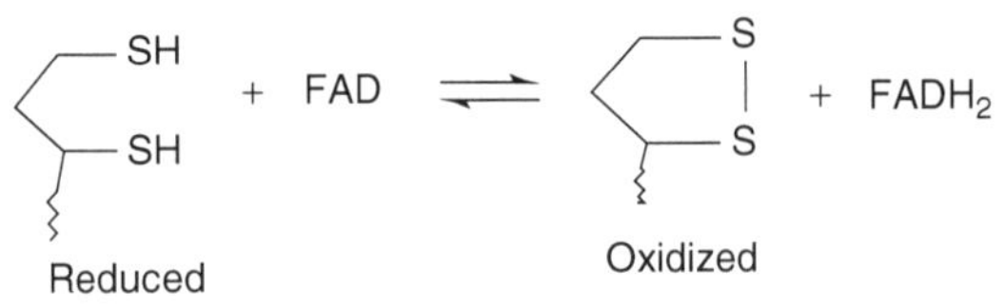

Reduced + FAD ⇌ Oxidized + FADH$_2$

5. FADH$_2$ is reoxidized by NAD$^+$.

$$FADH_2 + NAD^+ \leftrightarrow FAD + NADH + H^+$$

Acetyl CoA has a multiplicity of metabolic functions. One of these, its condensation with citrate, is the principal link between the tricarboxylic acid cycle and glycolysis. This complex is beautifully regulated. Pyruvate dehydrogenase is inhibited by elevated ATP, NADH, and acetyl CoA levels. In addition, this pathway is finely tuned by phosphorylation and dephosphorylation. Pyruvate dehydrogenase kinase phosphorylates the three serine residues in pyruvate dehydrogenase to inhibit the enzyme. Dephosphorylation of these residues by pyruvate dehydrogenase phosphatase reactivates pyruvate dehydrogenase.

The next step in the tricarboxylic acid cycle is depicted here.

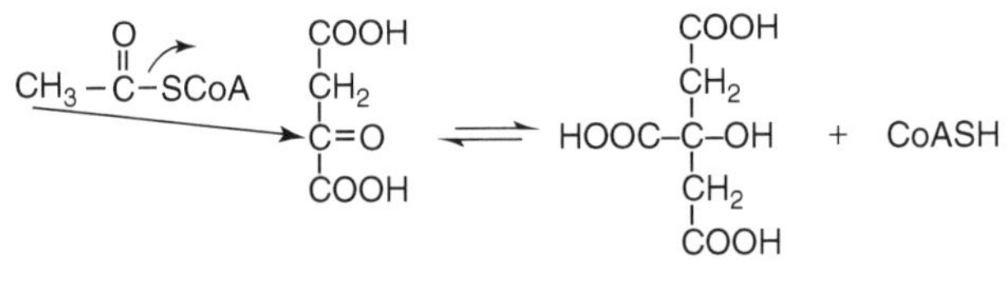

Citrate Synthase

Acetyl CoA enters the tricarboxylic acid cycle through its condensation with oxalacetate to form citrate. CoASH is also reformed in the reaction. The catalyzing enzyme is citrate synthase.

Aconitase

Aconitase mediates the dehydration of citrate to *cis*-aconitate and the subsequent hydration to isocitrate.

cis aconitate   Isocitrate

Isocitrate Dehydrogenase

Two isozymes of isocitrate dehydrogenase are found in mitochondria, one utilizes NAD^+ as a cofactor; the other $NADP^+$. $NADP^+$ differs from NAD by possessing a phosphate in the $2'$ position. The NAD^+-dependent enzyme is also present in the cytosol. Mg^{+2} or Mn^{+2} are necessary for enzyme activity.

α-Ketoglutarate provides another route for the entrance of carbon compounds into the tricarboxylic acid cycle. Glutamic acid and amino acids capable of conversion to glutamate (e.g., histidine, ornithine, proline, and glutamate) enter the pathway via transamination and/or oxidation to α-ketoglutarate.

α-Ketoglutarate Dehydrogenase

The conversion of α-ketoglutarate to succinate is accomplished by the coordinated action of an array of components similar to those involved in the oxidative decarboxylation of pyruvate (providing another "beautiful concept").

As depicted, the first step in the process is the reaction of α-ketoglutarate with thiamine pyrophosphate to form a hydroxyl derivative.

This is followed by the reaction of the thiamine derivative with lipoate (as shown).

Similarly, to the oxidative decarboxylation of pyruvate, the succinyl group is transferred to CoASH and subsequently lipoate is reoxidized by FAD in a manner similar to the oxidative decarboxylation of pyruvate.

The final step in the process is the oxidation of dihydrolipoate with the concomitant reduction of FAD.

$$FADH_2 + NAD^+ \leftrightarrow FAD + NADH + H^+$$

Succinyl CoA is also utilized for the biosynthesis of porphyrins. Its interaction with guanosine 5'-diphosphate results in the release of CoASH and the production of guanosine 5'-triphosphate (GTP) and succinate. The enzyme that catalyzes this reaction is succinyl-CoA synthetase.

$$Succinyl\ CoA + GDP + Pi \rightarrow Succinate + CoASH + GTP$$

The GTP produced may be utilized to produce ATP:

$$GTP + ADP \leftrightarrow GDP + ATP$$

Succinate Dehydrogenase

As shown subsequently, succinate is oxidized to fumarate by succinate dehydrogenase. The cofactor for this enzyme is FAD. Unlike the other tricarboxylic acid cycle enzymes, succinate dehydrogenase is tightly bound to the mitochondrial inner membrane. Succinate dehydrogenase is a unique enzyme as it is an iron–sulfur protein. It is composed of two subunits, each containing four atoms of iron and four atoms of sulfur; one of the units also contains one molecule of FAD.

$$\begin{array}{c} \text{COOH} \\ | \\ \text{HCH} \\ | \\ \text{HCH} \\ | \\ \text{COOH} \end{array} \quad + \quad \text{FAD} \quad \rightleftharpoons \quad \begin{array}{c} \text{H} \diagdown \underset{||}{\text{C}} \diagup \text{COOH} \\ \underset{\text{HOOC}}{\diagup} \text{C} \diagdown \text{H} \end{array} \quad + \quad \text{FADH}_2$$

Fumarase

As shown, fumarate is hydrated to form L-malate in a reaction catalyzed by fumarase. The reaction is stereospecific as only the L-isomer of malate is produced.

$$\begin{array}{c} \text{H} \diagdown \underset{||}{\text{C}} \diagup \text{COOH} \\ \underset{\text{HOOC}}{\diagup} \text{C} \diagdown \text{H} \end{array} \quad \xrightarrow{\text{HOH}} \quad \begin{array}{c} \text{COOH} \\ | \\ \text{HCH} \\ | \\ \text{HOCH} \\ | \\ \text{COOH} \end{array}$$

L-Malate Dehydrogenase

The oxidation of L-malate to oxalacetate is catalyzed by L-malate dehydrogenase. The cofactor is NAD⁺.

$$\begin{array}{c} \text{COOH} \\ | \\ \text{CH}_2 \\ | \\ \text{HOCH} \\ | \\ \text{COOH} \end{array} \quad + \quad \text{NAD}^+ \quad \rightleftharpoons \quad \begin{array}{c} \text{COOH} \\ | \\ \text{CH}_2 \\ | \\ \text{C}=\text{O} \\ | \\ \text{COOH} \end{array} \quad + \quad \text{NADH} + \text{H}^+$$

Pyruvate Carboxylase

An important enzyme involved in the tricarboxylic acid cycle, but not directly in the cycle, is pyruvate carboxylase. As you can see here, pyruvate carboxylase catalyzes the conversion of pyruvate to oxaloacetate.

$$\begin{array}{c} \text{CH}_3 \\ | \\ \text{C}=\text{O} \\ | \\ \text{COOH} \end{array} + \text{CO}_2 + \text{H}_2\text{O} \quad \xrightarrow[\quad]{\text{ATP} \quad \text{ADP}+\text{Pi}} \quad \begin{array}{c} \text{COOH} \\ | \\ \text{CH}_2 \\ | \\ \text{C}=\text{O} \\ | \\ \text{COOH} \end{array}$$

The enzyme contains four molecules of biotin and four atoms of manganese. Pyruvate carboxylase is an allosteric enzyme; its activity is enhanced by acetyl CoA. Thus, in situations of acetyl CoA excess, pyruvate supplies both the acetyl CoA and the oxaloacetate. The accumulation of acetyl CoA occurs during periods of active glycolysis and fatty acid catabolism.

At this stage of the metabolism of glucose, let us do a little bookkeeping.

- Glucose enters the cell and is phosphorylated to glucose 6-phosphate at the expense of one ATP.

Glycolysis

- The six-carbon glucose 6-phosphate enters the glycolytic pathway, where it is converted to 2 mol of the three-carbon compound, pyruvate.
- Also during glycolysis, 2 mol/mol glucose of NADH are formed.
- In addition, 4 mol ATP/mol glucose are formed, while 2 mol ATP/mol glucose are consumed for a net total production of 2 mol ATP/mol glucose.

Tricarboxylic Acid Cycle

- Preparatory to the tricarboxylic acid cycle, the pyruvate that has passed through the mitochondrial membrane from the cytosol is converted to acetyl CoA (2 mol/mol glucose, NADH (2 mol/mol glucose) and carbon dioxide (2 mol/mol glucose).
- In the tricarboxylic acid cycle, acetyl CoA is converted to carbon dioxide (4 mol/mol glucose), NADH (6 mol/mol glucose), $FADH_2$ (2 mol/mol glucose), and ATP (2 mol/mol glucose).
- Thus, from 1 mol of glucose, the tricarboxylic acid cycle (including the conversion of pyruvate to acetyl CoA) generates a total of 6 mol of carbon dioxide, 8 mol of NADH, 2 mol of $FADH_2$, and 2 mol of ATP.

Sum Total of Glycolysis and Tricarboxylic Acid Cycle

- From 1 mol of glucose, the glycolytic pathway and the tricarboxylic acid cycle together produce 6 mol carbon dioxide, 10 mol NADH, 2 mol $FADH_2$, and 4 mol ATP.

Summary

Both glycolysis and the tricarboxylic acid cycle do not produce much ATP; 2 mol per pathway. As explained earlier, the tricarboxylic acid cycle produces reducing units designated $[H:]^-$, which are transferred to and carried by NADH and $FADH_2$. These units can be viewed as two electrons traveling with a proton; a unit called a *hydride ion* (described earlier in this chapter). It is the reoxidation of NADH and $FADH_2$ by a process called *oxidative phosphorylation* (part of the electron transport system) that produces the bulk of the ATPs.

The electron transport system and oxidative phosphorylation will be the next topic in this chapter. Glycolysis occurs in the cytoplasm and the tricarboxylic acid cycle occurs in the mitochondria. The electron transport system and oxidative phosphorylation occur in the inner membrane of the mitochondria.

Summary Box 3.6

- The tricarboxylic acid cycle is the primary contributor for the formation of ATP.

- The tricarboxylic acid cycle produces hydride ions as the reducing agent to provide ATP.
- The hydride ions are transferred to NAD^+ and FAD to form NADH and $FADH_2$.
- Pyruvate dehydrogenase is a 60-protein subunit, three-enzyme, multicomponent complex that has two pentagon sides (dodecahedron).
- Pyruvate dehydrogenase converts pyruvate to acetyl CoA.
- Acetyl CoA subsequently is converted to citrate, *cis*-aconitic acid, isocitrate, α-ketoglutarate, succinate, fumarate, malate and oxaloacetate.
- The above-mentioned cycle produces 6 mol CO_2, 10 mol NADH, 2 mol $FADH_2$, and 4 mol ATP.

THE ELECTRON TRANSPORT SYSTEM AND OXIDATIVE PHOSPHORYLATION

The bulk of ATP formation is dependent upon the reoxidation of the NADH and $FADH_2$. The reoxidation of NADH and $FADH_2$ in turn is dependent upon electron transport by several compounds, to be described shortly, and ultimately the availability of molecular oxygen (O_2). The transfer of electrons is in a consecutive manner from the more electronegative components to the less electronegative components (more electropositive). At three points in the sequence, ATP is formed from ADP and inorganic phosphate in a process called *oxidative phosphorylation*. The components involved in electron transfer are collectively called the *respiratory chain* or *electron transport system*. The electron transport system is located in the inner membrane of the mitochondrion. The system is composed of the following:

- Flavin mononucleotide (FMN).

- Cytochromes b, c, c_1, a and a_3; a group of compounds similar in structure to that of heme and alternately called *heme b, c, c_1, a,* and *a_3*. Cytochrome a_3 is also called *cytochrome oxidase*.
- Coenzyme Q (CoQ) is also called *ubiquinone*.

n equals 10 in human liver

- Protein clusters of iron–sulfur proteins (FeS).

The above-mentioned categories of compounds undergo redox reactions in the following manner:

- Flavin mononucleotide undergoes redox reactions in a manner identical to FAD.
- The cytochromes undergo redox reactions through iron or copper ions as follows:

$$\underset{\text{Oxidized}}{Fe^{+3}} + e^- \rightarrow \underset{\text{Reduced}}{Fe^{+2}}$$

$$\underset{\text{Oxidized}}{Cu^{+2}} + e^- \rightarrow \underset{\text{Reduced}}{Cu^{+1}}$$

A representative structure of cytochrome is represented in Figure 3.14. As you can see, this is an iron-carrying cytochrome.

- CoQ undergoes redox reactions through quinol, and semiquinone and quinine forms.

| Quinone (oxidized) | Semiquinone (1 electron reduced) | Quinol (2 electron reduced) |

- Iron–sulfur proteins conduct electrons between the complexes that compose the chain, probably through reduction of iron.

Steps in the Electron Transport System

Refer to Figure 3.15.

- NADH enters the system at Complex 1 transferring two electrons to FMN; a reaction mediated by NADH-Q reductase (NAD dehydrogenase). The complex is composed of 46 proteins in addition to FMN, and approximately 9 Fe–S protein clusters.

$$NADH + H^+ \rightarrow FMNH_2 + NAD^+$$

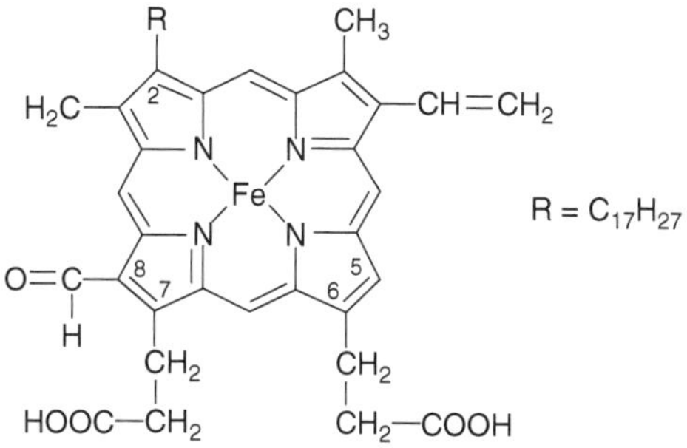

R = $C_{17}H_{27}$

Figure 3.14   A cytochrome.

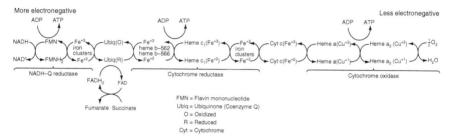

Figure 3.15   The electron transport system.

The pair of electrons is next passed to iron–sulfur protein clusters one at a time.

$$2\left(FMNH_2 + \left(Fe^{+3} - S\right)_{ox}\right) \rightarrow FMNH^{\cdot} + \left(Fe^{+2} - S\right)_{red} + H^+)$$

The FeS protein cluster is reoxidized by transferring its electron to the next iron–sulfur cluster in the system. The electrons are next transferred from the iron–sulfur protein clusters to CoQ. CoQ accepts two electrons/NADH and two protons/NADH to become $CoQH_2$.

$$CoQ + 2e + 2H^+ \rightarrow CoQH_2$$

- The next component in the chain is Complex II. It is composed of five proteins. As you may recall, succinate is oxidized to fumarate in the tricarboxylic cycle in a reaction mediated by succinate dehydrogenase (or succinate–CoQ reductase). Concomitantly, FAD is reduced to $FADH_2$. At Complex II, $FADH_2$ enters the chain and is oxidized to FAD by the transfer of electrons to iron–sulfur protein clusters and next to CoQ.
- The reduced CoQ ($CoQH_2$) transfers its electrons to cytochromes in Complex III. The enzyme in this complex is CoQ–cytochrome c reductase. It is composed of 11 proteins. Just as with iron–sulfur protein clusters, CoQ

carries two electrons but the recipient, the cytochromes, only carry one electron at a time. In this case, there are four cytochromes: cytochromes b_L (or b_{566}), bH (or b_{560}),[7] c_1, and c are involved in what is often called the Q *cycle*. The steps in this exquisite cycle are as follows:

1. Reduced CoQ ($CoQH_2$) binds to two Complex III proteins, an iron–sulfur protein (often called the *Rieske protein*) and cytochrome b. Oxidized CoQ binds to another site of Complex III.

2. $CoQH_2$ gives one electron to the iron–sulfur protein and one electron to cytochrome b_L.

$$CoQH_2 + FeS_{(oxid)} \rightarrow CoQH^{\cdot} + H^{+} + FeS_{(red)}$$

$$CoQH^{\cdot} + cyt\ b_{L(oxid)} \rightarrow CoQ + H^{+} + cyt\ b_{L(red)}$$

3. The reduced iron–sulfur protein cluster ($FeS_{(red)}$) transfers a single electron to oxidized cytochrome c_1, which in turn transfers an electron to oxidized cytochrome c (cyt $c_{(oxid)}$).

$$FeS_{(red)} + cyt\ c_{1(oxid)} \rightarrow FeS_{(oxid)} + cyt\ c_{1(red)}$$

$$cyt\ c_{1(red)} + cyt\ c_{(oxid)} \rightarrow cyt\ c_{1(oxid)} + cyt\ c_{(red)}$$

4. The reduced cytochrome b_L (cyt $b_{L(red)}$) transfers one electron to cytochrome b_H (cyt $b_{H(oxid)}$) and next to the second CoQ to form $CoQH^{\cdot}$. The proton that accompanies the electron is derived from glutamate, an amino acid that is located in the protein of Complex III.

$$cyt\ b_{L(red)} + cyt\ b_{H(oxid)} \rightarrow cyt\ b_{L(oxid)} + cyt\ b_{H(red)}$$

$$cyt\ b_{H(red)} + H^{+} + CoQ \rightarrow cyt\ b_{H(oxid)} + CoQH^{\cdot}$$

5. What happens to the $CoQH^{\cdot}$? A second cycle of the Q cycle supplies a second electron and another H^{+} to produce $CoQH_2$.

6. The $CoQH_2$ produced in step 5 is recycled (reoxidized) in step 2.

In summary, the products of the Q cycle are four protons (step 2, two cycles) that enter the intermembrane space of the mitochondrion. Their importance will be discussed in the oxidation phosphorylation section. Two protons enter Complex III from the matrix. Two molecules of cytochrome c are reduced.

- At Complex IV, two molecules of reduced cytochrome c are reoxidized in a reaction mediated by cytochrome c oxidase. It consists of 13 protein subunits, heme a and a_3, two copper atoms, and molecular oxygen (O_2). A single electron is transported from cytochrome c to the copper atom $A(Cu^{+2} \rightarrow Cu^{+1})$, then to heme a ($Fe^{+3} \rightarrow Fe^{+2}$), and then to the

[7]L and H denote lower and higher potential. B_H is also called b_{562} in the literature.

center, which contains both $heme_3$ $(Fe^{+3} \rightarrow Fe^{+2})$ and copper atom B $(Cu^{+2} \rightarrow Cu^{+1})$. Molecular oxygen (O_2) that diffuses into the cell oxidizes copper and reduces to H_2O.

$$O_2 + 4e^- + 4H^+ \rightarrow 2H_2O$$

The overall equation for Complex IV is

$$4 \text{ cyt } c \left(Fe^{+2}\right) + 8H^+ + O_2 \rightarrow 4 \text{ cyt } c \left(Fe^{+3}\right) + 2H_2O + 4H^+$$

Eight protons and one oxygen molecule are used to oxidize four reduced (Fe^{+2}) cytochrome c molecules to four oxidized (Fe^{+3}) cytochrome c molecules, two water molecules, and four protons. Thus, there is a net gain of four protons in the intermembrane space of the mitochondrion.

OXIDATIVE PHOSPHORYLATION (ATP SYNTHASE)

As stated earlier, the purpose of the electron transport system is the production of ATP. How does all of the foregoing biochemistry relate to ATP formation, since little mention is made of ATP as we described the electron transport system? The answer to that question rests in the protons (actually hydride ions) produced by the complexes that compose the electron transport system.

As depicted in Figure 3.16, the reoxidation of NADH and $FADH_2$ is instrumental in producing protons that are transferred from the matrix of the mitochondrion to the inner membrane space and then to the cytosol. Thus, a proton gradient (pH gradient) of 1.4 pH units is developed, with lower concentrations of protons in the matrix and higher concentrations of protons in the inner membrane space. Therefore, the mitochondrion is alkaline in the matrix and acid in the inner membrane space, and a membrane potential, which is negative in the matrix, is developed.

ATP is formed from the phosphorylation of ADP in the presence of ATP synthase (ATPase). The ATPase is contained in a spherical particle attached to the mitochondrial inner membrane by a stalk and base.

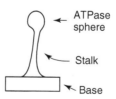

The ATPase is found at Complex I (the entry site for NADH), Complex III (the entry site for $FADH_2$), and Complex IV. The energy for the addition of phosphorus to ADP to produce ATP derives from the pH gradient. This hypothesis, first postulated by Peter Mitchel, is termed the *chemiosmotic theory* of oxidative phosphorylation. For this, Peter Mitchel earned the Nobel Prize.

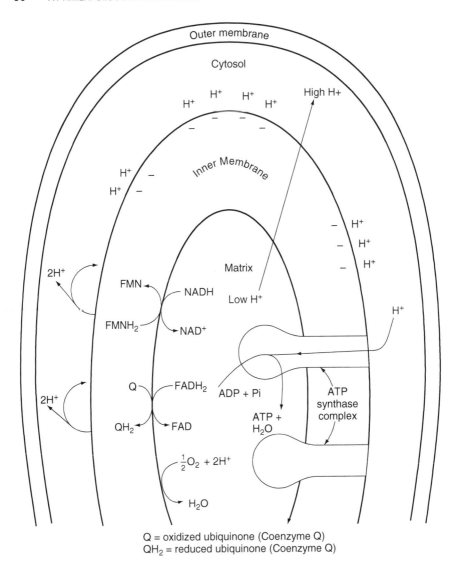

Figure 3.16   Chemiosmotic hypothesis.

It should be emphasized that NADH enters at Complex 1 and its electrons pass through ATPase sites at Complexes I, III, and IV, thus producing three ATPs. $FADH_2$ enters at Complex III and its electrons pass through ATPase sites at Complexes III and IV, thus producing two ATPs.

Electron transport is tightly coupled to phosphorylation of ADP to form ATP. The availability of ADP controls the rate of the oxidation. On the other hand, a decrease in oxygen reduces phosphorylation.

Shuttles

Before we total the sum of ATPs that derive from glucose, we must not forget the NADH that was produced in the glycolytic pathway by glyceraldehyde 3-phosphate dehydrogenase. Glycolysis occurs in the cytoplasm and ATP formation occurs in the mitochondrion. Furthermore, cytoplasmic NADH cannot pass through the inner mitochondrial membrane. The way around this is through shuttles. In the skeletal muscle and brain, the glycerol 3-phosphate shuttle operates and in the heart and liver, the malate-aspartate shuttle operates.

Glycerol 3-Phosphate Shuttle

On the cytoplasm side of the mitochondrial inner membrane, dihydroxyacetone phosphate is oxidized to glycerol phosphate by cytosolic glycerol 3-phosphate dehydrogenase. Concomitantly, NADH is converted to NAD^+. Mitochondrial membrane glycerophosphate dehydrogenase converts glycerol phosphate back to dihydroxylacetone while oxidizing FAD to $FADH_2$. This shuttle generates two ATPs.

Malate–Aspartate Shuttle

On the cytoplasm side of the mitochondrial membrane, oxaloacetate is oxidized to malate by cytosolic malate dehydrogenase. Concomitantly, NADH is converted to NAD^+. Oxaloacetate is derived from aspartate in a reaction mediated by cytoplasmic aspartate aminotransferase. Mitochondrial membrane malate dehydrogenase converts malate back to oxaloacetate by mitochondrial malate dehydrogenase, while NAD^+ is reduced to NADH. The oxaloacetate is converted to aspartate in a reaction catalyzed by mitochondrial aspartate aminotransferase. This shuttle generates three ATPs.

Moles ATP Produced by Oxidative Phosphorylation from 1 mol of Glucose

The tricarboxylic acid pathway produces 2 mol of ATP from succinyl-CoA synthetase, 8 mol NADH, and 2 mol $FADH_2$.

	Total Moles of ATP per mole of Glucose
Succinyl-CoA synthetase	2
NADH	24
$FADH_2$	4

Glycolysis produces 2 mol ATP directly. In addition, 2 mol of NADH produces either 4 mol ATP if occurring in the skeletal muscle or brain and 6 mol ATP if occurring in the heart or liver.

In summary, the total moles of ATP produced by oxidation of glucose is 36 or 38 depending on cell type.

Summary Box 3.7

- ATP production derives from the reoxidation of NADH and $FADH_2$.
- Reoxidation of NADH and $FADH_2$ occurs through the electron transport system.
- NADH enters at Complex I, transferring two electrons to FMN. These electrons are passed through Complexes I, II, III, and IV to O_2 to produce H_2O.
- $FADH_2$ enters the electron transport system at Complex II and is reoxidized to FAD while reducing an iron–sulfur protein. FADs electrons are transported through Complexes II, III, and IV to O_2 to produce H_2O.
- ATP synthase catalyzes the synthesis of ATP from ADP and inorganic phosphate. ATP synthase occurs at Complex I, III, and IV. Thus, 1 mol of NADH produces 3 mol of ATP and 1 mol of $FADH_2$ produces 2 mol of ATP.
- The energy source for the synthesis of ATP is the proton gradient across the matrix and the inner membrane of the mitochondrion. The pH of the matrix is about 1.4 units lower than the pH of the inner membrane. This theory is called the *chemiosmotic theory*.
- NADH formed by glycolysis in the cytoplasm is transported across the mitochondrion membrane by the glycerol 3-phosphate and the malate-aspartate shuttles.
- The glycerol 3-phosphate shuttle present in the skeletal muscle and brain produces $FADH_2$ in the mitochondrion and, therefore, 2 mol of ATP. The malate-aspartate shuttle present in the heart and liver produces NADH in the mitochondrion and, therefore, 3 mol of ATP.
- The total moles of ATP produced by glycolysis, the tricarboxylic acid cycle, electron transport, and oxidative phosphorylation is 36 if occurring in the skeletal muscle or brain and 38 if occurring in the heart or liver.

THE PHOSPHOGLUCONATE OXIDATIVE CYCLE

The phosphogluconate oxidative cycle (often called the *pentose phosphate shunt* or the *hexose monophosphate shunt*) is a secondary pathway for the oxidation of glucose 6-phosphate. As you can see from Figure 3.17, its principal products are ribose 5-phosphate, CO_2, and NADPH. Ribose 5-phosphate is the sugar found in nucleotides (and therefore nucleic acids), and the cofactors NAD^+, $NADP^+$, and FAD. NADPH is used for amino acid, fatty acid, and steroid biosynthesis. Unlike the tricarboxylic acid cycle, only one reducing agent is produced per mole glucose and no ATP molecules are formed. In fact, one ATP is required, if the phosphorylation of glucose to glucose 6-phosphate is included in the cycle.

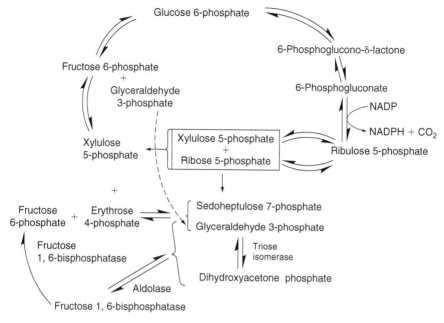

Figure 3.17 Phosphogluconate oxidative cycle.

The overall equation for the phosphogluconate oxidative pathway is

$$6\,\text{Glucose 6-phosphate} + 12\text{NADP}^+ \rightarrow 5\,\text{Glucose 6-phosphate}$$
$$+ 6\,\text{CO}_2 + 12\,\text{NADPH} + 12\text{H}^+ + \text{Pi}$$

STEPS IN THE PHOSPHOGLUCONATE OXIDATIVE CYCLE

Glucose 6-Phosphate Dehydrogenase; Lactonase

Glucose 6-phosphate dehydrogenase catalyzes the conversion of glucose 6-phosphate to 6-phosphoglucono δ-lactone, which is hydrolyzed to 6-phosphogluconate by lactonase.

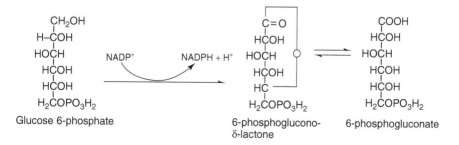

A deficiency of glucose 6-phosphate dehydrogenase leads to a defect in the erythrocyte membrane, which reduces its stability. The resulting hemolytic anemia can be severe.

6-Phosphogluconate Dehydrogenase; Ribose and Xylulose

6-Phosphogluconate dehydrogenase catalyzes the oxidative decarboxylation of 6-phosphogluconate through the intermediate, 3-keto-6-phosphogluconate to D-ribulose 5-phosphate. This pathway is especially active in the liver and adipose tissue.

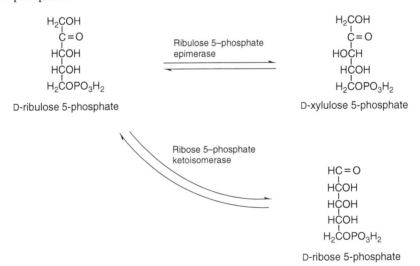

COOH	COOH	CO$_2$
HCOH	HCOH	+ H$_2$COH
HOCH	C=O	C=O
HCOH	HCOH	HCOH
HCOH	HCOH	HCOH
H$_2$COPO$_3$H$_2$	H$_2$COPO$_3$H$_2$	H$_2$COPO$_3$H$_2$
6-phosphogluconate	3-keto-6-phospho-gluconate	D-ribulose 5-phosphate

NADP$^+$ → Mn^{+2} → NADPH + H$^+$

This is the process that produces 5-carbon monosaccharides (pentoses) from 6-carbon monosaccharides (hexoses). As you will see in the subsequent text, ribulose isomerizes to D-ribose 5-phosphate, a major constituent of nucleotides (e.g., ATP, GTP, etc.), nucleic acids, and cofactors NAD$^+$, NADP$^+$, FMN, and FAD. Alternatively, ribulose 5-phosphate epimerizes to D-xylulose 5-phosphate.

H$_2$COH
C=O
HCOH
HCOH
H$_2$COPO$_3$H$_2$

D-ribulose 5-phosphate

Ribulose 5–phosphate epimerase

H$_2$COH
C=O
HOCH
HCOH
H$_2$COPO$_3$H$_2$

D-xylulose 5-phosphate

Ribose 5–phosphate ketoisomerase

HC=O
HCOH
HCOH
HCOH
H$_2$COPO$_3$H$_2$

D-ribose 5-phosphate

The catabolism of nucleic acids also utilizes this pathway. As the products of the degradation of the nucleic acids lead to products, that is, ribose 5-phosphate, which are utilized in other pathways, this portion of the phosphogluconate oxidative pathway is referred to as a salvage pathway. Salvage pathways allow the utilization of the products of catabolic metabolism.

Transaldolase

Transaldolase catalyzes the condensation of D-sedoheptulose 7-phosphate and D-glyceraldehyde 3-phosphate to form a 6-carbon ketose, fructose 6-phosphate, and a 4-carbon aldose, D-erythrose 4-phosphate. The reaction is similar to the aldolase-mediated reaction described in the section on the glycolytic pathway.

D-sedoheptulose 7-phosphate + D-glyceraldehyde 3-phosphate ⇌ D-fructose 6-phosphate + D-erthyrose 4-phosphate

Fructose 6-phosphate is isomerized to glucose 6-phosphate in a reaction catalyzed by the glycolytic pathway enzyme, phosphohexose isomerase.

Transketolase

The reaction of D-xylulose 5-phosphate and D-erythrose 4-phosphate to form D-fructose 6-phosphate and D-glyceraldehyde 3-phosphate is catalyzed by transketolase.

D-xylulose 5-phosphate + D-erthyrose 4-phosphate ⇌ (Transketolase) D-fructose 6-phosphate + D-glyceraldehyde 3-phosphate

The fructose 6-phosphate produced by this reaction is also converted to glucose 6-phosphate by phosphohexose isomerase.

The Fate of Glyceraldehyde 3-phosphate

Glyceraldehyde 3-phosphate is produced in two reactions of the phosphogluconate oxidative cycle. It is converted by enzymes of the glycolytic pathway,

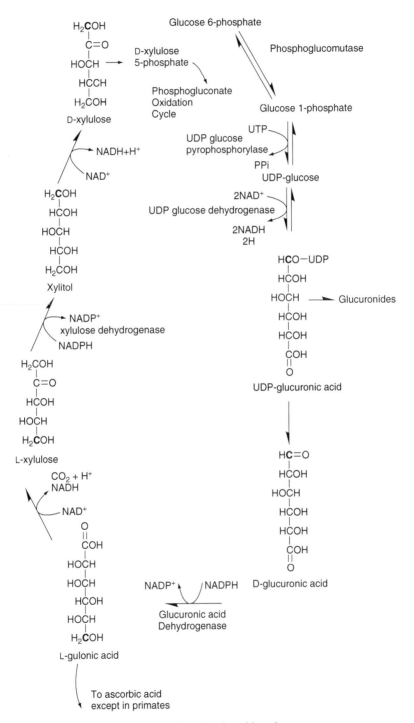

Figure 3.18   Uronic acid cycle.

namely, triose isomerase, aldolase, and fructose 1,6-bisphosphatase to fructose 6-phosphate, which is isomerized to glucose 6-phosphate.

The phosphogluconate cycle is complex. The enzymes of the pathway are located in the cytoplasm. For every 6 mol (or 36 carbon atoms) of glucose 6-phosphate that enter the cycle, 5 mol (30 carbon atoms) of glucose 6-phosphate leave and 1 mol (6 carbon atoms) of glucose 6-phosphate are converted to 6 mol (6 carbon atoms) of carbon dioxide. For each mole of carbon dioxide produced, two electrons are freed and are utilized to reduce 12 mol of NADP$^+$ to NADPH.

URONIC ACID PATHWAY

The uronic acid pathway is an alternative pathway for the metabolism of glucose 6-phosphate. The pathway provides a mechanism for the solubilization of drugs, steroid hormones, and bilirubin. UDP-glucuronic acid reacts with these substances to produce water-soluble glucuronides, which are eliminated by the kidneys.

Figure 3.18 shows the uronic acid pathway, which is sometimes referred to more specifically as the glucuronic acid oxidation pathway. The uronic pathway draws some interest because it involves the conversion of an L-isomer to the D form, that is, L-xylulose to D-xylulose. The product of the pathway, D-xylulose 5-phosphate, enters the phosphogluconate oxidative pathway. Carbon 1 of glucose is followed through the pathway by denoting it in bold-faced print.

Summary Box 3.8

- The phosphogluconate oxidative cycle is a secondary pathway in which glucose 6-phosphate is oxidized.
- An important product is ribose 5-phosphate; the sugar moiety for nucleosides, -nucleotides, nucleic acids, and the cofactors NAD$^+$, NADP$^+$, and FAD.
- In the cycle, the enzyme, 6-phosphogluconate dehydrogenase, catalyzes the oxidative decarboxylation of 6-phosphogluconate (a six-carbon sugar) to produce D-ribulose 5-phosphate (a five carbon sugar). D-Ribulose 5-phosphate isomerizes to D-ribose 5-phosphate.
- Transaldolase catalyzes the formation of a six-carbon ketose, fructose 6-phosphate and the four-carbon, D-erythrose 4-phosphate.
- Transketolase catalyzes the formation of fructose 6-phosphate and D-glyceraldehyde 3-phosphate from D-xylulose 5-phosphate and D-erythrose 4-phosphate.
- Glyceraldehyde 3-phosphate is converted to glucose 6-phosphate by enzymes of the glycolytic pathway.

- The phosphogluconate oxidative pathway produces 5 mol of glucose 6-phosphate and 6 mol of carbon dioxide and 12 mol of NADPH.
- The primary product of the uronic pathway is UDP-glucuronic acid.
- The uronic pathway also produces D-xylulose from L-xylulose.

HEXOSAMINE BIOSYNTHESIS PATHWAY

This pathway does not require ATP or any cofactors (NAD$^+$, FAD, etc.). As you can see from Figure 3.19, there are six steps to the pathway.

1. Glucose 6-phosphate isomerase converts glucose 6-phosphate to fructose 6-phosphate.

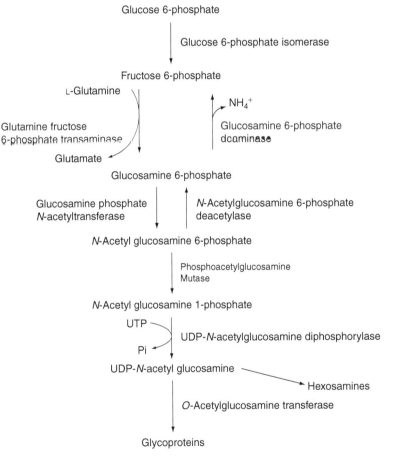

Figure 3.19   Hexosamine biosynthesis pathway.

2. Glutamine fructose 6-phosphate transaminase places the NH_4^+ group from glutamine on fructose 6-phosphate to form glucosamine 6-phosphate and L-glutamate. The reverse is catalyzed by glucosamine 6-phosphate deaminase.

3. Glucosamine 6-phosphate is acetylated by glucosamine phosphate N-acetyltransferase to produce N-acetylglucosamine 6-phosphate. The reverse reaction is catalyzed by N-acetylglucosamine 6-phosphate deacetylase.

4. N-Acetylglucosamine 6-phosphate is converted to the 1-phosphate form by phosphoacetylglucosamine mutase.

5. N-Acetylglucosamine 1-phosphate is coupled with UTP to form UDP-N-acetylglucosamine in a reaction mediated by UDP-N-acetylglucosamine diphosphorylase.

6. UDP-N-acetylglucosamine can form other hexosamines to glycosylate proteins or couple to protein serine and threonine groups in reactions catalyzed by O-acetylglucosamine transferases.

Summary Box 3.9

- The hexosamine biosynthesis pathway does not require ATP and cofactors.
- The initial step in the pathway converts glucose 6-phosphate to form fructose 6-phosphate.
- Glutamine is utilized to provide its ammonium group to fructose 6-phosphate.
- After acetylation, isomerization of the phosphate group from position 6 to the position 1, and coupling with UDP the product UDP-N-acetylglucosamine forms other hexosamines or couples to proteins.

Gluconeogenesis will be the last glucose metabolic pathway to be discussed in this chapter. It seems fitting to end the chapter with this pathway as it emphasizes the rule that we enunciated at the beginning of this chapter: a metabolic pathway is not simply reversed when going from product to starting substance.

THE STEPS OF GLUCONEOGENESIS

The steps in gluconeogenesis are shown in Figure 3.20. Contrast them with those shown for glycolysis (downward arrows).

Gluconeogenesis is described as follows:

1. Lactate is oxidized to pyruvate in a reaction catalyzed by lactate dehydrogenase (LD).

2. Pyruvate crosses the mitochondrial membrane and is converted to oxaloacetate in a reaction catalyzed by pyruvate carboxylase.

3. As previously described, the mitochondrial membrane is impermeable to oxaloacetate. How does the oxaloacetate reach the remaining enzymes of

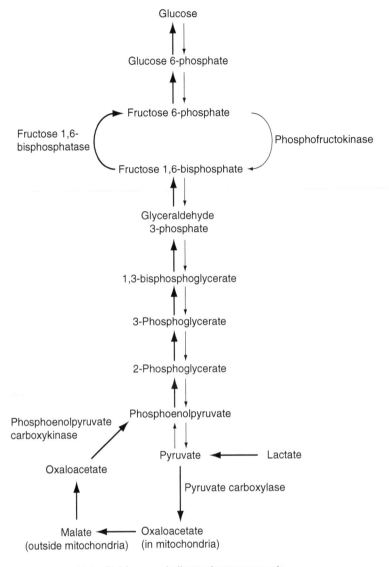

Note: **Bold** arrows indicate gluconeogenesis

Figure 3.20   Gluconeogenesis.

gluconeogenesis which are situated in the cytoplasm? Review the mechanism for the reoxidation of extramitochondrial NADH (malate–aspartate shuttle). Oxalacetate is converted to malate by mitochondrial malate dehydrogenase. The mitochondrial membrane is permeable to malate. When malate reaches the cytoplasm, it is reoxidized to oxaloacetate by a cytoplasm-located malate dehydrogenase.

4. Phosphoenolpyruvate carboxykinase, an enzyme unique to gluconeogenesis, catalyzes the conversion of oxaloacetate to phosphoenolpyruvate.

5. Glycolytic pathway enzymes convert phosphoenolpyruvate through the intermediates: 2-phosphoglycerate, 3-phosphoglycerate, 1,3-bisphosphate glycerate, glyceraldehyde 3-phosphate to fructose 1,6-bisphosphate.

6. Fructose 1,6-bisphosphatase and PFK provide the principal regulatory mechanism for gluconeogenesis. Fructose 1,6-bisphosphatase catalyzes the hydrolysis of fructose 1,6-bisphosphate to fructose 6-phosphate (gluconeogenesis). The reversal of this reaction, the phosphorylation of fructose 6-phosphate to fructose 1,6-bisphosphate, is a step in glycolysis. The phosphorylation is catalyzed by PFK.

   Fructose 1,6-bisphosphatase and phosphofructose kinase are positioned as traffic directors at the turnstile of the gluconeogenesis and glycolytic pathways, determining the direction of flow of the molecules. It is not surprising that AMP and ATP are effectors of the two enzymes. An increase in AMP levels reduces fructose 1,6-bisphosphatase activity and increases PFK activity. This is consistent with an inverse relationship between AMP and ATP levels. Thus, an increase in AMP levels signals a deficiency in ATP concentrations and leads to the activation of PFK (which favors glycolysis) and the inhibition of fructose 1,6-bisphosphatase (which reduces gluconeogenesis). ATP is an allosteric inhibitor of PFK. The inhibition of PFK by ATP reduces glycolysis when there are sufficient quantities of ATP.

7. Phosphoglucoisomerase, a glycolytic pathway enzyme, converts fructose 6-phosphate to glucose 6-phosphate and an enzyme unique to gluconeogenesis, glucose 6-phosphatase, carries out the final conversion to glucose.

Hormones such as epinephrine, cortisol, glucagon, and insulin play important roles in the regulation of glycolysis and gluconeogenesis. The rates of gluconeogenesis and glycolysis are closely regulated by blood glucose levels. The activities of hexokinase, PFK, and PK are increased after a meal rich in carbohydrates and reduced between meals. The enzymes of gluconeogenesis, glucose 6-phosphatase, fructose 1,6-bisphosphatase, and pyruvate carboxylase are reduced in activity after the ingestion of a carbohydrate-rich meal.

Summary Box 3.10

- Pyruvate, produced in the cytoplasm from lactate, crosses the mitochondrial membrane to be converted to oxaloacetate.
- Oxaloacetate uses the malate–aspartate shuttle to through the membrane into the cytoplasm as malate and then reoxidized to oxaloacetate.

- Afterwards, oxaloacetate is converted through several intermediate substances to fructose 1,6-bisphosphate.
- At this point, fructose 1,6-bisphosphate has two pathways that it may take: glycolysis or gluconeogenesis.
- Increased AMP reduces fructose 1,6-bisphosphatase activity and increases PFK activity favoring glycolysis. Increased ATP inhibits PFK and reduces glycolysis.
- Phosphoglucoisomerase converts fructose 6-phosphate to glucose 6-phosphate and glucose 6-phosphatase produces glucose.
- Epinephrine, cortisol, glucagon, insulin, and blood glucose levels regulate gluconeogenesis and glycolysis.

CONCLUSIONS

Glucose is a small molecule that is necessary for life. Its essentiality for life is evident from our description in this chapter of the many metabolic pathways that involve glucose. Glucose is the substance that produces the universal energy currency, ATP. In addition, glucose enters a pathway that produces ribose, which is the sugar contained in nucleosides, nucleotides, and nucleic acids and the cofactors, NAD^+, $NADP^+$, and FAD. Glucose is so important for life that it is stored as glycogen for situations where there is glucose starvation. Thus, a disorder in glucose metabolism has far-reaching consequences. What if glucose could not get through the cell membrane to undergo the metabolism described in this chapter? Next, we will describe how glucose enters the cell and how its metabolism is regulated.

QUESTIONS

Some research may be necessary to answer these questions.

3.1 What are the products of the hydrolysis of amylose and amylopectin with α-amylase?

3.2 During the incubation of saliva with starch, samples were taken at various intervals. These samples were reacted with iodine. What is observed? Explain.

3.3 Glucagon is a hormone that stimulates glycogenolysis. What effect would administration of this hormone have on blood glucose levels in a normal individual? In an individual with glycogen synthase deficiency?

3.4 What do you suppose is the effect of amylopectinosis on blood glucose? Why does amylopectin accumulate and not amylose?

3.5 Glucagon increases glycogenolysis in the liver; epinephrine acts in a similar manner in the muscle. Describe the mechanism of action of glucagon and epinephrine.

3.6 What does the formation of a Schiff base intermediate indicate about the conformation of fructose 1,6-bisphosphate? Suggestion: Look up Schiff bases in an organic chemistry textbook.

3.7 Each cycle of the tricarboxylic acid cycle produces:

____moles of ATP from each mole of acetyl CoA

____moles of ATP from each mole of pyruvate

____moles of ATP from each mole of glucose

____moles of ATP for each mole of glucose when the ATP produced by glycolysis are included.

3.8 Why do you suppose that there are no reported congenital defects involving deficiencies or absences of the tricarboxylic acid cycle enzymes?

3.9 Determine the moles of ATP produced by one cycle of the tricarboxylic acid pathway (exclude pyruvate oxidative decarboxylation). How many moles of ATP are there per mole of glucose?

3.10 How many moles of ATP are produced by the reoxidation of the NADH produced by glyceraldehyde 3-phosphate dehydrogenase? How many moles of ATP would be produced in this manner per mole of glucose entering the glycolytic pathway?

3.11 Explain the difference between isomerization and epimerization.

3.12 There four steps in the pathway where a metabolic defect results in an accumulation of L-xylulose. What are they?

GLOSSARY

Active transport   Transport of molecules across a membrane from low concentration to high concentration with the outlay of energy usually provided by conversion of ATP to ADP by ATPase.

Aerobic   Occurs in the presence of oxygen.

Aldehyde   A molecule that contains a $C=O$ functional group at an end.

Anaerobic   Occurs in the absence of oxygen.

ATP synthase   Enzyme that uses the energy produced by the electron transport system to generate ATP.

Carbohydrates   Group of compounds found in foods. Monomers (monosaccharides) are compounds that have an aldehyde or ketone group and several hydroxyl groups. Polymers (di- and polysaccharides) have glycoside linkages that bond several monosaccharides by their aldehyde or ketone groups to the hydroxyl group of the next monosaccharide forming a glycoside bond.

Cellulose   Polysaccharide composed of glucose units in $\beta(1\rightarrow4)$ linkage.

Coenzyme Q (CoQ, ubiquinone)   An organic component of the electron transport system. It is involved in redox reactions.

Coupling   Refers to a reaction that is endergonic receiving energy from an exergonic source such as ATP.

Cyclic AMP   A derivative of adenosine monophosphate that has its $3'$ and $5'$ groups linked by a phosphate (phosphodiester bond). It serves as a messenger to activate protein kinases and other substances.

Cytochromes   Heme containing proteins found in mitochondria that serve as electron carriers through the redox reactions of iron and copper ions.

Disaccharide   Two monosaccharides linked via a glycosidic bond.

Electron transport system   A chain of electron carriers arranged sequentially in order of increasing redox potential that transfer electrons from NADH and $FADH_2$ to O_2.

Facilitated transport   Transport of molecules across a membrane from a high to a low concentration by simple diffusion.

Fatty acid (or free fatty acid)   A molecule that has a carboxyl group attached to an unbranched aliphatic chain.

Flavin adenine dinucleotide (FAD)   A cofactor composed of riboflavin and ADP linked together by a phosphodiester bond.

Flavin mononucleotide (FMN)   A cofactor composed of riboflavin phosphorylated on the terminal carbon of the ribitol residue.

Free energy (ΔG)   Denotes the energy of a reaction that does useful work.

Gluconeogenesis   Metabolic pathway for the synthesis of glucose from lactate.

Glycerol 3-phosphate shuttle   A metabolic pathway for transferring electrons from NADH in the cytoplasm across the mitochondrial membrane into the matrix. Present in skeletal muscle and brain.

Glycogen   A glucose storage polysaccharide.

Glycogenesis   A metabolic pathway that produces glycogen from glucose units.

Glycogenolysis   A metabolic pathway that hydrolysis glucose monomers from glycogen.

Glycolysis   A metabolic pathway that converts glucose and other carbohydrates to pyruvate.

Glycosidic bond (linkage)   A bond between the aldehyde group on a monosaccharide and the hydroxyl group on another monosaccharide.

Heme   An organic compound with a porphyrin ring that has complexed an iron ion in its center.

Iron–sulfur proteins (clusters)   Clusters of proteins that have sequestered an array of iron and sulfur ions (Fe_xS_y) that function redox reactions in the electron transport system.

Ketone   A molecule that contains a C=O functional group in the interior.

Malate–aspartate shuttle   A metabolic pathway for transferring electrons from NADH in the cytoplasm across the mitochondrial membrane into the matrix. Present in heart and liver.

Monosaccharide   A polyhydroxyl aldehyde or ketone.

Nicotinamide adenine dinucleotide (NAD^+)   A cofactor composed of niacinamide ribose $5'$-phosphate and adenosine $5'$-phosphate linked by their phosphates (phosphodiester bond).

Nicotinamide adenine dinucleotide phosphate ($NADP^+$)   A phosphorylated derivative of NAD^+. Phosphate is bonded to the $2'$-position of adenosine.

Oxidative decarboxylation   Loss of electron(s) from a compound with concomitant loss of CO_2.

Oxidative phosphorylation   Formation of ATP from ADP catalyzed by ATP synthase and energized by the electron transport system.

Pentoses   A monosaccharide with five carbon atoms.

Phosphogluconate pathway (cycle)   The second oxidative metabolic pathway for glucose. It produces ribose $5'$-phosphate and NADPH.

Polysaccharide   More than two monosaccharides linked by glycosidic bonds.

Pyrophosphate

$$\text{--P}\overset{\overset{\displaystyle O^-}{|}}{\underset{\underset{\displaystyle O}{\|}}{}}\text{--O--P}\overset{\overset{\displaystyle O^-}{|}}{\underset{\underset{\displaystyle O}{\|}}{}}\text{--O}^-$$

Q-cycle   A part of the electron transport system that uses the oxidation of cytochrome *c* to produce four protons per cycle into the intermembrane space of the mitochondrion.

Rapoport–Luebering shunt   A metabolic pathway found in erythrocytes that converts 1,3-bisphosphoglycerate to 2,3-bisphosphoglycerate. The latter compound associates with hemoglobin to form a complex that has a greater affinity for oxygen than hemoglobin alone.

Redox reactions   Shorthand for oxidation–reduction reactions.

Synthase   An enzyme that catalyzes a reaction that joins molecules and does not necessitate ATP.

Synthetase   An enzyme that catalyzes a reaction that requires energy from ATP.

Tricarboxylic acid cycle   Metabolic pathway that converts acetyl CoA to CO_2, ATP, NADH, and $FADH_2$.

Ubiquinone   See coenzyme Q.

CHAPTER 4

REGULATION OF GLUCOSE METABOLISM

The human body maintains a delicate balance between glucose accumulation in the blood from diet, glycogenolysis, and gluconeogenesis and glucose removal from the blood due to transport into cells where it is metabolized producing ATP, the universal energy currency. Very little of the blood glucose is removed through the kidneys. The normal fasting range of blood glucose levels is a narrow 95–110 mg/dl. Anything that interferes with the removal process results in hyperglycemia and therefore causes diabetes.

As you may recall from Chapter 3, glucose in the circulation is transported across the cell membrane by transport proteins called *GLUTs* (glucose transporters) and *SGLTs* (sodium-glucose transporters). Intracellular glucose is phosphorylated at the 6-position producing glucose 6-phosphate. Glucose 6-phosphate is a substrate for five metabolic pathways. Insulin is obligatory for glucose transport. Thus, this chapter centers around the structure, biosynthesis, secretion, and functions of insulin. Once insulin has become attached to its receptor site on the cell membrane, it initiates a cascade of events that leads to transport of glucose across the cell membrane; the insulin signaling pathway. The insulin signaling pathway will be discussed in detail in this chapter.

Incretins are hormones secreted from the intestinal tract during the digestion of food. They cause the release of insulin. Incretins will be discussed later in this chapter. Other hormones that affect the release of insulin and regulate glucose metabolism are glucagon, epinephrine, somatotropin (growth hormone), somatostatin (SST), cortisol, adrenocorticotropic hormone (ACTH), and thyroid

Understanding Diabetes: A Biochemical Perspective, First Edition. Richard F. Dods.
© 2013 John Wiley & Sons, Inc. Published 2013 by John Wiley & Sons, Inc.

hormones. A brief review of Insulin-like growth factors (IGFs) will be included. Also covered in this chapter is a protein, FGF19, that is activated by bile acids and that stimulates glycogenesis by a distinctly different pathway than insulin.

Adenosine 5'-monophosphate-activated protein kinase (AMPK) increases glucose uptake into cells when ATP levels decrease and AMP levels increase. It maintains energy balance for the entire body. AMPK will be discussed at length in this chapter.

INSULIN

Structure

Insulin is synthesized as part of a large polypeptide with a molar mass of 11,500 Da in β-cells of the pancreatic islets of Langerhans. The human insulin precursor, preproinsulin, is depicted in Figure 4.1a. Preproinsulin is produced from ribosomes and mRNA located in the rough endoplasmic reticulum (RER). Notice that preproinsulin is a single chain protein. Proteolytic cleavage at site 1 occurs shortly after preproinsulin synthesis and produces proinsulin. The amino acid chain that is cleaved off is named the *presequence* or *signal protein* and consists of 24 residues. The amino acids that compose the presequence protein are primarily hydrophilic residues. Proinsulin also is a single chain polypeptide. It has a molar mass of 9000 Da.

Transport and Secretion of Insulin

Proinsulin is transported to the Golgi apparatus or Golgi body,[1] where it is packaged in a vesicle within the β-cell called the β-*granule*. Proteolytic cleavage at sites 2 and 3 produces insulin and C-peptide. The resulting insulin molecule consists of two polypeptide chains connected by two disulfide bonds derived from the cysteine residues (positions A7–B7 and A20–B19). The chain called the A-chain consists of 21 amino acids. It includes a third disulfide bond, A6–A11. The second chain called the B-*chain* consists of 30 amino acids. Refer to Figure 4.1b for the amino acid sequence of human insulin. Insulin has a molar mass of 6000 Da and C-peptide has a molar mass of 3000 Da.

---

[1]The Golgi body is composed of stacks of structures called *cisternae*. A human cell contains 40 to 100 stacks per Golgi body. The Golgi has four regions. Each region has specific enzymes that act on the macromolecule being transported. Preproinsulin is converted to proinsulin almost as soon as it is synthesized on the ribosomes of the RER (rough endoplasmic reticulum). Proinsulin is transported to the cis-Golgi region, which is found close to the RER. Proinsulin is transported from the cis site to the medial-Golgi region and next to the endo-Golgi region. The last region called the *trans-Golgi region* is where the proinsulin begins its conversion to insulin and where it is packaged into vesicles called β-*granules*. The transformation of proinsulin to insulin is completed in the β-granules. During the transfer to each succeeding region, the macromolecule is contained inside a vesicle. Microtubules serve as the conveyor.

(a)

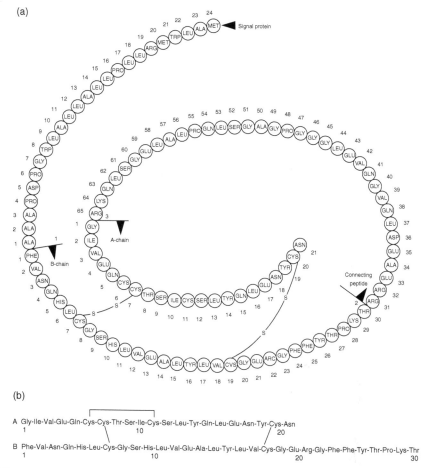

(b)

A Gly-Ile-Val-Glu-Gln-Cys-Cys-Thr-Ser-Ile-Cys-Ser-Leu-Tyr-Gln-Leu-Glu-Asn-Tyr-Cys-Asn
   1   10   20

B Phe-Val-Asn-Gln-His-Leu-Cys-Gly-Ser-His-Leu-Val-Glu-Ala-Leu-Tyr-Leu-Val-Cys-Gly-Glu-Arg-Gly-Phe-Phe-Tyr-Thr-Pro-Lys-Thr
   1   10   20   30

Figure 4.1   The amino acid sequence of (a) human preproinsulin and (b) human insulin. Modified by the author from Sures et al. (1).

The insulin molecule is complexed to zinc for storage within the molecule. Both insulin and proinsulin form a crystalline-like structure with zinc. Refer again to Figure 4.1. Zinc ion (Zn^{2+}) is complexed with the histidine residues at positions 5 and 10 of the insulin B-chain. Thus, a more stable structure is produced. Insulin, some proinsulin, and C-peptide are transported in a secretory vesicle called the *large dense core vesicle* (LDCV) by microfilaments and microtubules[2] to the periphery of the cell. This step is called the *recruitment step* (2).

[2]Microfilaments are thin, protein fibers composed of actin. Actin is a contractile protein, which, with the aid of microtubules, conveys the Golgi body vesicles from region to region and also transports the β-granule to the cell membrane for exocytosis. Microtubules are thin, cylindrical fibers composed of a protein called *tubulin*. They act as "tracks" for the movement of vesicles such as β-granules (LDCVs).

Fusion of the vesicles to the plasma membrane depends on protein complexes termed *SNAREs*. SNARE is an acronym derived from <u>S</u>oluble <u>N</u>-ethylmaleimide-sensitive factor <u>a</u>ttachment protein <u>r</u>eceptor. SNAREs mediate vesicle attachment to the plasma membrane. The SNARE complexes are composed of several proteins. The vesicle-SNARE (v-SNARE), located on the vesicle surface (specifically the LDCV) is composed of at least four proteins: syntaxin, synaptosomal-associated protein, molar mass 25 kDa (SNAP-25), vesicle associated membrane protein (VAMP2) and a calcium-binding component called *synaptotagmin* (SYT). The target-SNARE (t-SNARE), located on the plasma membrane, is composed of two of the above proteins, SNAP-25 and syntaxin, and has a third protein, mammalian uncoordinated 18 (Munc18). Munc18 binds with syntaxin to form a syntaxin–Munc18 complex. As shown in Figure 4.2, v-SNARE proteins located on the LDCV carrying the insulin, proinsulin and C-peptide, pairs its proteins with the proteins composing the t-SNARE at the plasma membrane. This step is called the *docking step*. Figure 4.2 was developed from three publications (3–5) and because of the rapid growth of information on this topic, the figure is likely to change to some extent by the time this book goes into print.

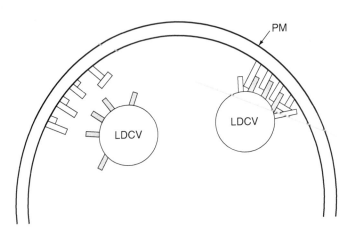

This Figure illustrates a proposed model for docking of the LDCV to the plasma membrane during insulin secretion.

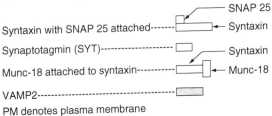

Syntaxin with SNAP 25 attached------- SNAP 25 / Syntaxin

Synaptotagmin (SYT)-------------------- Syntaxin

Munc-18 attached to syntaxin----------- Munc-18

VAMP2-------------------------------------

PM denotes plasma membrane

Figure 4.2 Docking of LDCV to the plasma membrane of a β-cell. A similar model depicts the docking phase of the vesicle carrying GLUT 4 during insulin-simulated GLUT 4 translocation from the cytoplasm to the cell surface.

Next, the complex of plasma membrane and LDCV undergoes priming. Priming involves alteration of membrane proteins by phosphatidylinositol transfer protein and type 1-phosphatidylinositol 4-phosphate 5′-kinase. This step is therefore dependent on ATP. Calcium ions, Ca^{2+}, are also obligatory for this step. Fusion of the LDCV and the plasma membrane is completed and a fusion pore or channel is formed. Equal quantities of insulin and C-peptide are released into the extracellular fluid as well as remaining proinsulin. The process is called *exocytosis* and is dependent on the contractile properties of microfilaments and microtubules. The vesicle membrane substances are retained and recycled for the formation of additional LDCVs. This process is called *endocytosis*.

The mechanism responsible for the extrusion of insulin, proinsulin, and C-peptide is also under active investigation. In resting β-cells, ATP-sensitive K^+ (K_{ATP}) channels are open and the cell is polarized. Potassium ion (K^+) flows from the β-cells to the extracellular fluid. Glucose metabolism alters the ATP/ADP ratio resulting in the closure of the K_{ATP} channels and depolarization of the β-cell and causes the opening of voltage-dependent Ca^{2+} channels (VDCCs). Movement of Ca^{2+} ions into the β-cell ensues. Movement of Ca^{2+} causes the extrusion (exocytosis) of the contents of the LDCV. The pore connecting the LDCV with the extracellular space is initially approximately 1.5 nm in diameter and widens to 12 nm. At the most dilated stage, the LDCV flattens and releases its contents (4–6).

Recently, another hypothesis has been developed for insulin secretion from the pancreatic islet β-cells. This alternative mechanism has been given the sexy name "kiss and run exocytosis." In the "kiss and run" mechanism, LDCVs fuse briefly with the plasma membrane and release the insulin, proinsulin and C-peptide, through a fusion pore. This model explains the fact that in the classical model retrieval of the LDCV membrane (endocytosis) is expected to be a slower process than is actually observed (7). A recent publication suggests that the "kiss and run" mechanism is a minor contributor to exocytosis occurring in only 6–8% of exocytotic events. According to the authors, the pore size in "kiss and run" is only 1.4–6 nm and, therefore, is too small for extruding the contents of the LDCV. They note that some vesicles at transient pores move away from the site and fail to secrete insulin (8).

Proinsulin has less than 10% of the physiological activity of insulin and C-peptide has no known activity. About 5% of proinsulin is removed on its pass through the liver and 50% is removed by the kidneys. Negligible amounts of C-peptide are removed by the liver and 70% is removed by the kidneys. Normally, release of insulin even under maximal stimulation involves only a small percentage of the total insulin packaged in vesicles. Interestingly, GLUT4 has a mechanism similar to the one just described in its translocation to the cell surface for attachment of glucose. GLUT4 is packaged in a vesicle containing, on its surface, v-SNARE proteins, transported to the plasma membrane where v-SNARE pairs with t-SNARE. Fusion occurs with the membrane, glucose from the extracellular fluid complexes with GLUT4, and through endocytosis the GLUT4-glucose complex enters the cell. As you may recall, GLUT4 is found predominantly in skeletal muscle, adipose cells, and cardiac muscle. GLUT2, located in the intestine, liver, and kidney, undergoes the same process. This exocytotic mechanism is insulin stimulated and probably pertains to all GLUTs.

Summary Box 4.1

- Insulin is synthesized in the β-cells of the pancreatic islets of Langerhans initially as a preproinsulin precursor.
- Immediately after synthesis preproinsulin is cleaved producing proinsulin.
- Proinsulin is transported to the Golgi apparatus where it is packaged in β-granules where it is cleaved in two positions along the polypeptide.
- While in the secretory granule (LDCV), the proinsulin is hydrolyzed to insulin and the insulin is complexed to zinc ion.
- v-SNARE is a complex of proteins located on the β-vesicle surface that binds to t-SNARE on the β-cell plasma membrane (a process called *docking*).
- Priming involves an alteration in the plasma membrane dependent on ATP and calcium ions, which leads to fusion of the vesicle with the plasma membrane.
- A fusion pore then opens and allows insulin, proinsulin, and C-peptide to be released into the extracellular space. This process is called *exocytosis*.
- The vesicle membrane substances are retained in the cell and are used to produce new vesicles. This is a process called *endocytosis*.
- Microfilaments are responsible for the transport of the vesicles to the plasma membrane with microtubules serving as the tracks for the transport.
- The actual mechanism for the secretion of insulin is under active investigation.
- An alternative mechanism for insulin exocytosis is called the "kiss and run" hypothesis in which LDCVs fuse to the plasma membrane and release insulin, proinsulin, and C-peptide through a fusion pore. Recent investigations suggest that "kiss and run" is a minor contributor to the exocytosis process.
- GLUT4 follows a process similar to that of insulin release from the cell.

INSULIN SIGNALING PATHWAYS

Insulin triggers the uptake of glucose; promotes glycogenesis, protein synthesis, and triacylglycerol (fatty acid) synthesis; and inhibits gluconeogenesis and triacylglycerol hydrolysis. It also promotes sodium transport and stimulates transcription, which leads to cell growth and apoptosis.[3] All of these effects are

---

[3]Apoptosis involves a programmed series of biochemical events that leads to cell death without releasing harmful cell debris.

accomplished by a cascade of elements leading from the initial signal in which insulin binds to the insulin receptor (IR) complex at the surface of the plasma membrane. Maps (9, 10) depicting the insulin signaling pathways appear as a smorgasbord of acronyms and in some cases strange sounding names. Some proteins have several different acronyms assigned by different authors. The cascade pathways are numerous, with some leading to the same proteins through several different paths. Reactions among elements of the pathways include phosphorylation–dephosphorylation (reversible phosphorylation) and binding of element to element (docking). Most elements that end in the letter k refer to a protein kinase and those that have an acronym starting with the letter p are phosphorylated.

To simplify matters rather than confront the reader with a difficult-to-interpret spider web of interrelated elements, I have teased out the principal pathways for the most important actions of insulin, namely, insulin binding to the IR protein, which leads to activation of Akt [also called *protein kinase B* (PKB)], gluconeogenesis, protein synthesis, glucose uptake, triacylglycerol synthesis (lipid synthesis), and triacylglycerol hydrolysis (lipolysis).

Akt Pathway

Figure 4.3 illustrates the Akt pathway that begins within the binding of insulin (I) to the IR and terminates with the activation of Akt (11). IR has two α-subunits and two β-subunits linked by disulfide bonds. The binding of insulin to the α-subunit activates tyrosine kinase, which is located on the β-subunits. Once activated, tyrosine kinase phosphorylates tyrosine residues on the β-subunits. The phosphorylated β-subunits phosphorylate insulin receptor substrates 1 (IRS-1) and insulin receptor substrates 2 (IRS-2). IRS-1 or IRS-2 binds (docks) to the p85 subunit of phosphatidyl inositol-3-kinase (PI3K), which leads to the activation of the p110 catalytic subunit. The activated PI3K phosphorylates phosphatidyl inositol (PIP) to produce phosphatidyl inositol-3-phosphate (PIP1), phosphatidyl inositol-4-phosphate to produce phosphatidyl inositol-3,4-bisphosphate (PIP2), and phosphatidyl inositol-4,5-bisphosphate resulting in phosphatidyl inositol-3,4,5-trisphosphate (PIP3). Subsequently, PIP3 activates phosphoinositide-dependent kinase-1 (PDK-1). Then, PDK-1 activates the serine/threonine kinase Akt (PKB).

In addition, mammalian target of rapamycin[4] complexes (mTORCs; mTORC1 and mTORC2), kinases involved in cell growth and cell division, are important components in the insulin signaling pathway. mTORC2 activates Akt and mTORC1 functions in a negative feedback loop that inhibits the initial signal from the insulin complexed to the receptors. The first step in the feedback is activation of mTORC1 by activated IRS 1 or 2. Active mTORC1 phosphorylates growth factor receptor-bound protein 10 (Grb10). Phosphorylated Grb10 inhibits tyrosine kinase and thus the activation of IRS 1 and IRS 2 (12, 13). This

---

[4]Rapamycin, an antibiotic, complexes with this kinase and inhibits it, hence the name mammalian target of rapamycin.

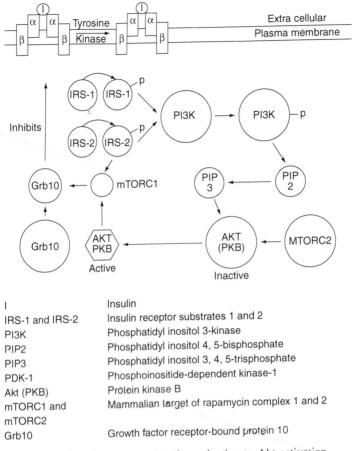

I	Insulin
IRS-1 and IRS-2	Insulin receptor substrates 1 and 2
PI3K	Phosphatidyl inositol 3-kinase
PIP2	Phosphatidyl inositol 4, 5-bisphosphate
PIP3	Phosphatidyl inositol 3, 4, 5-trisphosphate
PDK-1	Phosphoinositide-dependent kinase-1
Akt (PKB)	Protein kinase B
mTORC1 and mTORC2	Mammalian target of rapamycin complex 1 and 2
Grb10	Growth factor receptor-bound protein 10

Figure 4.3 Insulin-receptor site activation to Akt activation.

newly discovered negative feedback pathway provides an exquisite mechanism to regulate the response to insulin and IGFs.

Akt is not an acronym; it is a name used for a mouse strain developed for producing spontaneous thymic lymphomas. Akt is at the crossroads of insulin signaling. There are three Akt genes, Akt-1, Akt-2, and Akt-3. Akt-1 inhibits apoptosis (programmed cell death) and activates protein synthesis. Akt-2 is involved in glucose transport and other aspects of the insulin signaling pathway and Akt-3 is responsible for cell survival. Akt is activated by phosphorylation of threonine 30 and serine 473.

GLUT4 Translocation

As described in Chapter 3, glucose is cleared from the blood by insulin-stimulated transport dependent on glucose transporter proteins called *GLUTs*. GLUTs are packaged in vesicles that, in the presence of insulin, cycle from the interior of the cell to the plasma membrane much like a bucket parade during a fire; coming to the surface without glucose and then returning to the interior of the cell with glucose. Most reports on the mechanism by which GLUT is mobilized by insulin concerns GLUT4. Thus, our description will emphasize GLUT4.

Pathway 1   There are two insulin signaling pathways that are involved in GLUT transport. The initial insulin-stimulated pathway has been described as the Akt pathway. The first pathway is shown in Figure 4.4a. As shown in the figure, activated Akt and activated protein kinase C (PKC) are necessary for insulin-stimulated glucose transport.

Pathway 2   The second pathway is illustrated in Figure 4.4b. This pathway does not include the Akt pathway. As in the Akt pathway, this alternative pathway begins with insulin binding to the IR followed by phosphorylation of the IR by activated tyrosine kinase. Activated tyrosine kinase also concomitantly phosphorylates Casitas B-lineage lymphoma (Cbl) and adaptor-containing pleckstrin[5] oncogene (APS). Cbl binds to Cbl-associated protein (CAP). The complex then binds to lipid rafts via the protein flotilin. Lipid rafts are specialized regions in the plasma membrane, where there are more than the usual concentrations of cholesterol, and sphingolipids. They also contain caveoline and flotilin proteins. These regions that serve as receptor sites form hairpin loops, often called *caveolae*.

The complex of Cbl, CAP, and lipid raft (via flotilin) recruits SH2/SH3 adaptor protein (Crkll). Activation of Crkll is due to binding of Crkll SH2 regions to tyrosine phosphorylated Cbl. Also recruited with Crkll is guanyl nucleotide exchange factor (C3G). C3G activates a GTP-binding protein (GTPase) TC10, which complexes with ClP4/2. This complex probably is transported to the plasma membrane. To simplify the diagram, the complex is shown in the cytoplasm near the plasma membrane. Finally, a SNARE complex containing syntaxin 4-binding protein (Synip) is translocated from the plasma membrane to the cytoplasm. GLUT4 is sequestered in the vesicle and the vesicle is translocated to the surface of the plasma membrane. A glucose molecule is enshared and the vesicle is transported back to the cytoplasm of the cell (14). Other components and pathways of this ongoing story are likely to be discovered in the future. Some of the gaps you may identify in this story are present-day research projects.

Insulin-Stimulated Glycogenesis

As you may recall from Chapter 3, glucose surplus causes the augmentation of glucose incorporation into glycogen, a process called *glycogenesis*. The insulin signaling pathway explains the steps that occur in insulin-stimulated glycogenesis. Figure 4.5 illustrates this process.

---

[5]Pleckstrin is a region of 120 amino acids. Its name derives from the platelet leucocyte C kinase substrate.

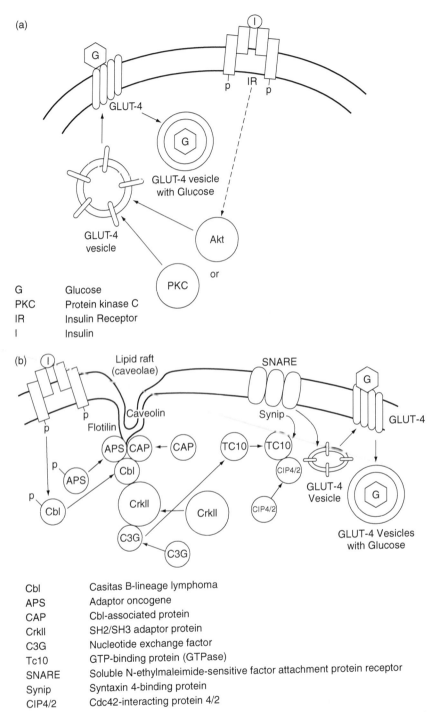

Figure 4.4 (a) Insulin-stimulated glucose uptake. (b) Insulin-stimulated glucose uptake. Alternate pathway.

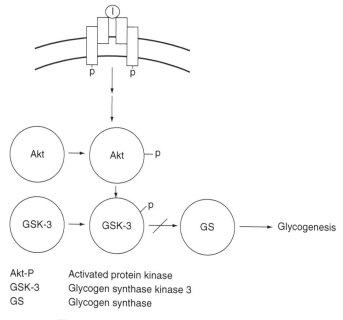

Akt-P   Activated protein kinase
GSK-3   Glycogen synthase kinase 3
GS   Glycogen synthase

Figure 4.5   Insulin-stimulated glycogenolysis.

The first part of the pathway is the activation of Akt (illustrated in Fig. 4.3). Akt then phosphorylates glycogen synthase kinase 3 (GSK-3), thus inhibiting its phosphorylation of glycogen synthase (GS). As you may recall (Chapter 3), GS phosphorylation results in a less active enzyme.

Insulin-Stimulated Inhibition of Gluconeogenesis

As shown in Figure 4.6, surplus levels of blood glucose results in insulin inhibition of gluconeogenesis. Refer to Figure 4.3 for the Akt activation pathway. Activated Akt phosphorylates and thereby inhibits the Forkhead transcription factor (Foxo1, previously FKHRL1), which in turn inhibits gluconeogenesis.

Insulin-Stimulated Protein Synthesis

As illustrated in Figure 4.7, a and b, there are two alternative pathways to insulin-stimulated protein synthesis. In both cases, refer to Figure 4.3 for the Akt activation pathway. In the first pathway, Akt activates glycogen synthase kinase-3 (GSK-3) by phosphorylation. GSK-3 activates eukaryotic translation initiation factor 2B (elF2B). ElF2B recruits ribosomes during the initial phase of translation. This action leads to protein synthesis.

In the second pathway, activated Akt phosphorylates mammalian target of rapamycin (complex) (mTOR or mTORC), a serine/threonine protein kinase.

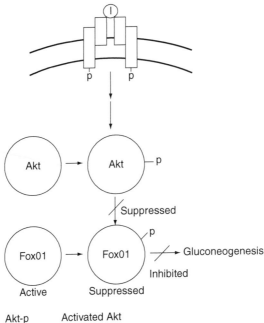

Figure 4.6   Insulin-stimulated inhibition of gluconeogenesis.

mTOR phosphorylates ribosomal protein S6 kinase 1 (p70S6K), another serine/threonine protein kinase. P70S6K phosphorylates S6 ribosomal protein and thereby induces protein synthesis at the ribosomes.

Insulin-Stimulated Lipogenesis (Fatty Acid Synthesis)

As can be seen in Figure 4.8, insulin has a stimulatory effect on lipid (fatty acid) synthesis. The Akt pathway (Fig. 4.3) initiates the stimulation. At basal levels of insulin, Akt phosphorylates and thus activates GSK-3, which next inhibits ATP-citrate lyase (CL) and thereby inhibits fatty acid synthesis. CL catalyzes formation of acetyl CoA, oxaloacetate, and ADP from citrate, ATP, CoA, and H_2O; a pivotal step in the synthesis of fatty acids.

$$\text{Citrate} + \text{ATP} + \text{CoA} + H_2O \rightarrow \text{Oxaloacetate} + \text{Acetyl CoA} + \text{ADP} + \text{Pi}$$

Insulin promotes lipogenesis by inhibiting the activity of GSK-3 and thereby enhancing CL activity. This results in increased lipogenesis.

Insulin-Inhibited Lipolysis (Fatty Acid Hydrolysis)

Lipolysis is inhibited by insulin. Starting from activated Akt (Fig. 4.3), as you may see from Figure 4.9, phosphodiesterase 3B (PDE3B) is activated by phosphorylation. PDE3B reduces the concentration of cyclic AMP and thereby causes

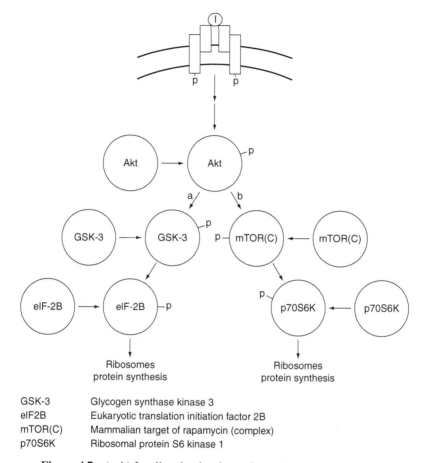

GSK-3	Glycogen synthase kinase 3
eIF2B	Eukaryotic translation initiation factor 2B
mTOR(C)	Mammalian target of rapamycin (complex)
p70S6K	Ribosomal protein S6 kinase 1

Figure 4.7 (a, b) Insulin-stimulated protein synthesis. Two pathways.

a decrease in the activity of an enzyme dependent on cyclic AMP for its activity, protein kinase A (PKA). PKA is also known as cAMP-dependent protein kinase. Decreased PKA activity results in decreased hormone sensitive lipase (HSL) and adipocyte triacylglycerol hydrolase (ATGL) activities. The function of HSL and ATGL is to hydrolyze adipose-cell-stored triacylglycerols (triglycerides) to free fatty acids.

Scaffold Proteins

How do the signaling pathway proteins find each other to bind and form multiprotein complexes? After all, intracellularly there are nearly a billion protein molecules to complex with and/or phosphorylate. The answer to this question is scaffold proteins—proteins that have sections (domains) that bind two or more proteins. Many of the proteins activated by tyrosine/serine phosphorylation (such

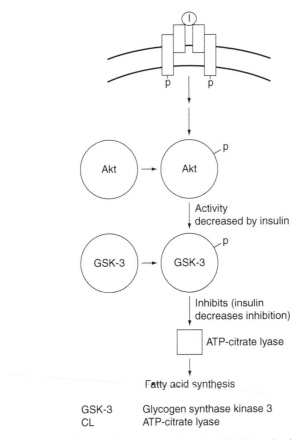

Figure 4.8   Insulin-stimulated lipid synthesis.

as PIP2, PIP3, and Akt) combine with proteins that possess domains of about 100 amino acids that preferentially bind to phosphotyrosine residues. These regions are called *Src homology 2* (SH2) domains. In addition, the proteins called *A-kinase anchoring proteins* (AKAPs) bind PKA, EPAC[6] (guanine nucleotide exchange factors), PP2B (calmodulin-dependent phosphatase), and phosphodiesterase (PD). AKAPs have isoforms (proteins with slightly different amino acid sequences but identical functions) that anchor to different cellular structures. AKAP18α anchors PKA to the Ca^{2+} channel, AKAP-350/450α to the K^+ channel, AKAP-350/450β to the Golgi apparatus. In addition, it has been recently published that a scaffold protein [e.g., kinase suppressor of RAS (KSR)], can serve both as a scaffold protein and as an enzyme (15). To learn more about these proteins, refer at references (16) and (17).

Additional insulin signaling routes will be described as the chapter advances.

---

[6]See later for one of the functions of EPAC.

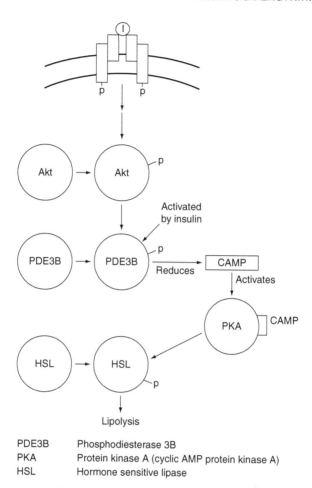

PDE3B	Phosphodiesterase 3B
PKA	Protein kinase A (cyclic AMP protein kinase A)
HSL	Hormone sensitive lipase

Figure 4.9 Insulin inhibition of lipolysis.

Summary Box 4.2

- The insulin receptor consists of two α-subunits and two β-subunits.
- Binding of insulin to the α-subunits activates tyrosine kinase located in the β-subunits.
- Tyrosine kinase phosphorylates tyrosine residues in the β-subunits.
- The β-subunits phosphorylate insulin receptor substrates, IRS-1 and IRS-2.
- IRS-1 or IRS-2 binds with phosphatidyl inositol-3-kinase (PI3K) activating the catalytic subunit of the enzyme, p110.
- The kinase phosphorylates PIP3.
- PIP3 activates PDK-1.

- PDK-1 activates Akt (also called *protein kinase B*).
- A negative feedback loop exists when IRS-1 and IRS-2 are activated. It consists of mTORC1 and Grb10. When activated by mTORC1, Grb10 inhibits the phosphorylation of the insulin-receptor complex, thus regulating the response to insulin.
- The above described pathway is called the *Akt pathway*.
- There are several insulin signaling pathways that emanate from the Akt pathway.
- The activation of the glucose transporter, GLUT is derived from two insulin signaling pathways. One derives from the Akt pathway and utilizes protein kinase B. The other is independent of the Akt pathway although it begins with insulin binding to the insulin receptor, and activation of tyrosine kinase.
- The pathway utilizes Cbl, APS, CAP, lipid rafts, Crkll, C3G, TC10, and CIP4/2. The final step in the pathway involves a SNARE complex that sequesters GLUT4, travels to the cell membrane, and picks up glucose.
- Initially, the Akt pathway is used in common for insulin-stimulated glycogenesis, protein synthesis, lipogenesis and insulin inhibition of gluconeogenesis, and lipolysis. But after Akt is activated, each metabolic pathway takes a different route utilizing different components.
- Scaffold proteins bind the components of the insulin signaling pathways and bring them in proximity to each other for enzyme action or to form multiprotein complexes.

THE INCRETIN HORMONES (INCRETINS)

Incretins[7] are peptide hormones produced by the intestinal mucosa in response to the ingestion of food. They stimulate insulin secretion from β-cells located in the islets of Langerhans. The incretins were discovered from the observation that intravenously administered glucose gave an insulin response that was 1/2 to 1/3 the response when an equal dose of oral glucose was administered. This discovery led to the isolation of two gastrointestinal tract peptides that stimulated insulin secretion—glucose-dependent insulinotropic polypeptide[8] (GIP) and glucagon-like peptide-1 (GLP-1).

The properties of GIP and GLP-1 are summarized in Table 4.1 and their amino acid sequences are presented in Figure 4.10 (18, 19). The 1-letter abbreviations for the amino acids are provided for your convenience in Table 4.2. As shown in Table 4.1 and Figure 4.10, GIP is secreted from the K-cells of the upper small intestine as a 42-amino acid polypeptide. GIP is initially synthesized as part of a 153-amino acid polypeptide called *preproGIP*. GIP receptors are located in

[7]Incretin is derived from intestine secretion insulin.
[8]Glucose-dependent insulinotropic polypeptide was originally named gastric inhibitory polypeptide.

TABLE 4.1   Properties of GIP and GLP-1 Compared

Property	GIP	GLP-1
Derived from	PreproGIP	Proglucagon
Number of amino acid residues	153	160
Amino acid residues	42	31
Secreted from	K-cells of upper small intestine	L-cells of lower intestine and colon
Receptors	GIPR located in pancreas, brain, GI tract, adipose cells, and bone	GLPR located in pancreas, brain, GI tract, and heart
Enzyme activated	Activates adenylyl cyclase	Activates adenylyl cyclase
Half-life	5 min	2 min
Hydrolysis enzyme	Dipeptidyl peptidase-4	Dipeptidyl peptidase-4

the pancreas, brain, GI tract, fat cells, and bone. GLP-1 is secreted from the L-cells in the lower intestine and colon as a 31-amino acid polypeptide. GLP-1 is synthesized as part of a larger polypeptide composed of 160 amino acids. This larger polypeptide, called *proglucagon*, is composed of five polypeptides: glicentin,[9] oxyntomodulin,[10] glucagon, GLP-1, and GLP-2.[11] GLP-1 receptor sites are located in the brain, pancreas, GI tract, and the heart. GIP and GLP-1 stimulate the β-cells of the pancreas to secrete insulin, increase β-cell proliferation, and decrease β-cell apoptosis. In addition, while GIP stimulates the α-cells to secrete glucagon, GLP-1 inhibits the secretion of glucagon. Both GIP and GLP-1 are rapidly hydrolyzed in the circulation by dipeptidyl peptidase-4 (DPP-4). The half-life of GIP is 5 min and that of GLP-1 is 2 min (20–22).

The promotion of insulin synthesis and secretion from β-cells and glucose transport into cells is illustrated in Figure 4.11. Attachment of GIP to GIP receptor (GIPR) and/or GLP-1 to GLP-1 receptor (GLP-1R) activates membrane-bound adenylyl cyclase to produce cyclic AMP (cAMP) from ATP. CAMP activates protein kinase A (PKA) and cAMP-regulated guanine nucleotide exchange factor (EPAC2 or cAMPGEF) by binding with them. PKA phosphorylates structural protein (SP), which crosses the nuclear membrane to stimulate transcription of preproinsulin to be eventually packaged as LDCVs (not depicted in the figure). As you may see in the figure, PKA also inhibits the K_v and K_{ATP} channels, thus preventing potassium ion from entering the extracellular fluid while activating the voltage-dependent calcium channel (VDCC) and thereby promoting the influx

[9]Glicentin is a 69 amino acid polypeptide whose function is not fully understood.
[10]Oxyntomodulin is a 37 amino acid polypeptide secreted from the colon. It contains a 29 amino acid sequence of glucagon. Oxyntomodulin stimulates adenylyl cyclase and has been shown to inhibit appetite, and gastric acid secretion. Receptor sites for oxyntomodulin have as yet not been found.
[11]GLP-2 stimulates intestinal mucosal growth, glucose transport in epithelial cells in the intestine, and nutrient absorption. It inhibits gastric emptying and acid secretion in the stomach, intestinal permeability, and bone cell death.

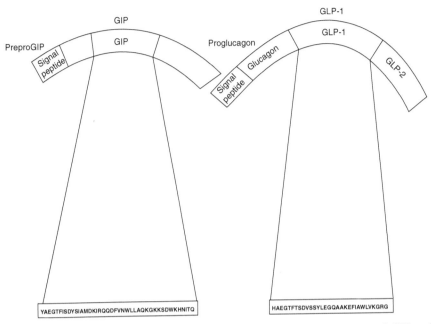

Figure 4.10 Precursors to GIP and GLP-1 and the amino acid sequence of GIP and GLP-1.

TABLE 4.2   Table of 1-Letter Abbreviations for Amino Acids

Amino Acids	1-Letter
Alanine	A
Arginine	R
Asparagine	N
Aspartic Acid	D
Cysteine	C
Glutamic Acid	E
Glutamine	Q
Glycine	G
Histidine	H
Isoleucine	I
Leucine	L
Lysine	K
Methionine	M
Phenylalanine	F
Proline	P
Serine	S
Threonine	T
Tryptophan	W
Tyrosine	Y
Valine	V

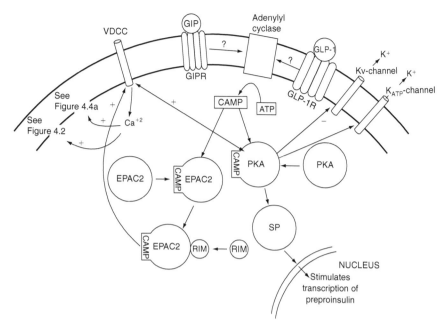

Figure 4.11 GIP and GLP-1 receptors' promotion of the insulin signaling pathway.

EPAC2 (CAMPGEF)	CAMP-regulated guanire nucleotide exchange factor
SP	Structure protein
RIM	Rab3-Interacting molecule
VDCC	Voltage dependent calcium channel

of calcium ion into the cell. As explained earlier in this chapter, this results in depolarization of the cell. The increase in Ca^{2+} ions results in extrusion of LDCVs from the cell and also promotes the transport of glucose into the cell.

EPAC2, when bound to cAMP, becomes activated to bind Rab3-interacting molecule (RIM). This complex also causes Ca^{2+}-dependent exocytosis of the LDCVs. Clearly, there is much research yet to be conducted before we fully understand the mechanisms described earlier. For example, how does the binding of GIP and GLP-1 to their respective receptor sites activate adenylyl cyclase? The full story for the stimulation of preproinsulin transcription is yet to be told. But that is what makes biochemistry fun!

The discovery of the incretin effect has led to a major advance in diabetes treatment. This part of the incretin story will be told in Chapter 10.

AMYLIN

Amylin is a 37-amino acid polypeptide that is packaged within the same β-granules as insulin. Amylin reduces insulin-stimulated glycogenesis in skeletal

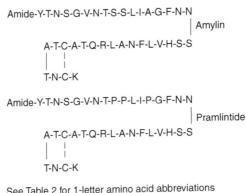

Amide-Y-T-N-S-G-V-N-T-S-S-L-I-A-G-F-N-N
| Amylin
A-T-C-A-T-Q-R-L-A-N-F-L-V-H-S-S
T-N-C-K

Amide-Y-T-N-S-G-V-N-T-P-P-L-I-P-G-F-N-N
| Pramlintide
A-T-C-A-T-Q-R-L-A-N-F-L-V-H-S-S
T-N-C-K

See Table 2 for 1-letter amino acid abbreviations
¦ Denotes a disulfide bond

Figure 4.12 Amino acid sequences of amylin and its mimic pramlintide.

muscle. Amylin follows the changes in insulin concentration, being low during fasts and increased after meals. Its amino acid sequence is shown in Figure 4.12 alongside its synthetic analog called *pramlintide* (pro-h-amylin). The amylin analog, pramlintide, will be discussed in Chapter 10.

Summary Box 4.3

- GIP, secreted from the upper small intestine, is a 42 amino acid polypeptide cleaved from a 153 amino acid polypeptide preproGIP.
- GLP-1, secreted from the lower small intestine and colon, is a 31-amino-acid polypeptide cleaved from a 160-amino-acid polypeptide called *proglucagon*.
- Proglucagon can be cleaved to five polypeptides, glicentin, oxyntomodulin, glucagon, GLP-1, and GLP-2.
- GIP receptors are located in the pancreas, brain, GI tract, fat cells, and bone. GLP-1 receptors can be found in the pancreas, brain, GI tract, and heart.
- GIP and GLP-1 increase insulin secretion, increase β-cell proliferation, and decrease β-cell apoptosis. GIP stimulates the secretion of glucagon, while GLP-1 inhibits glucagon secretion.
- Both GIP and GLP-1 are hydrolyzed by DPP-4.
- GIP and GLP-1 bind to their respective receptors and activate adenylyl cyclase.
- The cascade that follows results in glucose transport and insulin secretion.
- Amylin is a 37-amino-acid polypeptide that is secreted from the β-granule with insulin. It inhibits insulin stimulated glycogenesis.

OTHER HORMONES

Glucagon, epinephrine, somatotropin (previously called *growth hormone*), SST, cortisol, ACTH, and thyroid hormones are hormones that affect blood glucose levels. IGFs are proteins that have an amino sequence similar to insulin but do not affect blood glucose levels. We will briefly review their origins, structure and functions.

Glucagon

Glucagon is a 29-amino acid polypeptide (Fig. 4.13) that is synthesized in and secreted from α-cells located in the islets of Langerhans of the pancreas, from cells located in the gastrointestinal tract and cells in the central nervous system. It is synthesized as a larger polypeptide, proglucagon. You may remember proglucagon as the precursor to GLP-1. Interestingly, GIP stimulates glucagon secretion and GLP-1 inhibits secretion. Glucagon circulates in the blood stream until it passes through the liver, where hepatic enzymes cleave off the N-terminal histidine and serine residues, thus inactivating the hormone. Its half-life is approximately 5 min.

In Chinese philosophy, the concept of yin–yang describes opposites that are part of a system that are interconnected. Insulin and glucagon conform to the yin–yang philosophy. In accord with the yin–yang concept, glucagon's physiologic effects oppose those of insulin. As you may recall, insulin promotes energy storage by stimulating glycogenesis, lipogenesis, and protein biosynthesis. Glucagon stimulates glycogenolysis, gluconeogenesis, and lipolysis. By these reactions, it causes the mobilization of energy sources, glucose, and fatty acids. These effects are opposite to those of insulin, a yin–yang effect.

The activation of phosphorylase by glucagon (see Chapter 3) is illustrated in Figure 4.14. The pathway is initiated by glucagon binding to the glucagon receptor. The glucagon receptor consists of three subunits, α, β, and γ. When GR is inactivated guanosine 5'-diphosphate (GDP) binds to the α-subunit of the receptor. When activated by binding to glucagon the α-subunit exchanges the GDP for GTP. Next, the GTP-α complex binds to adenylyl cyclase which produces cyclic 3',5'-monophosphate (cAMP) from ATP. CAMP binds to protein kinase A (PKA), thus activating its phosphorylating activity. Activated PKA phosphorylates phosphorylase kinase (PK), consequently activating its phosphorylating activity. PK phosphorylates phosphorylase b (Phos-b) at each of its

H-S-Q-G-T-F-T-S-D-Y-S-K-Y-L-D-S-R-R-A-Q-D-F-V-Q-W-L-M-N-T

See Table 2 for 1-letter abbreviations of amino acid

Figure 4.13   Amino acid sequence of glucagon.

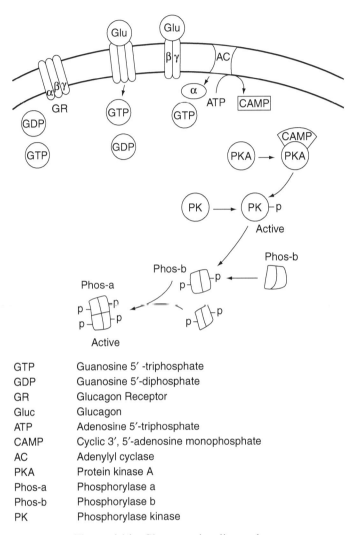

GTP	Guanosine 5′ -triphosphate
GDP	Guanosine 5′-diphosphate
GR	Glucagon Receptor
Gluc	Glucagon
ATP	Adenosine 5′-triphosphate
CAMP	Cyclic 3′, 5′-adenosine monophosphate
AC	Adenylyl cyclase
PKA	Protein kinase A
Phos-a	Phosphorylase a
Phos-b	Phosphorylase b
PK	Phosphorylase kinase

Figure 4.14   Glucagon signaling pathway.

two subunits. As you can see from the figure, the phosphorylated Phos-b dimer combines with a second dimer to form a tetramer, phosphorylase a (Phos-a). Phos-a hydrolyzes the $\alpha(1 \rightarrow 4)$ glycogen linkage, releasing glucose 1-phosphate units (glycogenolysis).

Summary Box 4.4

- Glucagon is a 29-amino-acid polypeptide synthesized and secreted from the α-cells of the islets of Langerhans. It is synthesized from proglucagon.

- Glucagon's activities are opposite to those of insulin, thus invoking the Chinese concept of yin yang.
- Phosphorylase is activated by the glucagon pathway. This pathway is initiated by the binding of glucagon to a three-protein subunit glucagon receptor complex.

Epinephrine

Epinephrine, depicted subsequently, is a member of the catecholamine family of organic compounds, which also includes dopamine and norepinephrine (NE).

The catecholamine hormones are often called the *fight or flight hormones*. They are synthesized from the amino acid tyrosine in the chromaffin granules of the adrenal medulla. The chromaffin granules are contained in adrenal medullary cells called *chromaffin cells*. The stepwise biosynthesis of dopamine, epinephrine, and NE is summarized here and illustrated in Figure 4.15:

- The amino acid tyrosine (T) enters the chromaffin cells.
- Tyrosine hydroxylase (TH) converts tyrosine to dihydroxphenylalanine (DOPA).
- Dopamine decarboxylase removes the carboxyl group from DOPA and produces dopamine.
- Dopamine is encapsulated in the chromaffin granule.
- Dopamine β-hydroxylase catalyzes the conversion of dopamine to NE.
- The NE can either be released from the granule to the cytoplasm of the chromaffin cell or exocytotically released from the granule and cell to the extracellular fluid.
- The NE that is released to the cytoplasm to converted to epinephrine (E) by phenylethanolamine-N-methyltransferase (PNMT). PNMT uses S-adenosylmethionine as a cofactor to supply the methyl group to NE. Release of NE to the cytoplasm is facilitated by the catecholamine-H⁺ exchanger, vesicular monoamine transporter 1 (VMAT1).
- E synthesized in the cytoplasm is encapsulated in a new population of chromaffin granules. VMAT1 also facilitates transport of E into chromaffin granules.
- Neural stimulation of adrenal medulla leads to fusion of the granules to the plasma membrane and a calcium-activated exocytotic release of the contents

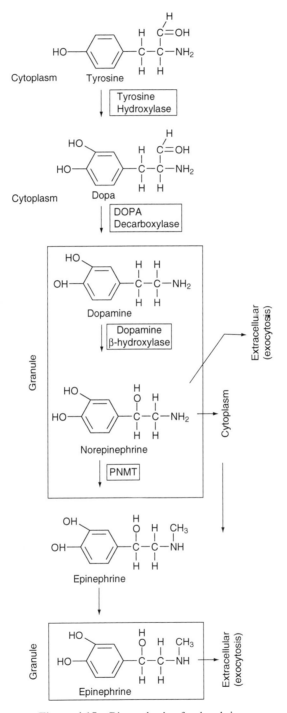

Figure 4.15   Biosynthesis of epinephrine.

of the granule. This mechanism occurs for both E and NE secretion from the chromaffin cell.

Epinephrine is a multifaceted catecholamine. It serves as a hormone and a transmitter of nerve impulses. It acts by binding to several receptors, including α and β receptors. By binding to α-receptors, epinephrine inhibits insulin secretion. It also stimulates glycogenolysis in hepatic and muscle cells and muscle glycolysis. By binding to β-receptors epinephrine causes glucagon and ACTH secretion. Triacylglycerol hydrolysis is increased in adipose cells by activation of lipoprotein lipase. Epinephrine causes an increase in blood glucose levels.

Epinephrine causes an elevation in glucose by a pathway that is similar to the glucagon signaling pathway (see Fig. 4.14), principally centered about the activation of adenylyl cyclase.

Summary Box 4.5

- Epinephrine is a catecholamine.
- Epinephrine is synthesized from tyrosine through DOPA and dopamine.
- Dopamine while in a chromaffin granule is converted to NE.
- NE can be released from the granule to the cytoplasm of the chromaffin cell or from both the granule and the cell.
- NE released to the cytoplasm is converted to epinephrine.
- Epinephrine synthesized in the cytoplasm is encapsulated in a second population of granules.
- Neural stimulation results in fusion of the granules to the plasma membrane and exocytotic release of epinephrine.
- Epinephrine inhibits insulin secretion and stimulates glucagon and ACTH secretion and lipolysis.
- Overall, epinephrine increases blood glucose levels.

Somatotropin (Growth Hormone)

Somatotropin (formally called *growth hormone*) is a polypeptide of 191 amino acids (Fig. 4.16) secreted from cells called *somatotropes* situated in the pituitary gland. Somatotropin has receptor sites in the muscle, adipose, hepatic, cardiac, kidney, brain and pancreatic cells. When bound to its receptor, somatotropin activates receptor-associated Janus kinase 2 (JAK2). Somatotropin signaling is outlined here and depicted in Figure 4.17.

- Two receptor subunits bind one somatotropin molecule.
- Each receptor subunits binds one JAK2 molecule.
- Each JAK2 molecule is phosphorylated on a tyrosine.

F-P-T-I-P-L-S-R-L-F^{10}-D-N-A-M-L-R-A-H-R-L^{20}-H-Q-L-A-F-D-T-Y-Q-E^{30}-F-E-E-

A-Y-I-P-K-E-Q^{40}-K-Y-S-F-L-Q-N-P-Q-T^{50}-S-L-C-

F-S-E-S-I-P-T^{60}-P-S-N-R-E-E-T-Q-Q-K^{70}-S- ┊

N-L-Q-L-L-R-I-S-L^{80}-L-L-I-Q-S-W-L-E-P-V^{90}- ┊

Q-F-L-R-S-V-F-A-N-S^{100}-L-V-Y-G-A-S-N-S- ┊

D-V^{110}-Y-D-L-L-K-D-L-E-E-G^{120}-I-Q-T-L-M-G- ┊

R-L-E-D^{130}-G-S-P-R-T-G-Q-I-F-K^{140}-Q-T-Y-S- ┊

K-F-D-T-N-S^{150}-H-N-D-D-A-L-L-K-N-Y^{160}-   ┊

G-L-L-Y-C-F-R-K-D-M^{170}-D-K-V-E-T-F-L-R-I-V^{180}-Q-C-

┊

R-S-V-E-G-S-C-G^{190}-F

- - - - - Denotes disulfide bonds

Figure 4.16   Amino acid sequence of somatotropin.

- The receptor proteins are also phosphorylated on tyrosine.
- Receptor protein and JAK2 form sites for several different signaling proteins, STAT1, STAT3, STAT5a, and STAT5b, the MAPK pathway and the phosphatidylinositol-3-kinase pathway (PIK3).
- STATs are phosphorylated on tyrosine by JAK2, dimerize, and then are transported to the nucleus, where they bind to DNA and effect transcription (the protein(s) affected depend on the targeted cell type).

Somatotropin affects protein, lipid, and carbohydrate metabolism. It stimulates protein biosynthesis, especially in muscle. Somatotropin causes an increase in the release of free fatty acids from triacylglycerols (lipolysis) and promotes fatty acid oxidation. This action leads to decreased glucose metabolism, thus reducing the uptake of glucose and conserving glycogen. Overall, somatotropin does not affect blood glucose levels. Administration of somatotropin increases insulin secretion, leading to hyperinsulinemia.

In summary, the effects of somatotropin in humans are that in periods of energy surplus it stimulates protein synthesis and when conditions provide an energy deficit somatotropin uses lipids for the body's energy needs, thus conserving protein.

Summary Box 4.6

- Somatotropin is a 191-amino-acid polypeptide secreted by the pituitary gland.
- Somatotropin receptors are found in muscle, adipose, hepatic, cardiac, kidney, brain, and pancreas.
- Somatotropin signals an increase in insulin secretion, lipolysis; stimulates protein biosynthesis; reduces glucose uptake; and decreases glycogenolysis.

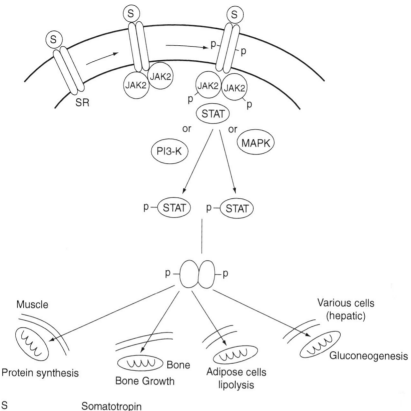

S   Somatotropin
SR   Somatotropin receptor
JAK2   Janus kinase 2
STAT   Signal transducers and activators of transcription (STAT 3, 5a, 5b)
PI 3-kinase Phosphoinositol 3-kinase
P   Phosphate
MAPK   Mitogen-activated protein kinase (pathway); also called ERK
            (extracellular response kinase). This signaling pathway takes a
            signal from the plasma membrane and communicates it to the DNA in
            the nucleus.

Figure 4.17   Somatotropin signaling pathway.

Somatostatin (SST)

SST is synthesized and released from pancreatic δ-cells. SST is synthesized as a large polypeptide, preprosomatostatin (116 amino acids) and is processed initially into a 28-amino acid prosomatostatin and finally a 14-amino acid polypeptide shown in Figure 4.18. Both the 14- and 28-amino acid polypeptides have physiological activity.

SST inhibits the release of glucagon, insulin, gastrin, and secretin. There are four principal receptor sites for somatotropin: the brain, GI tract, pancreas,

A-G-C-K-N-F-F-W

C-S-T-F-T-K

Denotes a disulfide
bond

Figure 4.18 Amino acid sequence of 14 amino acid somatostatin.

and pituitary. A total of five SSTRs have been characterized. SSTR2 mediates glucagon inhibition and SSTR5 regulates insulin secretion. SSTR1, SSTR3, and SSTR4 regulate neurotransmission, gastrointestinal motility, gastric acid secretion, intestinal absorption, pancreatic enzyme secretion, and cell proliferation. The 14-amino acid SST polypeptide is an inhibitor of glucagon secretion and the 28-amino acid polypeptide inhibits insulin secretion. But all five receptor sites tightly bind both 14- and 28-amino acid SST (23).

Summary Box 4.7

- Somatostatin is produced and secreted from the δ-cells of the pancreas.
- It is initially produced as a 116-amino-acid polypeptide, preprosomatostatin and processed to 14-amino-acid and 28-amino-acid peptides.
- Five receptor sites are available for its actions; neurotransmission, GI motility, secretion of gastric acid, intestinal absorption, pancreatic enzyme secretion, and cell proliferation.
- The 14-amino-acid peptide inhibits glucagon secretion, and the 28-amino-acid inhibits insulin secretion.

Cortisol

Cortisol (hydrocortisone) depicted subsequently is a steroid hormone synthesized and secreted from the zona fasciculata region of the adrenal gland. As illustrated in Figure 4.19, cortisol is regulated by the hypothalamus–pituitary–adrenal physiological control system (24). The release of cortisol is initiated by secretion of corticotropin-releasing hormone (CRH) from the hypothalamus. CRH promotes the release of ACTH from the pituitary. ACTH is transported by the circulation to the adrenal cortex, where it stimulates the secretion of cortisol.

Cortisol

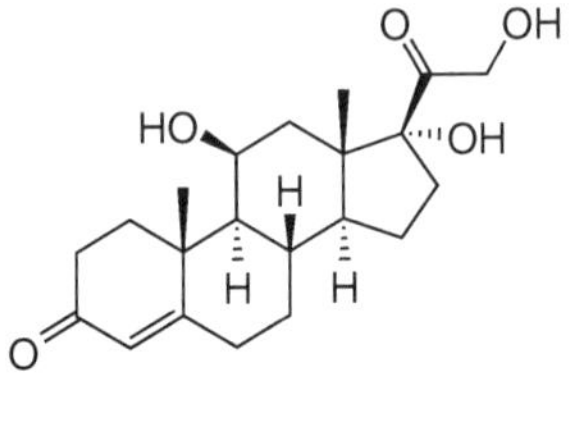

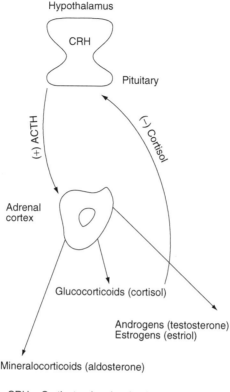

Figure 4.19   Hypothalamus–pituitary–adrenal physiological control system.

Cortisol increases gluconeogenesis and thus elevates blood glucose. It also activates glycogenesis and suppresses the immune system. Cortisol also inhibits translocation of glucose transporters. Cortisol follows a diurnal rhythm. In humans, it increases in the early morning and reaches peak levels between 6:00 AM and 8:00 AM. After 8:00 AM it slowly declines, reaching its lowest levels near midnight. Normally, the increase in cortisol secretion during morning hours is accompanied by inhibition of insulin secretion, resulting in unchanged glucose levels.

Summary Box 4.8

- Cortisol is synthesized in the adrenal gland.
- It is regulated by the hypothalamus–pituitary–adrenal physiological control system.

- ACTH is the principal stimulator of the release of cortisol.
- Cortisol increases gluconeogenesis and glycogenesis. It also inhibits cycling of glucose transporters.
- Cortisol follows a diurnal rhythm.

Adrenocorticotropin

As pictured in Figure 4.20, adrenocorticotropin (ACTH) is a polypeptide consisting of 39 amino acids. As illustrated in Figure 4.21, ACTH is produced in the pituitary gland from a large 285-amino acid precursor polypeptide named *pro-opiomelanocortin* (POMC). POMC produces a family of hormones. In the anterior pituitary (adenohypophysis), POMC is cleaved to ACTH and β-lipotropin (β-LPH). ACTH can be further cleaved to α-melanocyte stimulating hormone (α-MSH) (amino acids 1–13) and corticotropin-like intermediate lobe peptide (CLIP). β-LPH (amino acids 42–134) is cleaved to γ-LPH (amino acids 42–101) and β-endorphin. An alternative pathway for β-LPH is hydrolysis to β-MSH

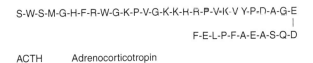

S-W-S-M-G-H-F-R-W-G-K-P-V-G-K-K-H-R-P-V-K-V Y-P-D-A-G-E
F-E-L-P-F-A-E-A-S-Q-D

ACTH   Adrenocorticotropin

Figure 4.20   Amino acid sequence of ACTH.

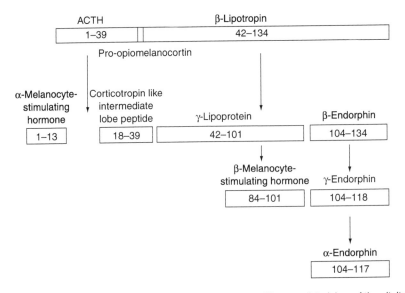

Pro-opiomelanocortin is synthesized in the anterior and intermediate lobes of the pituitary

Figure 4.21   Origins of ACTH.

(amino acids 84–101) and γ-endorphin (amino acids 104–118). Γ-Endorphin can be converted to α-endorphin by losing one amino acid (104–117).

Return to Figure 4.19 to study the regulation of ACTH. ACTH binds to receptor sites and activates adenylyl cyclase, thus promoting the synthesis of cyclic AMP. In adipose cells, this leads to the activation of lipoprotein lipase and lipolysis. ACTH stimulates the synthesis of steroids in the adrenals by accelerating the formation of pregnenolone from cholesterol. ACTH binds with the melanocortin 2 receptor (MC2-R) in pancreatic β-cells. The binding of ACTH with MC2-R activates protein kinase A (PKA). Activation of PKA leads to increased cytosolic Ca^{2+} concentrations. These effects cause a significant insulin secretory response.

Summary Box 4.9

- ACTH, a 39-amino-acid peptide, synthesized in the pituitary gland from a 285 amino acid protein, POMC.
- POMC is cleaved to form ACTH, β-LPH, γ-LPH, α-MSH, CLIP, and α-, β- and γ-endorphins depending on the specific lobe of the pituitary cleavage occurs.
- ACTH binds to its receptor site and activates adenylyl cyclase.
- ACTH stimulates lipolysis and synthesis of steroids.
- ACTH binds with MC2-R in pancreatic β-cells activating PKA leading to increased Ca^{+2} concentrations resulting in insulin secretion.

Thyroid Hormones

The hypothalamus–pituitary–thyroid physiological control system illustrated in Figure 4.22 governs the production of 3,5,3'-triiodothyronine (T_3) and 3,5,3',5'-tetraiodothyronine (T_4, also known as *thyroxine*). T_3 is believed to be the physiologically active hormone. Thyrogobulin (5000 amino acids; 660,000 Da) is the precursor to T_3 and T_4. The synthesis of the thyroid hormones is pictured in Figure 4.23 and outlined here (25):

- Stimulated by the thyroid stimulating hormone (TSH), iodine is covalently bound to the 120 tyrosine residues found in thyroglobulin.
- Iodination of thyroglobulin, catalyzed by thyroid peroxidase, produces two tyrosine derivatives: 3-monoiodotyrosine (MIT) and 3,5-diiodotyrosine (DIT).
- While still attached to thyroglobulin, two DIT molecules form a bond to form the precursor to T_4 and one DIT molecule links to one MIT molecule to form the precursor to T_3.
- Enzymes hydrolyze T_4 and T_3 from thyroglobulin.
- In the liver and kidneys, T_3 is produced from deiodination of T_4.

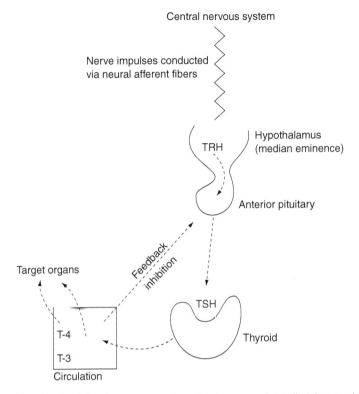

TSH = Thyroid-stimulating hormone; a polypeptide hormone also called thyrotropin
TRH = Thyrotropin-releasing hormone; a polypeptide hormone

Figure 4.22   Hypothalamus–pituitary–thyroid physiological control system.

- T_4 and T_3 are bound to two binding proteins, thyroxine-binding globulin (TBG) and thyroxine-binding prealbumin (TBPA).
- Unbound T_3 is physiology active.

   Thyroid hormones affect a wide range of metabolic pathways, including insulin signaling, gluconeogenesis, glycogenolysis, lipogenesis, cell proliferation, and apoptosis. Thyroid receptors (TRs) are expressed by two genes, α and β. T_3, the physiologically active thyroid hormone, binds to TRs and activates or inhibits metabolic pathways. Akt (PKB) synthesis is inhibited by T_3. Refer to Figures 4.3 and 4.4a to understand how inhibition of Akt and PKC expression results in reduction of GLUT4 glucose transport and therefore increases blood glucose. Look at Figure 4.6 to understand how Akt inhibition results in increased gluconeogenesis. Look at Figure 4.8 to understand how inhibition of Akt results in increased lipogenesis and Figure 4.9 to understand the increase in lipolysis and the increase in cyclic AMP that occurs.

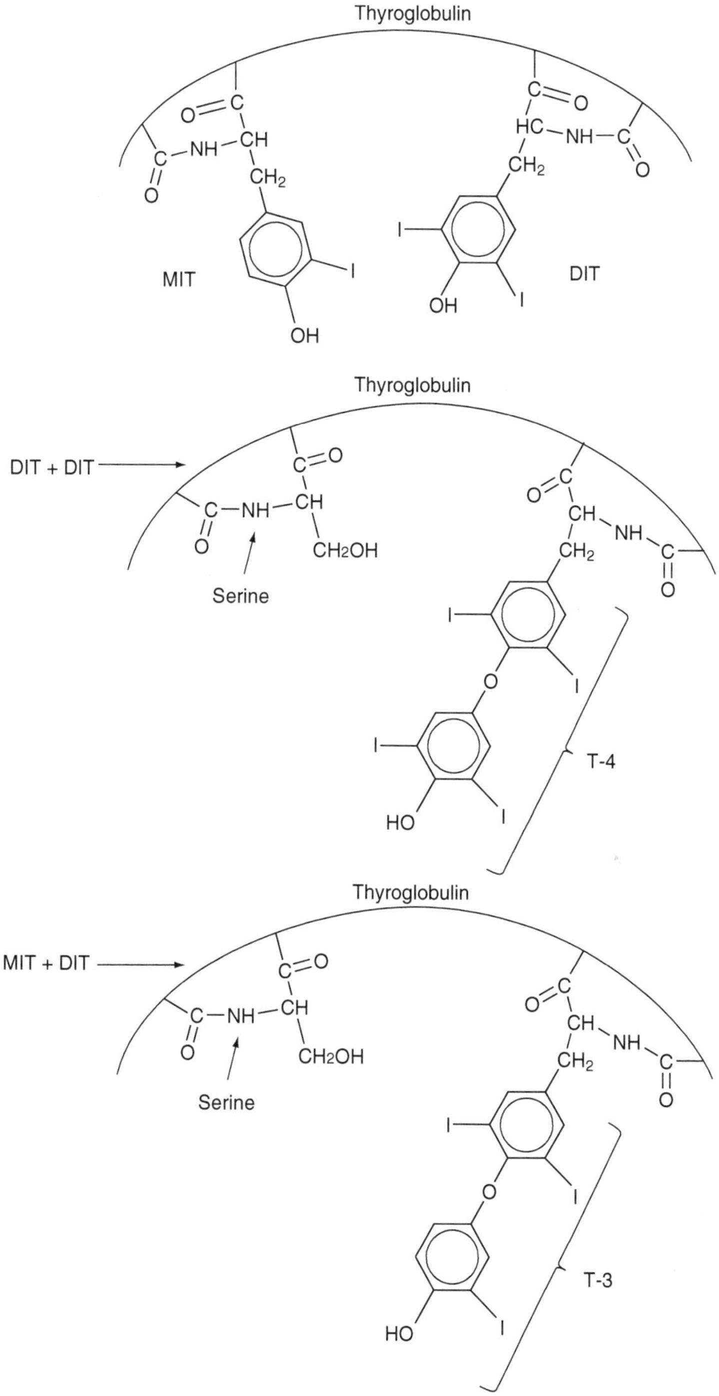

Figure 4.23 Synthesis of thyroid hormones.

Summary Box 4.10

- Two thyroid hormones are synthesized on thyroglobulin, T_3 and T_4.
- T_3 is the physiologically active thyroid hormone.
- Thyroid hormones affect several metabolic pathways.
- Thyroid hormones inhibit glucose transport and augment gluconeogenesis.

Insulin-Like Growth Factor (IGF)

IGFs (IGF-1 and IGF-2) are polypeptides that are similar to insulin in sequences of amino acids, but despite this similarity do not have a direct effect on glucose metabolism. IGFs are synthesized and secreted from the liver when stimulated by somatotropin. The function of these polypeptide hormones is promotion of cell proliferation, inhibition of apoptosis, and to effect neural development and early developmental growth. It has been reported that although IGFs and insulin have structurally and functionally similar receptors and share some of the same signaling molecules, they have separate and distinctly different roles in the cell. Both IGFs and insulin, when bound to their respective receptors, activate tyrosine kinase and thence activate the Akt pathway. But beyond this, IGFs and insulin apparently take different paths leading to different responses from the cell (26).

FIBROBLAST GROWTH FACTOR 19

When bile acids come in contact with cells of a section of the small intestine called the *ileum*, a signal is activated for the synthesis and secretion of a substance called *fibroblast growth factor 19* (FGF19). FGF19 inhibits synthesis of bile acids and stimulates both liver fatty acid oxidation and protein synthesis. Similar to insulin, FGF19 also stimulates liver glycogenesis. But, interestingly, FGF19 uses a different signaling system from insulin (27). Compare Figures 4.5 and 4.24 to see the distinctly different glycogenesis signaling pathways for FGF19 and insulin.

Summary Box 4.11

- Although IGF has an amino acid sequence similar to insulin, it does not have a direct on glucose metabolism.
- IGF activates tyrosine kinase and the Akt pathway.
- IGF then takes different pathways than insulin to promote cell proliferation, inhibit apoptosis, and effect neural development and promotion of developmental growth.
- Fibroblast growth factor (FGF19) is stimulated for release from the small intestine by bile acids.

- It inhibits synthesis of bile acids and stimulates fatty acid oxidation and protein synthesis.
- FGF19 also stimulates glycogenesis but through a different pathway than does insulin.

ADENOSINE 5′-MONOPHOSPHATE-ACTIVATED PROTEIN KINASE

AMPK is an enzyme that is involved in numerous metabolic pathways. It inhibits fatty acid oxidation and lipogenesis. AMPK stimulates synthesis of GLUT4 and

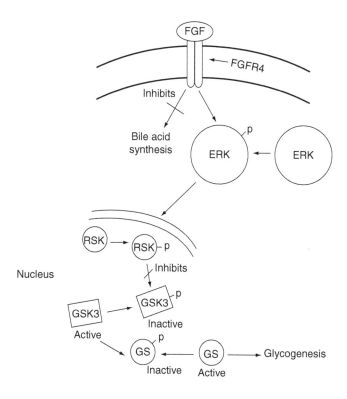

ERK	Extracellular signal-regulated protein kinase
FGF19	Fibroblast growth factor19
FGFR4	Fibroblast growth factor receptor
RSK	Ribosomal S6 kinase
GSK3	Glycogen synthase kinase 3
GS	Glycogen synthase

Figure 4.24 Fibroblast growth factor 19 inhibition of bile synthesis and stimulation of glycogenesis.

cellular glucose uptake and inhibits gluconeogenesis. It has been implicated in the regulation of food intake by affecting the hypothalamus. As the principal topic of this chapter is regulation of glucose metabolism, we will focus on that to the exclusion of other pathways.

AMPK is a three-subunit (heterotrimeric) enzyme composed of α, β, and γ polypeptides. The α-subunit has kinase activity and the β and γ-subunits provide regulation. The regulatory nature of the β and γ-subunits is very interesting. The β-subunit serves as the scaffold for the α- and γ-subunits. It also contains an amino acid loop that binds to glycogen. The γ-subunit contains four cystathionine beta synthase (CBS) domains, two of which (the Bateman domains) bind AMP and ATP and thereby recognizing shifts in the ratio of AMP to ATP. When both the Bateman domains are occupied by AMP, the enzyme undergoes a conformational change, which positions the kinase site of the α-subunit in a more exposed position to phosphorylation and activates the enzyme fivefold. The kinase site becomes activated an additional 100-fold when it is phosphorylated at threonine-172 by an upstream kinase.[12] When ATP occupies the Bateman domains, it positions the kinase so that AMPK is more susceptible to dephosphorylation by a phosphatase.

The Bateman domains serve as sensitive sensors for AMP and ATP levels. AMP and ATP are in competition for these binding sites. When AMP levels increase and ATP decreases, such as in muscle during contraction, AMP binds to the γ-subunit site, thus activating AMPK fivefold and rendering it susceptible to kinase phosphorylation, which increases its activity 100-fold. When ATP levels increase and AMP levels decrease, as in muscle relaxation, ATP replaces AMP in the Bateman domains, thus rendering AMPK susceptible to phosphatase dephosphorylation and reorienting the enzyme to the inactive conformation. What a beautiful mechanism!

There are three known kinases that phosphorylate AMPK. The most widely expressed AMPK kinase is LKB1, also called *serine/threonine kinase 11* (STK11) (28). Ca²⁺-calmodulin-dependent kinase kinase β (CaMKKβ) acts in the brain and its activity is dependent on Ca^{2+} release. The third kinase is transforming growth factor β-activated kinase (TAK1). Less is known about this kinase than LKB1 and CaMKKβ kinases. Recently, TAK1 has been reported as a regulator of muscle cell formation.

AMPK acts as a sensitive sensor for any biological stress (specifically an energy deficit such as during exercise) that results in a reduction of ATP and an increase in the AMP/ATP ratio. As stated earlier, an increase in the AMP/ATP ratio concomitant with LKB1 (or CaMKKβ or TAK1) phosphorylation of AMPK results in full AMPK activity. Phosphorylated AMPK is the most active form of AMPK.

---

[12]Upstream and downstream refer to the order of molecular events. For example, reception sites on plasma membranes are upstream to a molecule activated by the attachment of a hormone to the reception site. The activated molecule is downstream to the receptor site–hormone complex.

Activated AMPK stimulates glucose uptake due to increased translocation of GLUT4 to the plasma membrane. The transported glucose enters glycolysis and next the tricarboxylic acid pathway for the production of ATP, thereby lowering the AMP/ATP ratio. Activated AMPK also inhibits the expression of the enzymes involved in gluconeogenesis (29, 30).

Summary Box 4.12

- AMPK inhibits fatty acid oxidation, lipogenesis, and gluconeogenesis and stimulates GLUT4 synthesis and glucose uptake.
- AMPK is composed of three proteins, α, β, and γ. The β-subunit serves as the scaffold, the γ-subunit contains two domains: the Bateman domains, which bind AMP and/or ATP, and the α-subunits, which contains the catalytic region of the enzyme.
- ATP and AMP compete for sites on the Bateman domains; when AMP concentrations increase, AMP binds, activating the enzyme fivefold. AMP binding makes AMPK susceptible to phosphorylation and thus increasing activation by 100-fold. When ATP concentrations raise, the Bateman domains bind ATP, and if the kinase is phosphorylated, AMPK becomes susceptible to phosphatase activity and thus inactivation.
- There are three kinases that phosphorylate the AMP-activated AMPK. They are LKB1 (also called *STK11*), CaMKKβ, and TAK1.
- Active AMPK increases recycling of GLUT4 and therefore increases glycolysis and the tricarboxylic acid cycle, thereby resulting in an increase of ATP. In addition, gluconeogenesis is inhibited.

GLOSSARY

Analog   A compound that is similar to a natural occurring substance.

Catecholamine   A compound containing a $-NH_2$ group that is derived from tyrosine. Examples: dopamine, epinephrine, and norepinephrine.

Chromaffin cells   Neuroendocrine cells principally found in the medulla section of the adrenal gland which stain a brown-yellow color with chromium salts (hence the name).

Cleave   Pertains to the hydrolysis and removal of a portion of a polypeptide chain by peptidases.

Depolarization   Reversal of the separation of charge on the cell membrane rendering it more neutral.

Disulfide bonds   A covalent bond between two sulfur atoms on two cysteine molecules in one or more polypeptide chains.

Endoplasmic reticulum   A folded membrane network within the cytoplasm of cells that is involved in synthesis, transport in and out of cells, and modification of compounds. The rough endoplasmic reticulum (RER) has ribosomes attached to it.

Eukaryotic   Cells that have a nucleus containing the DNA.

Extracellular fluid   Fluid outside cells.

Half-life   Half of the period time that it takes a protein to undergo digestion by proteolytic enzymes.

Hydrophilic   Region of a protein that has several ionic groups and therefore attracts water molecules.

Intravenous   Administration of substances directly into a vein.

Polarity   Development of a separation of charge on a cell membrane.

Proliferation   The multiplication of cells.

Receptor sites   A complex of molecules on the cell surface that recognizes a specific molecule and when bond to the molecule sets off a signaling mechanism leading to physiological changes in the cell.

Translation   Messenger-RNA-directed synthesis of proteins from amino acids.

Translocation   Transport of molecules or vesicles from one place in the cell to another.

Vesicle   A small sac found within the cytoplasm of a cell, which is surrounded by a membrane and contains molecules of biological importance.

REFERENCES

1. Sures I, Goeddel DV, Gray A, Ullrich A. Nucleotide sequence of human preproinsulin complementary DNA. Science 1980;208:57.

2. Dods RF. *Pathophysiology for Chemists*. Washington: American Chemical Society; 1993.

3. Lang J. Molecular mechanisms and regulation of insulin exocytosis as a paradigm of endocrine secretion. Eur J Biochem 1999;259:3.

4. Takahashi N, Kishimoto T, Nemoto T, et al. Fusion pore dynamics and insulin granule exocytosis in the pancreatic islet. Science 2002;297:1349.

5. MacDonald PE, Rorsman P. The ins and outs of secretion from pancreatic beta-cells: control of single-vesicle exo- and endocytosis. Physiology 2007;22:113. DOI: 10.1152/physiol.00047.2006.

6. Pittman IV, I, Philipson LH, Steiner DF. Insulin biosynthesis, secretion, structure, and structure-activity relationships, Chapter 3. Endotext.com; 2004. Available at http://www.endotext.org/diabetes/diabetes3_new/diabetes3.htm. Accessed December, 2010.

7. Valtora F, Meldolesi J, Fesce R. Synaptic vesicles: is kissing a matter of competence? Trends Cell Biol 2001;11:324. DOI: 10.1016/S0962-8924(01)02058-X.

8. Kasai H, Hatskeyama H, Ohno M, Takahashi N. Exocytosis in islet beta cells. In: Islam S, editor. *Advances in Experimental Medicine and Biology*. Vol. 654. Berlin: Springer; 2010. p 305

9. Insulin Receptor Pathway. SABiosciences. Available at http://www.sabiosciences. com/pathway.php?sn=Insulin_Receptor_Pathway. Accessed December, 2010.

10. Insulin Receptor Signaling (IRS). Cell Signaling Technology. Available at http://www .cellsignal.com/reference/pathway/Insulin_Receptor.html. Accessed December, 2010.

11. Dods RF. Diabetes Mellitus. In: Pesce AJ, Kaplan LA, editors. Volume 38. *Clinical Chemistry: Theory, Analysis, Correlation*. 5th ed. St. Louis: Mosby; 2010. p 729.

12. Yu Y, Yoon S-O, Poulogiannis G, et al. Phosphoproteomic analysis identifies Grb10 as an mTORC1 substrate that negatively regulates insulin signaling. Science 2011;332:1322. DOI: 10.1126/science.1199484.

13. Hsu PP, Kang SA, Rameseder J, et al. ThemTOR-regulated phosphoproteome reveals a mechanism of mTORC1-mediated inhibition of growth factor signaling. Science 2011;332:1317. DOI: 10.1126/science.1199498.

14. Chang L, Adams RD, Saltiel AR. The TC10-interacting protein CIP4/2 is required for insulin-stimulated Glut4 translocation in 3T4L1 adipocytes. Proc Natl Acad Sci U S A 2002;99:12835. DOI: 10.1073/pnas.202495599.

15. Shi F, Lemmon MA. KSR plays CRAF-ty. Science 2011;332:1043.

16. Scott JD, Pawson T. Cell signaling in space and time: where proteins come together and when they're apart. Science 2009;326:1220. DOI: 10.1126/science.1175668.

17. Good M, Zalatan JG, Lim WA. Scaffold proteins: hubs for controlling the flow of cellular information. Science 2011;332:680. DOI: 10.1126/science.1198701.

18. Jornvall H, Carlquist M, Kwauk S. Amino acid sequence and heterogeneity of gastric inhibitory polypeptide (GIP). FEBS Lett 1981;123:205.

19. Lopez LC, Frazier ML, Su C-J. Mammailian pancreatic preproglucagon contains three glucagon-related peptides. PNAS 1983;80:5485.

20. Seino Y, Fukushima M, Yabe D. GIP and GLP-1, the two incretin hormones: similarities and differences. J Diabetes Invest 2010;1:8. DOI: 10.1111/j2040-1124 .2010.00022x,2010.

21. Shibasaki T, Takahashi H, Miki T. Essential role of Epac2/rap1 signaling in regulation of insulin granule dynamics by cAMP. PNAS 2007;104:19333. DOI: 10.1073 /pnas.0707054104.

22. Holz GG. Perspectives in diabetes. Epac: a new cAMP-binding ptrotein in support of glucagon-like peptide-1 receptor-mediated signal transduction in the pancreatic beta-cell. Diabetes 2004;53:5.

23. Strowski MZ, Parmar RM, Blake AD, et al. Somatostatin inhibits insuklin and glucagon secretion via two receptor subtypes: an in vitro study of pancreatic islets from somatostatin receptor 2 knockout mice. Endocrinology 2000;141:111.

24. Dods RF. *Section I The integrators and communicators: hormone physiological control systems. Pathophysiology*. Washington: American Chemical Society; 1993. p 369.

25. Dods RF. The integrators and communicators: hormone physiological control systems. In: *Pathophysiology for Chemists*. Washington: American Chemical Society; 1993. p 362–364.

26. Dalle S, Ricketts W, Imamura T, et al. Insulin and insulin-like growth factor 1 receptors utilize different G protein signaling components. J Biol Chem 2001;276:15688.

27. Kir S, Beddow SA, Samuel VT. FGF19 as a postprandial, insulin-independent activator of hepatic protein and glycogen synthesis. Science 2011;331:1621.

28. Shaw RJ, Lamia KA, Vasquez D. The kinase LKB1 mediates glucose homeostasis in liver and therapeutic effects of metformin. Science 2005;310:1642.

29. Towler MC, Hardie DG. AMP-activated protein kinase in metabolic control and insulin signaling. Circ Res 2007;100:328.

30. Gruzman A, Babai G, Sasson S. Adenosine monophosphate-activated protein kinase (AMPK) as a new target for antidiabetic drugs: a review on metabolic, pharmacological and chemical considerations. Rev Diabet Stud 2009;6:13.

CHAPTER 5

GLUCOSE METABOLISM GONE WRONG

The best laid schemes o' mice and men Gang aft a-gley

—Robert Burns (1759–1796)

For every reason there is a good reason and then there is a real reason

—Anonymous

You may recall from Chapter 2 that Aretaeus, Sushruta, and Ibn Sina all reported the presence of two types of diabetes. But it took until Wrenshall and colleagues in 1952 (1) to conclusively report that diabetes took two different forms. GA Wrenshall extracted insulin from the pancreas of young diabetic corpses and found them to weigh significantly less than those extracted from the corpses of nondiabetic subjects of the same age and sex. But, in addition, he did the same investigation in elderly diabetics and found the difference from nondiabetic deceased subjects to be rather small. J Bornstein and RD Lawrence conducted a similar experiment, except that they measured the plasma insulin content and recognized the consequence of their finding.

> We have now investigated the available plasma insulin content of two different types of human diabetes... The first type is generally young and is characterized not only by hyperglycemia but by rapidly developing ketosis[1] and severe weight

[1]Ketosis refers to the formation of acetoacetate, β-hydroxybutyrate, and acetone in quantities in diabetics.

Understanding Diabetes: A Biochemical Perspective, First Edition. Richard F. Dods.
© 2013 John Wiley & Sons, Inc. Published 2013 by John Wiley & Sons, Inc.

loss, and requires insulin to live. The second type consists largely of middle-aged obese females with similar grades of hyperglycaemia and glycosuria, but with *no ketosis* and no important loss of weight. Their diabetes is easily controlled by low (carbohydrate) diets without insulin, particularly if weight is reduced.

Bornstein and Lawrence assayed for plasma insulin levels in these two groups of patients, and

It seems clear from these results that available plasma insulin is absent and present respectively in these two types of diabetes. (2)

In a subsequent paper, Bornstein and Lawrence use the terms *growth onset* to describe diabetics under the age 22 with "practically no insulin" extractable and "maturity onset" in "older patients in whom, independent of duration, the extractable insulin is about 50% of that of the normal controls" (3). "Growth onset" eventually changed to "juvenile onset." Another naming system developed on the basis of the need for injectable insulin in order to maintain the well-being of the individual. Juvenile diabetes became known as *insulin-dependent diabetes mellitus* (IDDM) and maturity onset diabetes changed to noninsulin-dependent diabetes mellitus (NIDDM).

In 1979, the nomenclature for diabetes changed once again. The National Diabetes Data Group classified IDDM as type 1 diabetes (T1D) and NIDDM as type 2 diabetes (T2D) (4). In 1980, the World Health Organization (WHO) adopted this naming system (5). With that as the background, we shall proceed to discuss the pathophysiology of diabetes.

As you may recall from Chapter 1, Footnote 3, T1D results from lack of insulin secretion because of β-cell destruction by an immune response. Thus, you would expect a significant decrease in pancreatic β-cell mass in T1D.

As you may recall from Chapter 4, insulin causes an increased glucose flux into cells, and increased glycogenesis, protein synthesis, and fatty acid production (lipogenesis). Gluconeogenesis and triacylglycerol hydrolysis is impeded. Thus, in T1D where there is an insufficient production of insulin we should expect decreased glucose metabolism and therefore increased levels of glucose in the blood, reduced glycogenesis with increased glycogenolysis[2] adding more glucose to the blood, decreased protein synthesis, decreased fatty acid production, increased gluconeogenesis again putting more glucose into the blood, and increased triacylglycerol production. Decreased insulin levels result in increased glucagon levels (the yin–yang effect). Glucagon stimulates glycogenolysis, gluconeogenesis, and triacylglycerol hydrolysis (lipolysis). By these reactions, glucagon causes the mobilization of energy sources, glucose, and fatty acids. In T2D there is a surplus of insulin, resulting in decreased glucagon levels. One would expect this relationship to produce all of the effects listed earlier for insulin, except that glucose levels are well above normal. This

[2]Claude Bernard who discovered glycogen is proved to be partially correct in defining diabetes as a disease of the liver.

is what we would expect for these pathways for low insulin (type 1) and for high insulin (type 2) concomitant with elevated glucose levels. A T2D hallmark is insulin resistance. Insulin resistance is the condition found in T2D that results in a given concentration of insulin, resulting in lowered rather than normal biological responses.

Summary Box[3] 5.1

- Wrenshall extracted insulin from the pancreas of nondiabetics and diabetics and found that insulin from diabetics weighed significantly less than from nondiabetics.

- Bornstein and Lawrence measured plasma insulin content of two types of diabetics: young, hyperglycemic, ketotic, weight loss, require insulin to maintain life and middle-aged, hyperglycemic, nonketotic, obese, no loss of weight, glycosuric and found plasma insulin as missing from the former and present in the latter. Bornstein and Lawrence labeled the former growth onset and the latter as maturity-onset diabetics.

- Growth onset changed to juvenile onset. Juvenile onset changed to insulin-dependent diabetes mellitus and subsequently was renamed type 1 diabetes mellitus. Maturity onset changed to noninsulin-dependent diabetes mellitus and subsequently was renamed type 2 diabetes mellitus.

- T1D is caused by an immune response that destroys β-cells, the pancreatic cells that produce insulin.

- Low insulin levels should result in decreased glucose metabolism, decreased glycogenesis, decreased fatty acids, increased glycogenolysis, increased gluconeogenesis, increased triacylglycerol production, and, consistent with the yin–yang effect, increased glucagon. Increased glucagon levels should result in increased glycogenolysis, increased gluconeogenesis, and increased triacylglycerol hydrolysis. The increase in glycogenolysis, gluconeogenesis, and reduced glucose metabolism promote above normal blood glucose. High insulin levels, on the other hand, should promote the opposite of the effects listed. Nonetheless, when insulin levels are high, insulin resistance results in above normal blood glucose.

The Bottom Line

There are two types of diabetes, initially called growth onset and maturity onset, then juvenile onset, and maturity onset, then IDDM and NIIDM and presently T1D and T2D. T1D is an immune response disease that destroys β-cells.

---

[3]For the Summary Box in this chapter, the principal salient concepts will be denoted by bold print at the bottom of the box and labeled "The Bottom Line."

PANCREATIC β-CELL MASS

Pancreatic β-cell mass is determined by β-cell reproduction, changes in numbers of β-cells through differentiation of exocrine pancreatic duct epithelial cells, and changes in size and programmed β-cell death (apoptosis). In humans, reproduction of β-cells from existing β-cells (mitosis) and production of new β-cells from epithelial cells significantly increases immediately after birth and outstrips apoptosis. Through adolescence, all three of the aforementioned processes decline equally. Pancreatic β-cell mass stays constant until about 40 years of age, when it decreases slightly owing to an increase in apoptosis and decreases in reproduction and the production of new cells from pancreatic epithelial cells.

Since 1952, numerous investigations have substantiated Wrenshall's studies. T1D is characterized by progressive β-cell destruction. By the time T1D symptoms (thirst, fatigue, polyuria, and weight loss) have appeared, 70–80% of β-cells have been destroyed. β-Cell destruction is by an autoimmune attack. The autoimmune attack manifests itself as an inflammatory reaction (insulitis). Apoptosis is the principal route for T1D β-cell death (6, 7).

Relative β-cell volume and mass are decreased in both obese and normal weight individuals with T2D. However, T2D decrease in β-cell mass is about 25–50%. The principal defect in T2D is similar to T1D—β-cell apoptosis. However, the specific routes for β-cell death are different and distinct.

Glucose and free fatty acid (FFA) toxicities induce synthesis of interleukin-1 beta (IL-1β) in both T1D and T2D. IL-1β regulates β-cell inflammation in T2D (8). In order to prevent inflammatory disease, the human body maintains a balance between proinflammatory and antiinflammatory regulatory T-cells. In both T1D and T2D, proinflammatory T-cells that produce chronic inflammation are predominant (9). The sequence of events that purportedly (7, 10) occurs in T1D is as follows:

- Elevation of IL-1β and TNF-α[4] in immune cells
- Infiltration by activated macrophages
- Infiltration of T-cells (especially $CD8^+$ and $CD4^+$)
- β-Cells produce procaspase-3[5] which is converted to caspase-3
- β-Cells undergo apoptosis.

---

[4]TNF-α is the acronym for tumor necrosis factor-alpha, an inflammatory substance secreted by immune cells that effect other cells in various ways (cytokine).

[5]Procaspase-3 is the precursor to caspase-3 (cysteine-aspartic acid protease-3). Caspase-3 plays a major role in the execution phase of apoptosis.

In T2D, the following events are postulated (7, 10):

- High glucose/high FFAs/leptin[6] result in the production of reactive oxygen species (ROS).[7]
- TXNIP[8] is activated.
- NLRP3 inflammasome is activated by TXNIP.
- NLRP3 inflammasome promotes the production of IL-1β from pro-IL-1β.
- β-Cells die either by apoptosis or by necrosis.[9]
- Scavenger macrophages remove the necrotic β-cells.

A characteristic of T2D is deposition of amyloid in the beta islets. A component of amyloid is islet amyloid polypeptide (IAPP) (also known as *amylin*). IAPP activates caspase-1, which is an integral component of the inflammasome complex, to cleave pro-IL-1β to its active form, IL-1β (11, 12). IAPP forms β-sheet fibrils in the beta islet cells in T2D. Amyloid deposits are not unique to T2D but are found in at least 20 other diseases and conditions including Alzheimer's, Parkinson's, and Huntington's diseases. For more information on amyloid deposits, refer to Reference 13.

It has been also suggested that IRS-2 plays a pivotal role in the maintenance of β-cell mass. Decreased IRS-2 promotes β-cell apoptosis and increased IRS-2 promotes β-cell reproduction and new β-cell formation. Thus, suppression of IRS-2 possibly is linked to increased apoptosis in T2D and could find a place in the bulleted events listed earlier (14).

A connection between obesity and T2D has been reported (15) involving protein kinase C zeta (PKCζ). In obese individuals, PKCζ causes adipose cells to secrete interleukin-6 (IL-6), which travels to the liver promoting inflammation and insulin resistance. IL-6 may contribute to the inflammation found in T2D (16).

Summary Box 5.2

- Pancreatic β-cell mass is regulated by β-cell reproduction, differentiation of epithelial cells, size changes, and apoptosis. Pancreatic β-cell mass increases after birth, stays constant from adolescence through about 40 years of age, and then declines slightly.

---

[6]Leptin is a protein that is synthesized primarily in adipose cells. Leptin regulates metabolism.

[7]ROS is the name given to free radicals produced in all cells. Free radicals possess one unpaired electron. The precursor to the free radical is a superoxide anion ($O_2{}^-$). ROS is neutralized by antioxidants, but when this balance is disrupted surplus ROS molecules are produced, causing damage to proteins, nucleic acids, and other macromolecules. ROS and its relation to diabetes will be presented in the mitochondrion section of this chapter.

[8]TXNIP is the acronym for thioredoxin-interacting protein.

[9]Necrosis is cell death caused by enzyme degradation as opposed to apoptosis, which is genetically destined cell death.

- T1D patients have greatly reduced pancreatic β-cell mass (70–80%). Obese and T2D patients have about a 25–50% loss in pancreatic β-cell mass.
- In both T1D and T2D, interleukin-1β plays a prominent role in the loss of pancreatic β-cell mass. Nonetheless, the pancreatic β-cell mass loss takes different pathways in T1D and T2D.
- In T1D, infiltration by immune cells initiates the destruction of the β-cells. In T2D, hyperglycemia, high FFAs, and production of leptin promote reactive oxygen as the initiating step and the activation of NLRP3 inflammasome plays an important role in the process.
- In both T1D and T2D, apoptosis of β-cells occurs. In T2D, β-cell necrosis also is involved in the loss of pancreatic β-cell mass.
- IRS-2 is important in maintaining pancreatic β-cell mass. Secretion of interleukin-6, promoted by PKCζ, causes pancreatic inflammation and insulin resistance in T2D.

The Bottom Line

T1D subjects have only 20–30% of normal β-cell mass, while T2D subjects have 50–75% of normal β-cells. Apoptosis and necrosis processes destroy the β-cells. Interleukin 1β and NLRP3 inflammasome play a role in the destruction of β-cells.

GLUCOSE TRANSPORT AND HEXOKINASE

You may wish to review the information presented on the properties and functions of GLUT4 presented in Chapter 4.

T1D presents with elevated blood glucose levels and hypoinsulinemia. Without sufficient quantities of insulin, one would expect diminished glucose transport. And, indeed, investigations studying glucose metabolism in T1D individuals bear this out. The rate-limiting defect for glucose intake in skeletal muscle of T1D subjects is at the glucose transport step (17, 18).

On the other hand, T2D is a polygenic disease that results in the absence of a proportionate glucose response to insulin (termed *insulin resistance*). Thus, T2D presents itself with hyperglycemia accompanied by hyperinsulinemia. The pathology underlying T2D immediately coming to mind would be a defect in the transport of glucose into the cell. The inability of insulin to affect the uptake of glucose into skeletal muscle and fat cells of animals and humans has been shown by numerous investigators as a principal characteristic of T2D. As shown in Figures 4.3 and 4.4, there are several sites where the defect could be present. Reviewing Chapter 4, the defect can occur at the following sites:

- Receptor site for insulin
- Activation of IRS1/IRS2

- Activation of PI3K
- Conversion of PIP2 to PIP3
- Activation of Akt/AS160[10]
- Activation of GLUT4 vesicle
- Expression of GLUT4 and/or vesicle
- GLUT4 vesicle transport to the surface of the cell
- GLUT4 vesicle complex formation with glucose
- Transport of GLUT4 vesicle containing glucose into the cell
- Hexokinase phosphorylation of glucose to glucose 6-phosphate.

Although genetic defects have been reported for the insulin receptor site, they are very rare and therefore are only an infrequent cause of T2D (19). Several investigators report that GLUT4 is expressed normally in T2D but there is evidence of reduced targeting and trafficking[11] (20). Insulin-promoted activation of Akt and AS160 is normal in diabetic subjects. But PI3K has been reported as diminished by 50% and 39% in IRS-1 and IRS-2-associated PI3K activities, respectively (21). AS160 and TBC1D1, as substrates for Akt phosphorylation, are required for GLUT4 vesicle trafficking (22). AS160 activity is increased by insulin in normal individuals but not in T2D persons (an effect called *uncoupling*) (23, 24). Insulin-activated AS160 phosphorylation was reduced nearly 40% in T2D subjects (25).

Hexokinase II (HKII) is decreased in the skeletal muscle of T2D persons. The expression of HKII is increased in the muscle and fat cells of healthy individuals when hyperinsulinemia has been induced. This does not occur in T2D. In other words, glucose 6-phosphate formation from glucose is not as readily produced in T2D as in healthy or obese subjects.

From this, one can conclude that the principal T2D defect occurs at the link between AS160 activation of the GLUT4 vesicle that shuttles glucose between the surface of the cell where the glucose is picked up and the interior of the cell where the glucose is phosphorylated to glucose 6-phosphate (this is the step in glucose metabolism called *translocation*).

Summary Box 5.3

- The principal defect in T1D is at the glucose transport step(s). Thus, T1D exhibits hyperglycemia and hypoinsulinemia.

---

[10] AS160 stands for Akt *substrate* of 160 kDa. AS160 is a substrate for phosphorylation by Akt and is required for insulin-stimulated translocation of the GLUT4 vesicle.

[11] Targeting is the modification of a molecule or organelle so that it can be translocated to a specific cellular location. Trafficking is the transportation of a molecule or organelle from one place to another in the cell.

- T2D is a disease that involves several genes (polygenic), which result in insulin resistance. Insulin resistance is the absence of a proportionate decrease in glucose in response to insulin. Thus, T2D exhibits hyperglycemia and hyperinsulinemia.
- Mutations at the insulin receptor site are rare causes of T2D.
- GLUT4 is expressed normally in T2D but there is evidence of reduced targeting and trafficking.
- Activation of Akt and AS160 are normal in T2D.
- PI3K is reduced. AS160 is normally increased by insulin but is uncoupled from this activity in T2D.
- HKII is low in T2D. HKII activity is increased by insulin normally but this does not occur in T2D.
- It is therefore concluded that the T2D defect occurs at the AS160 activation of the GLUT4 vesicle.

The Bottom Line

The principal defect in T1D is at glucose transport. T1D exhibits high glucose and low insulin. T2D is a polygenic disease that results in a disproportionate response of insulin for blood glucose levels (called *insulin resistance*). This results in high glucose and high insulin levels. The principal defect in T2D occurs at the AS160 activation of the GLUT4 vesicle.

GLYCOGEN SYNTHESIS AND BREAKDOWN

You may wish to review glycogenesis, glycogenolysis, and gluconeogenesis pathways presented in Chapter 3.

During fasting, normal individuals derive about 75% of their blood glucose from liver glycogenolysis, $45 \pm 6\%$ and gluconeogenesis, $55 \pm 6\%$ (26). Researchers have shown (27) that T1D patients have both reduced liver glycogen synthesis and breakdown. Liver glycogen levels increase slowly during the day in healthy subjects. After an 800 kcal meal in the afternoon (5 pm), normal individuals showed a 29% higher liver glycogen increase than in poorly controlled T1D subjects. Both net glycogen synthesis and net glycogen breakdown were lower in T1D subjects than in healthy controls. Glycogen synthesis is 74% lower in diabetics than in healthy subjects. Diabetic individuals infused with insulin for the short-term (24-h) have an increase of 23% in total glycogen. This value was lower than that in healthy subjects despite the fact that short-term normalization of blood glucose levels (normoglycemia) had been achieved.

Overnight fasting resulted in a slowly decreased concentration of liver glycogen in healthy persons. The maximum decrease was lower by almost twofold in T1D than in normal subjects and was increased in short-term insulin-infused

T1D subjects (brought to normal blood glucose levels) above that of noninfused T1Ds but below that of normal individuals.

Further experimentation by the same investigators (28) showed that long-term normoglycemia resulted in "peak hepatic glycogen synthesis and breakdown in type 1 diabetic patients that were identical to those of nondiabetic humans." Diminished glycogenesis in poorly controlled T1D is expected from the earlier observation that T1D individuals have markedly reduced insulin levels. Glucose is a negative allosteric inhibitor of glycogen phosphorylase (29). Therefore, a decrease in glycogenolysis is expected from hyperglycemia inhibition of glycogen phosphorylase.

In T2D subjects, muscle glycogen synthesis was almost 50% reduced than in healthy subjects. If the defect were in glycogen synthase, glucose 6-phosphate would be higher in T2D than in healthy individuals (Why? Refer to Chapter 3). When similar concentrations of blood glucose and blood insulin were established in healthy subjects and T2D subjects, a greater increase in glucose 6-phosphate was found in the healthy subjects than in the T2D subjects. Glucose 6-phosphate formation was blunted in T2D subjects. If HKII activity was reduced relative to glucose transport, an increase in intracellular glucose should occur. If glucose transport is defective in T2D, then intracellular glucose and glucose 6-phosphate should decrease in T2D subjects. Intracellular glucose concentrations were significantly lower in T2D subjects. This reasoning and experimental results support a similar earlier experiment and suggest that reduced glucose transport plays an important role in the occurrence of T2D (30, 31).

You may recall from Chapter 3 that glycogen synthase is inhibited by phosphorylation by GSK-3, while dephosphorylation and thereby activation of glycogen synthase is promoted by phosphoprotein phosphatase. Several studies have demonstrated that protein phosphorylase is not defective in T2D subjects. Researchers have determined that GSK-3 protein levels and activity are elevated in T2D muscle in lean and obese patients. Thus, increased GSK-3 activity may contribute to the reduced activity of glycogen synthase in T2D (32).

GLYCOGEN CYCLING

Glycogen cycling is defined as the simultaneous synthesis and breakdown of glycogen. The rate of glycogen synthesis is defined as the sum of two processes: formation of glycogen from glucose 6-phosphate derived from noncarbohydrate substances such as lactate, pyruvate, glycerol, and alanine and glucose 6-phosphate derived from phosphorylation of glucose (see Fig. 5.1). The rate of glycogen breakdown is the rate of formation of glucose 6-phosphate from glycogen.

There are at least four procedures for glycogen cycling measurements. These four procedures are briefly described subsequently. Reference 33 illustrates how each of these methodologies determines glycogen cycling. In the first method, net hepatic glycogenolysis is estimated by measuring blood glucose that has been

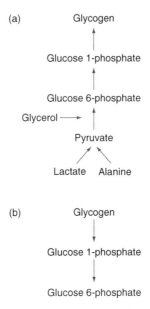

Figure 5.1 Glycogen cycling: (a) synthesis and (b) breakdown.

labeled in the C-2 and C-5 positions by the ingestion of deuterated water (^2H_2O). The rate of net glycogenolysis is determined from the decline in liver glycogen measured by NMR. In this procedure, glucose 6-phosphate is assumed to be converted to glycogen and glucose only.

In the second method, in order to measure the rate of glycogenesis, [(1-^{13}C)]glucose is infused into the subject. The label ends up on C-1 of the glucose residues in glycogen. NMR is used to measure the glycogen binding of the label. The rate of glycogenesis is measured from the intensity of the ^{13}C signal. NMR next measures the rate of disappearance of the label during an unlabeled glucose infusion. This method assumes that glucose 6-phosphate is derived from the breakdown of glycogen and is not reversibly converted to glycogen.

The third method involves the conversion of galactose to glucose. In this procedure, acetaminophen and a trace amount of [2-^3H, 6-^{14}C]galactose is administered. [2-^3H, 6-^{14}C]Galactose is converted to 6-^{14}C-glucose in the following manner: [2-^3H, 6-^{14}C]galactose is phosphorylated to [2-^3H, 6-^{14}C]galactose 6-phosphate and then binds an uridyl group from uridine diphosphate glucose (UDP-glucose) to form UDP-[2-^3H, 6-^{14}C]galactose (Refer to Chapter 3 for details.) UDP-[2-^3H, 6-^{14}C]galactose replaces UDP-glucose and incorporates into the growing glycogen molecule. In the next step, epimerization changes [2-^3H, 6-^{14}C]galactose to [2-^3H, 6-^{14}C]glucose in the glycogen molecule. During glycogenolysis, [2-^3H, 6-^{14}C]glucose 1-phosphate is formed. [2-^3H, 6-^{14}C]Glucose 1-phosphate is then converted to [2-^3H, 6-^{14}C]glucose 6-phosphate. Next, [2-^3H, 6-^{14}C]glucose 6-phosphate is converted to [6-^{14}C]fructose 6-phosphate. Notice that in this conversion the ^3H is lost, while the ^{14}C is retained. The reversibility of these steps results in the formation of [6-^{14}C]glucose 1-phosphate, which is reincorporated into the glycogen molecule. This method depends on galactose being incorporated into glycogen before conversion to glucose. The $^{3}H/^{14}C$ ratio in acetaminophen glucuronide excreted in the urine to that of the [2-^{3}H, 6-^{14}C]galactose that was administered measures the rate of the synthesis of glycogen produced from glucose 1-phosphate (glycogenesis). This method assumes that ^{3}H is removed completely and glucose 6-phosphate that is not converted back to glycogen.

In the fourth method, glycogenesis is measured by the dilution of the label from [2-^{3}H, 6-^{14}C]galactose as it is converted to acetaminophen glucuronide. Glycogenolysis is measured from the quantity of label found in blood glucose. The assumption in this assay is that galactose is converted to glycogen and then to glucose.

Researchers (34) using a combination of these methods determined that poorly controlled T1D fasting gluconeogenesis and glycogen cycling are increased when compared to controls and that intravenous insulin infusion restores these differences to near normal values. Also reported in this research are the following:

- Liver glycogen synthesis was 70% lower in poorly controlled T1D (T1Dp) than in healthy subjects. T1D subjects treated by continuous subcutaneous insulin infusion (T1Di) had near normal glycemia and had glycogen synthesis values comparable to healthy subjects.
- During fasting, T1Dp subjects had 42% higher endogenous glucose production than healthy subjects. T1Dp subjects were 25% higher than T1Di.
- Gluconeogenesis was 67% higher in T1Dp than in healthy subjects. T1Dp subjects were 74% higher than T1Di subjects.
- Glycogen phosphorylase flux[12] in T1Dp was twofold higher than net glycogenolysis. T1Di was comparable to healthy subjects.
- Glycogen cycling in T1Dp accounted for 47% of glycogen phosphorylase flux.

A principal conclusion derived from this research is that hepatic glucose metabolism in T1D is not irreversibly altered and is restorable by insulin infusion.

Summary Box 5.4

- T1D subjects have decreased liver glycogen synthesis and breakdown.
- After a meal, normal subjects have a higher increase in their liver glycogen than T1D subjects. Net glycogen synthesis and net glycogen was lower in T1D than in normal subjects.

---

[12]Glycogen phosphorylase flux refers to the rate of the production of glucose 1-phosphate from glycogen catalyzed by glycogen phosphorylase. In general, the term *flux* refers to the rate of a single reaction occurring in one direction as though the reaction occurred in isolation.

- T1D subjects infused with insulin short-term had an increase in total glycogen, which was lower than that for normal subjects treated in the same way.
- Overnight fasting normal subjects had a decrease in liver glycogen greater than that of T1D subjects. Short-term insulin infusion resulted in a further glycogen decrease but still less than in normal subjects.
- Long-term normal blood glucose values (sustained by insulin infusions) resulted in liver glycogen synthesis and breakdown that was equivalent to T1D subjects.
- Glycogen synthesis in T2D subjects was reduced relative to normal subjects.
- When normal subjects were subjected to the same glucose and insulin levels as T2D subjects, glucose 6-phosphate levels were found to be higher in the normal subjects. If the defect occurred in HKII, increased intracellular glucose would be expected. If the defect occurred in glucose transport in T2D subjects, intracellular glucose and glucose 6-phosphate would be expected to decrease. Intracellular glucose was lower inT2D subjects, suggesting that the defect occurred at glucose transport.
- GSK-3 is increased in T2D. As GSK-3 inhibits glycogen synthase, it may contribute to the decreased activity of the latter in T2D.
- Poorly controlled T1D fasting gluconeogenesis and glycogen cycling are increased and insulin infusion restores them to normal. It is concluded that liver glucose metabolism is not irreversibly altered in T1D.

The Bottom Line

After a meal or short-term infusion with insulin, T1D subjects have decreased liver glycogen synthesis relative to normal subjects. Fasted T1D subjects have decreased liver glycogen that is less than in normal subjects. However, long-term insulin infusion resulted in liver glycogen synthesis and breakdown that was equivalent to that in normal subjects. T1D gluconeogenesis and glycolysis is increased above normal but insulin infusion restores them to normal. In T2D subjects, glycogen synthesis was reduced. The principal defect is believed to be at glucose transport. In addition, GSK-3 is increased in T2D and may contribute to decreased glycogen synthesis.

GLUCONEOGENESIS AND GLYCOGENOLYSIS

Many investigators have reported excessive liver glucose production in T2D subjects. In T2D patients, hepatic glucose was increased threefold and gluconeogenesis was increased nearly threefold. Substances that feed into the gluconeogenesis pathway such as blood lactate and alanine were also increased (35). T2D subjects are subject to an excessive and prolonged increase in blood glucose after ingesting a meal (postprandial). After a meal, liver glycogen was reported (36) as lower by close to twofold in T2D subjects when compared to healthy subjects. Gluconeogenesis was increased about 1.5 times in T2D subjects and net liver glycogenolysis was decreased by about twofold as compared to healthy subjects. The increased production of gluconeogenesis in T2D accounted for 88% of the total glucose produced as compared to 70% in healthy individuals.

After a 15-h fast, assays for gluconeogenesis and glycogenolysis were conducted on a population of lean and obese healthy individuals and lean and obese T2D patients (37). Gluconeogenesis flux was found in this population to have increasing values as follows: normal lean, normal obese, T2D obese, and T2D lean subjects. Glycogenolysis flux was found to have increasing values as follows: normal obese = T2D obese, T2D lean, and normal lean individuals. Thus, gluconeogenesis was increased in obesity and more markedly in T2D subjects and glycogenolysis flux was reduced only in obesity (either normal or T2D). These results suggest that obesity is a precursor to T2D as increased gluconeogenesis would increase blood glucose concentrations. In fact, these studies demonstrate that about 90% of the increase in liver glucose production is attributable to increased gluconeogenesis.

Another characteristic feature of T1D and T2D is excessive postprandial hyperglycemia. Researchers reported (38) that T2D subjects had an increased postprandial glucose release than did normal individuals primarily because of increased gluconeogenesis and glycogenolysis, and the ingested glucose derived from the meal. The ingested glucose that was released into the blood was similar in normal individuals and T2D individuals. Postprandial gluconeogenesis and glycogenolysis in T2D subjects were nearly twice that of normal individuals. In conclusion, the excess glucose found after meals in T2D persons is largely due to excessive gluconeogenesis and glycogenolysis.

In another study (39), insulin administration, which restored glucose levels to normal, reduced gluconeogenesis in both normal and T2D subjects by about 20%. Gluconeogenesis remained abnormally elevated in T2D. Glycogenolysis was almost completely inhibited in both groups. From these studies, one may conclude that the defect in gluconeogenesis and glycogenolysis is not affected by returning glucose levels to normal.

Studies have shown (40) that gastric emptying in response to postprandial hyperglycemia is defective in individuals with T1D. Thus, glucose derived from meals appears in the blood of persons with T1D more rapidly than in healthy individuals and adds to postprandial hyperglycemia. We will discuss gastric emptying in greater detail in Chapter 8 "Complications."

Summary Box 5.5

- Liver glucose production and gluconeogenesis is excessive in T2D.
- After a meal, T2D subjects exhibit an excessive and prolonged increase in blood glucose and glycogen was lower than in normal subjects. Glycogenolysis was decreased.
- After a 15-h fast, gluconeogenesis is lowest in normal lean subjects, with normal obese being intermediate, and T2D obese and T2D lean subjects being highest.
- Glycogenolysis was decreased only in normal obese and T2D obese individuals.
- T1D and T2D have abnormally postprandial glucose levels.
- Insulin infusion that restored glucose to normal decreased gluconeogenesis in both normal and T2D subjects, while glycogenolysis was nearly completely inhibited.
- Gastric emptying is abnormal in T1D and results in glucose appearing early in the blood.

The Bottom Line

Blood glucose in T2D lean and T2D obese patients is increased by excessive gluconeogenesis and glycogenolysis, while normal obese subjects also show an increase. Obesity is suggested by these results as a precursor to T2D. T1D and T2D show excessive postprandial hyperglycemia. Insulin administration to T2D subjects demonstrates that the gluconeogenesis defect is reversible.

GLYCOLYSIS, GLUCOSE OXIDATION, AND PYRUVATE DEHYDROGENASE

Insulin infusion that resulted in normal glucose levels did not affect the activity of hexokinase I (HKI) or HKII. You may remember that HK is decreased in T2D patients. This suggests that reduced HK activity is not due to elevated glucose levels. It is likely that the effect on HK activity is at the translation level of HK production. The irreversibility of the defect in HK upon treatment with insulin suggests that the defect is an inherited one. Thus, insulin resistance in T2D is likely composed of a hereditary defect and an acquired defect such as hyperinsulinemia, hyperglycemia, increase in fatty acids, and so on (41).

In order to give you an idea of how these types of metabolic experiments are conducted, I will describe the methods and results of an experiment with respect to glycolysis, glycogenesis, and glucose oxidation in normal and T2D subjects. The studies described (42) subsequently utilized 3-[^{3}H]glucose to measure glycolytic flux. At the triose phosphate isomerase step of glycolysis, ^{3}H on the

3-hydroxyl (OH) group of glucose is transferred to form ^3H_2O. U-$[^{14}C]$glucose[13] metabolized to $^{14}CO_2$ is used to measure glucose oxidation (primarily, the tricarboxylic acid (TCA) and oxidative phosphorylation pathways). Knowing the quantity of glucose infused into the subject, the quantity of ^3H_2O (glycolysis), and the quantity of $^{14}CO_2$ expelled, the glycolytic flux, glucose oxidation, and glycogenesis were determined.

Researchers in this study determined the aforementioned parameters in control subjects (nondiabetic) who had blood glucose levels of 5.1 mM (normal levels equivalent to approximately 100 mg/dl) and a clamp[14] that provided moderate hyperinsulinemia, 258 pM (~172 pM is normal). T2D patients were compared to control subjects under three different sets of conditions:

1. Insulin was infused into the T2D patient until normal blood glucose (5.2 mM) was obtained. Insulin levels were close to that of normal subjects at 269 pM.

2. At the start of insulin infusion, glucose was infused into the T2D patients at a rate equal to that of control subjects during the last hour of insulin infusion. The glucose levels reached a plateau (14.9 mM) and remained constant till the completion of the study. Insulin levels were close to normal at 259 pM.

3. During insulin infusion, glucose values were permitted to drop to near normal values (5.5 mM). Insulin levels were at 1650 pM.

The following summarizes just one aspect of this research: the absolute rates of glycogen formation, glycolysis, glucose oxidation, and nonoxidative glycogenesis in control subjects and T2D subjects in the last hour of clamp studies and the conclusion drawn from this data.

In study 1, T2D subjects had glucose and insulin concentrations that were equal to control subjects. T2D subjects exhibited glycolysis that was nearly 64% lower than control subjects. In study 2, where the rate of glucose utilization was equal between T2D subjects and normal subjects, glycolysis remained the same for T2D subjects as in study 1. In the third study, glycolysis was insignificantly greater in T2D than in control subjects.

In study 1, T2D subjects exhibited significantly lower glucose oxidation (43%) than control subjects. In study 2, T2D subjects showed an increase from the results in study 1 but the glucose oxidation results remained lower (69%) than in control subjects. In study 3, T2D subjects showed a further increase but glucose oxidation still remained lower (86%) than in control subjects.

Glycogenesis was determined from the sum of the glycolytic flux and glucose oxidation subtracted from the total glucose infused into the subject. In study

[13]The U designates uniformly labeled glucose, that is, all six carbons are ^{14}C.

[14]Clamp studies are performed using a catheter inserted into the anticubital vein for the infusion of insulin or glucose. A second catheter is inserted into a wrist vein for blood sampling. The clamp technique will be discussed in greater detail later in this chapter.

1, T2D subjects exhibited significantly lower glycogenesis (25%) than control subjects. In study 2, T2D subjects exhibited a significant increase in glycogenesis, bringing it to control levels. In study 3, T2D subjects exhibited glycogenesis near control subjects.

In study 1, T2D subjects exhibited nonoxidative glycolysis 2.6 times higher than controls. In study 2, T2D subjects exhibited nonoxidative glycolysis that was nearly three times controls. In study 3, nonoxidative glycolysis increased to nearly four times the values found in control subjects.

The authors concluded[15] that glucose uptake was reduced and the rates of glycolysis, glycogenesis, and glucose oxidation were significantly reduced in the T2D subjects who had their glucose normalized by moderate insulin infusion. This is consistent with reports that glycogen synthase activity is decreased in T2D subjects. Nonoxidative glycolysis and lipid oxidation is increased. The defect in glycogenesis in T2D subjects is about twice that of the defect in total body glycolysis. In terms of percentage of total body glucose infusion, glycogenesis corresponds to 47% and glycolysis to 57% in normal subjects. In T2D subjects, this ratio was 26% and 74%, respectively.

The decrease in glucose oxidation concomitant with the increase in nonoxidative glycolysis suggests a defect at pyruvate dehydrogenase (PD) [the connection between glycolysis and the TCA pathway]. This conclusion is consistent with the reports of reduced PD activity in T2D subjects and their healthy children (43).

As you may recall from Chapter 3, pyruvate dehydrogenase kinase-4[16] (PDK4) inhibits PD by phosphorylating it. After a meal, when insulin levels are elevated, PDK4 is decreased and therefore PD activity is increased. Blood glucose levels are decreased and pyruvate is directed into the TCA pathway. When in the fasting state, the situation is just the opposite and pyruvate enters the gluconeogenesis pathway and glucose levels are increased.

PDK4 activity is increased in T2D and therefore PD is inactivated (44). In T1D, the decreased insulin levels increase PDK4 transcription and PD is decreased. Increased PDK4 activity has been reported in rats fed high fat meals and obese rats (45, 46). Glucocorticoids have been reported to increase PDK4. The increase in PDK4 is mediated by Forkhead transcription factor (Foxo1) (refer to Fig. 4.3) through a decrease in AkT phosphorylation (47).

Researchers (48) used ^{31}P magnetic resonance spectroscopy to investigate glycolysis and glucose oxidation in T1D subjects. The study was designed to determine the difference in the energetic properties of exercised muscle (ankle dorsiflexor) in normal and well-controlled T1D subjects. The difference in pH of the muscle at rest and the pH at the end of exercise was significantly different between normal subjects and T1D subjects. Glycolytic flux began earlier during exercise and reached a higher peak in T1D subjects than in normal individuals. Reduced glucose oxidation occurred in T1D muscles than in normal individuals.

---

[15]Referral to Chapter 3 might be helpful in interpreting these conclusions.
[16]There are four isoforms of PDK: PDK1, PDK2, PDK3, and PDK4.

These results are consistent with the studies of glycolysis and glucose oxidation in T2D and obesity described earlier (in Reference 41).

PD and PDK are located in the mitochondrion and serve as the intermediate system between glycolysis and glucose oxidation. In the next section, we will discuss defects in the mitochondrion found in T1D and T2D.

Summary Box 5.6

- Hexokinase is reduced in T2D. Insulin infusion bringing glucose levels to normal does not affect HKI or HKII. Irreversibility of the hexokinase defect suggests that it is a genetic defect and occurs at the level of translation.
- Experiments utilizing a variety of conditions (glucose and insulin levels) determined that the rates of glycolysis, glycogenesis, and glucose oxidation were lower in T2D subjects than in normal subjects under all conditions of the studies. Glycogenesis was lower in T2D subjects than in control subjects in study 1 but was at control levels in studies 2 and 3. This is consistent with reports that glycogen synthase is reduced in T2D. T2D subjects exhibited nonoxidative glycolysis that was higher than normal subjects under all conditions of the study. The authors concluded that PD was defective in T2D subjects.
- PDK4 activity is increased in T2D and therefore PD is inhibited. The decreased insulin levels in T1D increase PDK4 production and result in decreased PD activity.

The Bottom Line

Reduced hexokinase activity in T2D is likely caused by an genetic defect at the transcription level. A study utilizing insulin infusion reported increased nonoxidative glycolysis and decreased glucose oxidation in T2D subjects as compared to nondiabetic subjects. The conclusion of the authors was that PD was defective. The underlying defect may be due to increased PDK4 levels inT2D. PDK4 inhibits PD. T1D subjects with reduced insulin levels have increased PDK4 production and reduced PD activity.

MITOCHONDRIAL DEFECTS

Tricarboxylic Acid Pathway and Oxidative Phosphorylation

UCPs   As we learned in Chapter 3, the TCA pathway is linked to the electron transport system and oxidative phosphorylation by NADH and $FADH_2$. Uncoupling proteins (UCPs) 1–5 are present on the mitochondrial inner membrane. UCPs regulate the pumping of protons across the inner membrane of the mitochondrion and thus provide the energy for the synthesis of ATP. UCPs therefore regulate the coupling of the electron transport system to oxidative phosphory-lation. UCP1 and UCP3 are found in skeletal muscle tissue. UCP4 and 5 are located in the brain, while UCP2 is widely distributed. UCP2 decreases the production of ROS (see Footnote 6) in mitochondria, regulates insulin secretion and fatty acid metabolism. In β-cells, increased ROS production has been reported to activate UCP2 and therefore increase mitochondrial proton leakage.[17] This results in decreased ATP production and inhibits insulin secretion. In addition, mice lacking the UCP2 gene and bred for obesity (termed *ob/ob mice*) were found to have increased insulin secretion in response to glucose (49).

The ATP/ADP ratio is an important sensing unit for insulin secretion. When the ATP/ADP ratio increases, the ATP-sensitive K^+ channel of the mitochondrial inner membrane closes. Subsequently, the Ca^{+2} channel opens and the influx of Ca^{+2} triggers the secretion of insulin. Inhibition of UCP2 decreases the formation of ATP, thus decreasing the ATP/ADP ratio and increasing secretion of insulin (50, 51).

There have been several studies in humans relating the polymorphisms[18] in the UCP locus[19] to T2D. But these studies have produced results that are contradictory to each other. In conclusion, we can say that UCP2 may contribute to the pathophysiology of T2D, but this remains to be conclusively determined.

Sirt3   The sirtuin[20] (Sirt or Sir) group of NAD-dependent protein deacetylases is located in the mitochondria. They were initially identified as regulators of the life span of yeast, worms, flies, and mice. A polymorphism of Sirt3 has been reported as increasing the life span of humans to 100 years. Sirt3 has been found to be reduced in T1D and T2D. This reduction in the expression of Sirt3 results in a decreased TCA pathway, increased ROS, and activation of c-Jun N-terminal kinase (JNK). The activation of JNK results in increased serine and decreased tyrosine phosphorylation of IRS-1 and therefore results in insulin resistance (Fig. 5.2) (52).

More recently, T2D mice were found to revert to normal when fed nicoti-namide mononucleotide (NMN) (53). The rationale for this reversal derives from the fact that NMN is an intermediate in the biosynthesis of NAD^+ (Figure 5.3). As you can see from Figure 5.3, nicotinamide is converted to NMN by nicotinamide phosphoribosyltransferase (Nmnat). NMN is next converted to NAD^+ by nicoti-namide acid adenylyltransferase (54). The significance of this metabolic pathway is that SIRT1[21] is dependent on NAD^+ for its deacetylase activity. Refer to Figure 5.2 to see how activation of Sirt1 activates the insulin signaling response. The Sirt association with diabetes is likely to be explored further in the near future.

---

[17]Proton leakage refers to transport of protons to sites other than ATP synthase.
[18]Polymorphisms refer to specific mutations occurring in a gene.
[19]Locus refers to a specific location in a gene.
[20]Sirt is the acronym for silent mating type information regulation 2 homolog.
[21]There are seven known human Sirts.

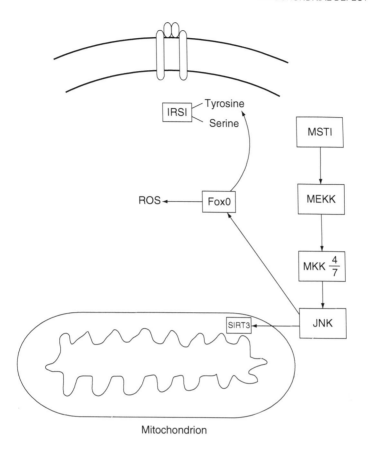

MST1	Macrophage stimulating 1
MEKK (also called MAP3K)	Mitogen-activated protein kinase kinase kinase
MKK $\frac{4}{7}$	Mitogen-activated protein kinase kinase
Fox0	Forkhead translation factor
JNK	c-Jun N-terminal kinase
IRS1	Insulin receptor substrate 1

Figure 5.2   c-Jun N-terminal kinase signaling pathway.

Lipids (Free Fatty Acids)   During the past decade, there has been growing evidence that T2D individuals possess higher concentrations of FFAs[22] than healthy individuals. The lipid is stored within the skeletal muscle cell as lipid droplets called *intramyocellular lipids* (IMCLs). Skeletal muscle and liver are the principal insulin-reacting tissues that maintain normal glucose levels. Accumulation of FFA is suspected to cause the insulin resistance exhibited by T2D individuals

---

[22]FFAs are unesterified fatty acids.

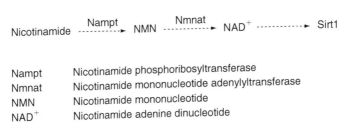

Nampt	Nicotinamide phosphoribosyltransferase
Nmnat	Nicotinamide mononucleotide adenylyltransferase
NMN	Nicotinamide mononucleotide
NAD^+	Nicotinamide adenine dinucleotide

Figure 5.3 Metabolic pathway for the biosynthesis of NAD^+.

(55–57). In fact, insulin resistance occurs nearly 20 years before T2D is diagnosed. Children of patients with T2D exhibit insulin resistance long before they are diagnosed with T2D. The insulin resistance they exhibit is associated with increased intramyocellular fatty acid deposits (58). The term used for the impairment of insulin secretion by increased FFA and their deposition in skeletal muscle and liver is *lipotoxicity*.

Increased FFA levels cause an increase in the intramyocellular levels of long chain fatty acyl-CoAs[23] and diacylglycerol[24] (59, 60). Long chain fatty acyl-CoAs and diacylglycerol activate PKC-θ. Refer to Figures 4.3 and 4.4a for the following description of the long chain fatty acyl-CoAs and diacylglycerol suppression of insulin-stimulated glucose transport. Activated PKC-θ inhibits IRS-1 by phosphorylating serine-307. Deactivating IRS-1 disrupts the remainder of the insulin signaling pathway. Inactivated IRS-1 does not activate PI3K. Inactive PI3K does not activate Akt via PIP3. Inactivated Akt does not activate the insulin-stimulated GLUT4 vesicle to be transported to the plasma membrane to pick up glucose. Glucose accumulates in the circulation.

Malonyl-CoA ($HOOCCH_2COSCoA$) is a short chain fatty acyl-CoA derivative that has been implicated in T2D (61). It inhibits carnitine palmitoyltransferase-1, which catalyzes the mitochondrial β-oxidation of fatty acids. This results in the accumulation of long chain fatty acyl-Cos and could, in a manner similar to diacylglycerol, cause disruption of the insulin signaling pathway at the glucose transportation level. This model has been challenged by several reports in the literature (62).

T2D is also characterized by increased FFA oxidation in the mitochondrion, a reduced TCA cycle and synthesis of ATP. A current hypothesis explaining these events is that increased FFA oxidation activates peroxisome-proliferator-activated receptor alpha (PPARα),[25] leading to the expression of several genes that are in the FFA oxidation pathway. Concomitantly, the TCA cycle is reduced as GLUT4 transport is inhibited by one of the mechanisms described earlier (61).

[23]Long chain fatty acyl-CoAs are long chain fatty acids bonded to coenzyme A.
[24]Diacyl glycerol consists of two fatty acids bonded to glycerol.
[25]PPARα is a nuclear receptor located in muscle, fat cells, liver, and many other organs that regulates gene translation.

Summary Box 5.7

- UCPs regulate the passage of proteins across the inner membrane of the mitochondrion.
- UCP2 is known to decrease the production of ROS, thus regulating insulin secretion and fatty acid metabolism. On the other hand, increased ROS production has been shown to activate UCP increasing proton leakage, decreasing ATP production, and decreasing insulin secretion.
- Several human studies suggest that polymorphisms in the UCP gene result in T2D. There some studies that contradict this finding.
- Sirtuin (Sirt or Sir) is reduced in T1D and T2D resulting in decreased TCA pathway, increased mitochondrial ROS and activation of JNK. Activation of JNK results in insulin resistance.
- NMN is converted to NAD^+. Sirt1 is dependent on NAD^+ for its activity. Sirt activates the insulin signaling pathway.
- T2D patients have high levels of FFAs, which result in long chain fatty acyl-CoAs and diacylglycerol.
- This activates PKC-θ, which next inhibits IRS-1.
- Malonyl-CoA has been shown to inhibit carnitine palmitoyltransferase-1, which is involved in mitochondrial β-oxidation of fatty acids and could disrupt the insulin signaling pathway.
- T2D subjects have increased FFA oxidation while GLUT4 transport is inhibited.

The Bottom Line

UCPs uncouple electron transport from the synthesis of ATP (oxidation phosphorylation) by causing the leakage of protons from the mitochondrion. The ATP/ADP ratio is a sensing unit for insulin secretion. When ATP increases, ATP/ADP increases and the Ca^{+2} channel opens and insulin is secreted. Polymorphisms in the UCP gene have been related to T2D. Sirt3 is reduced in T1D and T2D. This results in activation of JNK. JNK inhibits IRS-1 and results in insulin resistance. NMN is an intermediate in the synthesis of NAD^+, which is necessary for SIRT1 activity and its subsequent activation of the insulin signaling response. T2D patients have high levels of FFAs. FFAs are expected to cause insulin resistance. Three fatty acids, long chain fatty acyl-CoA, diacylglycerol, and malonyl-CoA have been implicated in the disruption of the insulin signaling pathway. T2D is characterized by increased fatty acid oxidation and reduced TCA cycle and therefore reduced ATP synthesis.

HEXOSAMINE BIOSYNTHESIS PATHWAY

Refer to Chapter 3 and Figure 3.19 for the following section.

The hexosamine biosynthesis pathway (HBP) senses blood glucose blood levels. If the levels increase above normal, the HBP reduces insulin-stimulated glucose transport. Glutamine fructose 6-phosphate transaminase (GFPT) activity is greatly elevated in the skeletal muscle of T2D individuals. Elevated GFPT levels lead to elevated UDP-*N*-acetylglucosamine (UDP-Glc-Nac). Next, UDP-Glc-Nac is linked by oxygen to cytoplasmic and nuclear proteins (O-linked glycosylation). This reaction is catalyzed by *O*-acetylglucosamine transferase (OGT). The protein residues that undergo O-linked glycosylation are serine and threonine.

(Refer to Figure 4.3 for the following.) OGT has a phosphatidyl inositol-3,4,5-trisphosphate (PIP3) binding site. PIP3 transports OGT from the nucleus to the plasma membrane, where OGT catalyzes O-linked glycosylation of a binding protein called *mammalian target of rapamycin complex 2* (mTORC2).[26] O-Linked glycosylation of mTORC2 activates gluconeogenesis in the liver. The manner in which O-linked glycosylation of mTORC2 affects gluconeogenesis is interesting and is probably one of the beautiful concepts in biochemistry. mTORC2 increases the production of glucose 6-phosphatase. Glucose 6-phosphatase is normally decreased in its expression by insulin. AMPK (see Chapter 4) phosphorylates the serine residue in mTORC2 and thus deactivates the binding protein. O-Linked glycosylation attaches to the same serine that is normally phosphorylated and thereby deactivated. O-linked glycosylation of the serine of mTORC2 prevents the deactivation of mTORC2 and gluconeogenesis continues although blood glucose levels are high. The added glucose produced by gluconeogenesis is added to the already high glucose levels (63–65).

Summary Box 5.8

- The HBP senses blood glucose levels and regulates insulin-stimulated glucose transport.
- GFPT activity is increased in T2D subjects and that leads to increased UDP-Glc-Nac. UDP-Glc-Nac undergoes O-linked covalent glycosylation to proteins. A protein that it is glycosylated is mTORC2. The glycosylated mTORC2 increases glucose 6-phosphatase production. The same serine that is glycosylated is the one that, when phosphorylated by AMPK, deactivates mTORC2. Serine glycosylation results in an activated mTORC2 despite elevated glucose levels. Glucose 6-phosphate is decreased and gluconeogenesis is increased.

---

[26] As usual, binding proteins have more than one name. mTORC2 is also known as *CREB regulated transcription coactivator 2* (CRTC2).

> *The Bottom Line*
>
> In normal persons, the HBP is a sensor of glucose levels; if glucose levels are high, insulin-stimulated glucose transport is increased; if low, transport is reduced. T2D subjects have GFPT levels that are increased. Through a series of steps including O-linked glycosylation of mTORC2, glucose 6-phosphatase is increased in activity and gluconeogenesis is thereby activated. Glucose from gluconeogenesis is added to already high levels found in the T2D subject.

TECHNIQUES USED IN THE INVESTIGATIONS

To conclude this chapter, it seemed useful and interesting to present in some detail the techniques used in the aforementioned investigations. Many of them are clever and imaginative.

Hyperinsulinemic-Euglycemic Clamp

The hyperinsulinemic-euglycemic clamp is used to assess enzymes, glycogen cycling, insulin signaling components, and other substances involved in glucose metabolism under controlled glucose and insulin conditions. The clamp procedure[27] was first described in the literature in 1979 (66).

In Animals   The animal, under anesthesia, is catheterized at the left common carotid artery and right jugular vein. The catheters are connected to pumps containing insulin and glucose solutions. Blood is collected from the arterial catheter and assayed for glucose while insulin is infused into the jugular vein catheter. When glucose levels decrease (which occurs immediately at the start of the insulin infusion), glucose (D50[28]) is infused through the jugular vein catheter. The glucose levels are kept at normal fasting levels throughout most experiments, hence the term euglycemic. However, if the experiment demands higher or lower blood glucose levels than normal, they can be maintained through the infusion of lesser or more glucose. If the animal is a rodent, the blood for sampling for insulin and glucose is obtained from the tail. In many experiments, radioactive tracers are infused as well into the jugular vein.

In Humans   In humans, the technique is similar except that the glucose is infused by a catheter in one arm and the insulin is infused by a catheter into the other. The hyperinsulinemic-euglycemic clamp technique is used to determine insulin resistance and this aspect of the technique will described in Chapter 9.

---

[27]The clamp designation derives from the fact that the glucose is clamped or, in other words, held at a constant level.

[28]D50 is 25 g of glucose in 50 ml water. The D stands for dextrose, an older name for glucose.

Vastus Lateralis Muscle Biopsy

The vastus lateralis muscle is located laterally in the outside frontal part of the thigh. Needle biopsy (about 100–250 mg) is often utilized with the hyperinsulinemic-euglycemic clamp technique to determine the effect of insulin and glucose levels on insulin-stimulated glucose uptake, glycolysis, glucose oxidation, and glycogenolysis (67).

Summary Box 5.9

- The hyperinsulinemic-euglycemic clamp is used to assess glucose metabolism under controlled conditions.
- In animals, two catheters are used. One is used to infuse insulin and/or glucose into the jugular vein, the other to draw blood samples from the carotid artery. If a rodent is the subject, blood can also be drawn from the tail vein.
- In humans, one catheter is placed in one arm to infuse insulin or glucose and in the other arm a second catheter is placed to draw blood.
- Often, in conjunction with the hyperinsulinemic-euglycemic clamp, biopsy of the vastus lateralis muscle is used to determine effects of various levels of insulin and glucose on various parameters of glucose metabolism.

The Bottom Line
The hyperinsulinemic-euglycemic clamp and vastus lateralis muscle have been effectively used in the investigations presented in this chapter.

GLOSSARY

Allosteric effector   In addition to the active site, regulatory enzymes possess sites that bind small molecules. These sites are called *allosteric sites*. The substance that binds to the allosteric site is called the *modulator*. The modulator reversibly binds to the allosteric site and changes the conformation of the enzyme. The modulator can inhibit or accelerate the reaction taking place at the active site.

Epimerization   Epimers differ in the configuration about a specific carbon. Epimerization refers to converting one epimer to another.

Epimers   Epimers are two carbohydrates that differ only by the configuration around one carbon. Glucose and galactose differ only in configuration around carbon 4.

Epithelial cells   Cells forming a protective layer on bodily surfaces, internal cavities, ducts, and organs.

Euglycemic   Normal glucose levels.

Exocrine cells (pancreatic) Cells in the pancreas that produce digestive enzymes.

Expression Expression refers to the transcription (formation of mRNA from a gene) or translation (synthesis of a protein from mRNA).

Glucose oxidation Refers to glucose oxidation via the tricarboxylic acid cycle and electron transport. Most investigations do not include the phosphogluconate oxidative cycle (2–3% of glucose).

Hepatic Pertaining to the liver.

Nonoxidative glycolysis Glycolysis as discussed in Chapter 3.

Postprandial Postprandial refers to "after eating a meal."

Triacylglycerols Often called *triglycerides*; triacylglycerols is a category of lipids usually stored in adipose (fat) cells.

REFERENCES

1. Wrenshall GA, Bogoch A, Ritchie RC. Extractable insulin of pancreas; correlation with pathological and clinical findings in diabetic and non diabetic cases. Diabetes 1952;1:87.

2. Bornstein J, Lawrence RD. Two types of diabetes mellitus with and without available plasma insulin. Brit Med J 1951;1:732.

3. Bornstein J, Lawrence RD. Plasma insulin in human diabetes mellitus. Brit Med J 1951;1:1541.

4. National Diabetes Data Group. Classification and diagnosis of diabetes mellitus and other categories of glucose intolerance. Diabetes 1979;28:1039.

5. World Health Organization. WHO Expert Committee on Diabetes Mellitus: second report. World Health Organ Tech Rep Ser 1980;646:1.

6. Butler AE, Janson J, Bonner-Weir S, et al. Beta-Cell deficit and increased Beta-cell apoptosis in humans with type 2 diabetes. Diabetes 2003;52:102.

7. Cnop M, Welsh N, Jonas J-C, et al. Mechanisms of pancreatic beta-cell death in type 1 and type 2 diabetes. Diabetes 2005;54(Suppl. 2):S97.

8. Donath MY, Boni-Schnetzler M, Ellingsgaard H, et al. Islet inflammation impairs the pancreatic beta-cell in type 2 diabetes. Physiology 2009;24:325. DOI: 10.1152/physiol.00032.2009.

9. Jagannathan-Bogdan M, McDonnell ME, Shin H, et al. Elevated proinflammatory cytokine production by a skewed T cell compartment requires monocytes and promotes inflammation in type 2 diabetes. J Immunol 2011;186:1162. DOI: 10.4049/jimmunol.1002615.

10. Schroder K, Zhou R, Tschopp J. The NLRP3 inflammasome: a sensor for metabolic danger? Science 2010;327:296.

11. Dinarello CA, Donath M, Mandrup-Poulsen T. Role of IL-1[beta] in type 2 diabetes. Curr Opin Endocrinol Diabetes Obes 2010;17:314.

12. Masters SL, Dunne A, Subramanian SL, et al. Activation of the NLRP3 inflammasome by islet amyloid polypeptide provides a mechanism for enhanced IL-1beta in type 2 diabetes. Nat Immunol 2010;11:897. DOI: 10.1038/ni.1935.

13. Laganowsky A, Liu C, Sawaya MR, et al. Atomic view of a toxic amyloid small oligomer. Science 2012;335:1228.

14. Rhodes CJ. Type 2 diabetes-a matter of beta-cell life and death. Science 2005;307:380.

15. Lee SJ, Kim JY, Nogueiras R, et al. PKCzeta-regulated inflammation in the non-hematopoietic compartment is critical for obesity-induced glucose intolerance. Cell Metab 2010;12:65. DOI: 10.1016/j.cmet.2010.05.003.

16. Kristiansen OP, Mandrup-Poulsen T. Interleukin-6 and diabetes: the good, bad, or the indifferent? Diabetes 2005;54(Suppl 2):S114.

17. Yki-Jarvinen H, Sahlin K, Ren JM, et al. Localization of rate-limiting defect for glucose disposal in skeletal muscle of insulin-resistant type 1 diabetic patients. Diabetes 1990;39:157.

18. Kahn BB, Rosen AS, Bak JF, et al. Expression of GLUT1 and GLUT4 glucose transporters in skeletal muscle of humans with insulin-dependent diabetes mellitus: regulatory effects of metabolic factors. J Clin Endocrin Metab 1992;74:1101. DOI: 10.1210/jc74.5.1101.

19. Pessin JE, Saltiel AR. Signaling pathways in insulin action: molecular targets of insulin resistance. J Clin Invest 2000;106:165.

20. Garvey WT, Maianu L, Zhu JH, et al. Evidence for defects in the trafficking and translocation of GLUT4 glucose transporters in skeletal muscle as a cause of human insulin resistance. J Clin Invest 1998;101:2377.

21. Kim Y-B, Nikoulina SE, Ciaraldi T, et al. Normal insulin-dependent activation of Akt/protein kinase B, with diminished activation of phosphoinositide 3-kinase, in muscle in type 2 diabetes. J Clin Invest 1999;104:733.

22. Sakamoto K, Holman GD. Emerging role for AS160? TBC1D4 and TBC1D1 in the regulation of GLUT4 traffic. Am J Physiol Endocrinol Metab 2008;295:E29. DOI: 10.1152/ajpendo.90331.2008.

23. Miinea CP, Sano H, Kane S, et al. AS160, the Akt substrate regulating GLUT4 translocation, has a functional Rab GTPase-activating protein domain. Biochem J 2005;391:87.

24. Karlsson HKR, Ahlsen M, Zierath J, et al. Insulin signaling and glucose transport in skeletal muscle from first-degree relatives of type 2 diabetic patients. Diabetes 2006;56:1283.

25. Karlsson HK, Zierath JR, Krook A, et al. Insulin-stimulated phosphorylation of the Akt substrate AS160 is impaired in skeletal muscle of type 2 diabetic subjects. Diabetes 2005;54:1692.

26. Peterson KF, Price T, Cline GW, et al. Contribution of net hepatic glycogenolysis to glucose production during the early postprandial period. Am J Phys Endo Metab 1996;270:E186.

27. Bischof MG, Krssak M, Krebs M, et al. Effects of short-term improvement of insulin treatment and glycemia on hepatic glycogen metabolism in type 1 diabetes. Diabetes 2001;50:392.

28. Bischof MG, Bernroider E, Krssak M, et al. Hepatic glycogen metabolism in type 1 diabetes after long-term near normoglycemia. Diabetes 2002;51:49.

29. Petersen KF, Laurent D, Rothman DL, et al. Mechanism by which glucose and insulin inhibit net hepatic glycogenolysis in humans. J Clin Invest 1998;101:1203.

30. Cline GW, Petersen KF, Krssak M, et al. Impaired glucose transport as a cause of decreased insulin-stimulated muscle glycogen synthesis in type 2 diabetes. N Engl J Med 1999;341:240.

31. Shulman GI. Cellular mechanisms of insulin resistance. J Clin Invest 2000;106:171.

32. Nikoulina SE, Ciaraldi TP, Mudaliar S, et al. Diabetes 2000;49:263.

33. Landau B. Methods for measuring glycogen cycling. Am J Physiol Endocrinol Metab 2001;281:E413.

34. Kacerovsky M, Jones J, Scmid AI, Barosa C, et al. Postprandial and fasting hepatic glucose fluxes in long-standing type 1 diabetes. Diabetes 2011;60:1752.

35. Consoli A, Nurjhan N, Capani F, et al. Predominant role of gluconeogenesis in increased hepatic glucose production in NIDDM. Diabetes 1989;38:550. DOI: 10.2337/diabetes.38.5.550.

36. Magnusson I, Rothman DL, Katz LD, et al. Increased rate of gluconeogenesis in type II diabetes mellitus: A 13C nuclear magnetic resonance study. J Clin Invest 1992;90:1323.

37. Gastaldelli A, Baldi S, Pettiti M, et al. Influence of obesity and type 2 diabetes on gluconeogenesis and glucose output in humans. Diabetes 2000;49:1367. DOI: 10.2337/diabetes.49.8.1367.

38. Woerle HJ, Szoke E, Meyer C, et al. Mechanisms for abnormal postprandial glucose metabolism in type 2 diabetes. Am J Physiol Endocrinol Metab 2006;290:E67. DOI: 10.1152/ajpendo.00529.2004.

39. Gastaldelli A, Toschi E, Pettiti M, et al. Effect of physiological hyperinsulinemia on gluconeogenesis in nondiabetic subjects and in type 2 diabetic patients. Diabetes 2001;50:1807.

40. Woerle HJ, Albrecht M, Linke R, et al. Impaired hyperglycemia-induced delay in gastric emptying in patients with type 1 diabetes deficient for islet amyloid polypeptide. Diabetes Care 2008;31:2325. DOI: 10.2337/dco7.2446.

41. Pratipanawatr T, Cusi K, Ngo P, et al. Normalization of plasma glucose concentration by insulin therapy improves insulin-stimulated glycogen synthesis in type 2 diabetes. Diabetes 2002;51:462.

42. Del Prato S, Bonadonna RC, Bonora E, et al. Characterization of cellular defects of insulin action in type 2 (Non-insulin-dependent) diabetes mellitus. J Clin Invest 1993;91:484.

43. Mostert M, Rabbone I, Piccinini M, et al. Derangements of pyruvate dehydrogenase in circulating lymphocytes of NIDDM patients and their healthy offspring. J Endocrinol Invest 1999;22:519.

44. Jeoung NH, Harris RA. Role of pyruvate dehydrogenase kinase 4 in regulation of blood glucose levels. Korean Diabetes J 2010;34:274.

45. Wu P, Sato J, Zhao Y, et al. Starvation and diabetes increase the amount of pyruvate dehydrogenase kinase isoenzyme 4 in rat heart. Biochem J 1998;329:197.

46. Holness MJ, Kraus A, Harris RA, et al. Targeted upregulation of pyruvate dehydrogenase kinase (PDK)-4 in slow-twitch skeletal muscle underlies the stable modification of the regulatory characteristics of PDK induced by high-fat feeding. Diabetes 2000;49:775.

47. Puthanveetil P, Wang Y, Wang F, et al. The increase in cardiac pyruvate dehydrogenase kinase-4 after short-term dexamethasone is controlled by an Akt-p38-Forkhead box other factor-1 signaling axis. Endocrinology 2010;151:2306.

48. Crowther GJ, Milstein JM, Jubrias SA, et al. Altered energetic properties in skeletal muscle of men with well-controlled insulin-dependent (type 1) diabetes. Am J Physiol Endocrinol Metab 2003;284:E655. DOI: 10.1152/ajpendo.00343.2002.

49. Krauss S, Zhang C-Y, Scorrano L, et al. Superoxide-mediated activation of uncoupling protein 2 causes pancreatic beta cell dysfunction. J Clin Invest 2003;112:1831.

50. Zhang C-Y, Baffy G, Perret P, et al. Uncoupling protein-2 negatively regulates insulin secretion and is a major link between obesity, beta cell dysfunction and type 2 diabetes. Cell 2001;105:745.

51. de Souza BM, Assmann TS, Kliemann LM, et al. The role of uncoupling protein 2 (UCP2) on the development of type 2 diabetes mellitus and its chronic complications. Arg Bras Endocrinol Metab 2011;55:239.

52. Jing E, Emanuelli B, Hirschey MD, et al. Sirtuin-3 (Sirt3) regulates skeletal muscle metabolism and insulin signaling via altered mitochondrial oxidation and reactive oxygen species production. Proc Natl Acad Sci U S A 2011;108. DOI: 10.1073/pnas.1111308108.

53. Yoshino J, Mills KF, Yoon MJ, et al. Nicotinamide mononucleotide, a key NAD+ intermediate, treats the pathophysiology of diet- and age-induced diabetes in mice. Cell Metab 2011;14:528. DOI: 10.1016/j.cmet.2011.08.014.

54. Revollo JR, Grimm AA, S-i I. The NAD biosynthesis pathway mediated by nicotinamide phosphoribosyltransferase regulates Sir2 activity in mammalian cells. J Biol Chem 2004;279:50754. DOI: 10.1074/jbc.M408388200.

55. Schrauwen-Hinderling VB, Hesselink MK, Schrauwen P, et al. Intramyocellular lipid content in human skeletal muscle. Obesity (Silver Spring) 2006;14:357.

56. Schrauwen-Hinderling VB, Kooi ME, Hesselink MK, et al. Impaired in vivo mitochondrial function but similar intramyocellular lipid content in patients with type 2 diabetes mellitus and BMI-matched control subjects. Diabetologia 2007;50:113.

57. Phielix E, Mensink M. Type 2 diabetes mellitus and skeletal muscle metabolic function. Physiol Behav 2008;94:252.

58. Petersen KF, Dufour S, Befroy D, et al. Impaired mitochondrial activity in the insulin-resistant offspring of patients with type 2 diabetes. N Engl J Med 2004;350:664. DOI: 10.1056/NEJMoa031314.

59. Yu C, Chen Y, Cline GW. Mechanism by which fatty acids inhibit insulin activation of insulin receptor substrate-1 (IRS-1)-associated phosphatidylinositol 3-kinase activity in muscle. J Biol Chem 2002;277:50230. DOI: 10.1074/jbc.M200958200.

60. Samuel VT, Petersen KF, Shulman GI. Lipid-induced insulin resistance: unravelling the mechanism. Lancet 2010;375:2267. DOI: 10.1016/S0140-6736(10)60408-4.

61. Prentki M, Joly E, El-Assad W, et al. Malonyl-CoA signaling, lipid partitioning, and glucolipotoxicity. Diabetes 2002;51(Suppl 3):S405. DOI: 10.2337/diabetes.51. 2002.S405.

62. Abel ED. Free fatty acid oxidation in insulin resistance and obesity. Heart Metab 2010;48:5.

63. Buse MG. Hexosamines, insulin resistance, and the complications of diabetes: current status. Am J Physiol Endocrinol Metab 2006;290:E1. DOI: 10.1152/ajpendo. 00329.2005.

64. Dentin R, Hedrick S, Xie J, et al. Hepatic glucose sensing via the CREB coactivator CRTC2. Science 2008:1402.

65. Yang X, Ongusaha PP, Miles PD, et al. Phosphoinositide signalling links O-GlcNAc transferase to insulin resistance. Nature 2008;451:964. DOI: 10.1038/nature06668.

66. DeFronzo RA, Tobin JD, Andres R. Glucose clamp technique: a method for quantifying insulin secretion and resistance. Am J Physiol Endocrinol Metab 1979;237:E214.

67. Holck P, Porksen N, Nielsen MF, et al. Effect of needle biospy from the vastus lateralis muscle on insulin-stimulated glucose metabolism in humans. Am J Physiol Endocrinol Metab 1994;267:ES44.

CHAPTER 6

CLASSIFICATION SYSTEM FOR DIABETES MELLITUS

The first step in wisdom is to know the things themselves; this notion consists in having a true idea of the objects; objects are distinguished and known by classifying them methodically and giving them appropriate names. Therefore, classification and name-giving will be the foundation of our science.

—Carolus Linnaeus Systema Naturae (1735)

As stated at the outset of this textbook, diabetes is an array of diseases that have a common symptom, abnormally high blood glucose levels. Diabetes is chronic,[1] often debilitating, costly and nontransmissible. In the early chapters of this book, we have looked exclusively at type 1 diabetes (T1D) and type 2 diabetes (T2D). In this chapter, we will review the characteristics of T1D and T2D and then we will discuss other categories of diabetes.

It should be emphasized that although we have often defined categories of diabetes, it is difficult to decide which category a specific individual with his or her own set of symptoms falls in. An example is a woman who first shows symptoms of diabetes during pregnancy and is classified as a gestational diabetic but retains her diabetic condition after childbirth and is then determined to have T2D (1).

Although the American Diabetes Association (ADA) (2) classifies diabetes into four clinical classes, T1D, T2D, other types of diabetes due to other causes,

---

[1]A chronic disease or condition is one that lasts at least 3 months. In the case of diabetes mellitus, it usually lasts a lifetime.

Understanding Diabetes: A Biochemical Perspective, First Edition. Richard F. Dods.
© 2013 John Wiley & Sons, Inc. Published 2013 by John Wiley & Sons, Inc.

and gestational diabetes, we have added several distinctly defined additional categories. As you can see from Table 6.1, the additional classes are latent autoimmune diabetes of adulthood (LADA, often also called *type 1.5*); hybrid diabetes (T1D and T2D combined, occasionally called *type 1.5*); idiopathic diabetes; prediabetes, which includes impaired glucose tolerance (IGT) and impaired fasting glucose (IFG); and the statistical risk classes, Previous abnormality of glucose tolerance (PrevAGT) and potential abnormality of glucose tolerance (PotAGT). We will end the discussion with a description of the metabolic syndrome, which plays a major role in T2D.

T1D

As previously related, T1D is an autoimmune disease in which the immune system attacks pancreatic β-islet cells. T1D is found for 5–10% of those with diabetes. Destruction of β-islets results in insulinopenia. Often, 70–80% of the β-islet cells are destroyed before detection of the illness. This condition is best defined by the presence in the blood of specific β-islet cell autoimmune antibodies. This form of diabetes results in exactly what you would expect if insulin is not present to activate glucose transport into cells, activate glycogen synthase (and therefore glycogenesis), inhibit glycogenolysis, inhibit lipolysis, promote protein synthesis, and suppress gluconeogenesis. Review Chapters 3 and 4 for a more specific presentation. T1D is often diagnosed at an early age, less than 40 years of age, which resulted previously in naming it *juvenile diabetes*.

LATENT AUTOIMMUNE DIABETES (LADA) OR TYPE 1.5

LADA or type 1.5 was originally discovered in patients diagnosed as T2D. Autoimmune antibodies are the hallmark of T1D. Therefore, the approximately 10% of individuals initially diagnosed as T2D, who were found to contain these antibodies, were probably better identified as a subset of T1D. LADA patients are past the age of 30 years, with slow progression of β-cell death. These diabetics, unlike T1D patients, are not initially insulin requiring. It may take as long as 12 years until β-cell failure occurs in LADA patients, whereas in T1D patients it has taken place by the time the condition is diagnosed. When β-cell failure occurs, the LADA patient has to be treated with insulin. LADA does have some attributes of T2D. For example, insulin resistance is found in LADA patients, especially those who are obese (3, 4).

The type of autoantibody can predict when β-cell failure will occur in patients with LADA (4, 5). If both glutamic acid decarboxylase antibody (GADA) and islet cell antibody (ICA) are found, it is likely that β-cell failure will occur within 5 years. If one of the antibodies is found, β-cell failure will occur after 5 years.

TABLE 6.1 Classification of Diabetes and Other Categories of Glucose Intolerance

Class	Pathophysiology	Characteristics
T1D*	An autoimmune response to pancreatic β-islet cells results in insulin deficiency	Patients are dependent on insulin for survival Usually occurs before 40 years of age 70–80% of β-islet cells are destroyed Lymphocytic infiltration of islet cells is observed Presence of autoimmune antibodies Likely to show complications such as: neuropathy, nephropathy, angiopathy, cataracts and ketoacidosis (complications will be discussed in Chapter 8)
LADA	A slowly progressing autoimmune response to pancreatic β-islet cells	Patients are not dependent on insulin until later in the progression of the disease Usually occurs after the age of 30 years Progressive destruction of β-islets cells (complete β-islet cell failure does not occur until up to 12 years from diagnosis) Presence of autoimmune antibodies Insulin resistance present especially in obese subjects Complications occur such as described for T1D especially in advanced LADA Individuals are not as dependent on insulin as T1D
T2D*	A polygenic disease in which insulin resistance is a central characteristic	Weight loss and/or medications are for some the only necessary recommendations Not likely to show complications Most prevalent form of diabetes The majority of T2D persons are overweight or obese Heredity plays a major role in T2D
Hybrid	T1D associated with T2D in children	Insulin is needed Presence of autoimmune antibodies Presence of insulin resistance Majority tend to be obese Diabetes medications for insulin resistance Healthy eating habits and exercise help

(*continued*)

TABLE 6.1 (*Continued*)

Class	Pathophysiology	Characteristics
Idiopathic Diabetes (T1b)	No known pathophysiology	Various degrees of insulin deficiency No evidence of autoimmune antibodies Episodes necessitating insulin replacement therapy
Secondary*	Diabetes caused by various secondary conditions	Includes: genetic defects of β-islet function (MODY 1–8); genetic defects in insulin action (lipoatrophic diabetes); diseases of the exocrine pancreas (pancreatitis); endocrinopathies (Cushing's syndrome, pheochromocytoma); drug or chemically caused diabetes (glucocorticoids, nicotinic acid); infections; uncommon forms of immune-mediated diseases (anti-insulin receptor antibodies); other genetic syndromes sometimes associated with diabetes (Down syndrome, Klinefelter syndrome) Accounts for 1–5% of diabetes cases
Prediabetes	Impaired glucose tolerance (IGT) and Impaired fasting glucose (IFG)	IGT is defined as a glucose value of 140–199 mg/dl 2 h after a glucose load. IFG is defined as a fasting blood glucose levels of 100–125 mg/dl after an 8-h fast. IGT and/or IFG usually progresses to T2D
Gestational* (GDM)	Diabetes derived from pregnancy	Criteria to diagnose GDM is in dispute Neonates have Increased birth weight Increased birth weight results in risk of instrument delivery or vaginal delivery (shoulder dystocia) Increased risk of low blood glucose, jaundice, high red blood cell mass, low blood calcium, and magnesium
Statistical risk diabetes	Previous abnormality of glucose tolerance Potential abnormality of glucose tolerance	Individuals who previously had diabetes mellitus, and gestational diabetes and IGT whose fasting blood levels have returned to normal fall into the Previous abnormality of glucose tolerance category. Potential abnormality of glucose tolerance is a category that includes individuals who have blood islet cell antibodies, being an identical twin or sibling of a T1D patient, being a child of a T1D patient, being a relative of a T2D patient, having gestational diabetes, being a member of an ethnic or racial group which has a high incidence of diabetes and being obese.

*One of the four clinical classifications cited by the American Diabetes Association.

T2D

As previously presented, T2D is a polygenic disease that accounts for approximately 90–95% of the total diabetics. T2D patients exhibit both hyperglycemia and hyperinsulinemia. Insulin resistance is the hallmark of this type of diabetes. Most patients with T2D are obese. For some, the distribution of weight plays a role. Increased body fat accumulated predominantly in the abdominal region predisposes a person who is not overweight to T2D. Weight loss measures including exercise and dieting can reverse the diabetes. T2D concerns not only weight. Simply being overweight will not cause a person to develop T2D. Heredity plays a major role in the development of T2D. Thus, thin people also develop T2D.

T2D patients usually do not need insulin and generally do not show the complications associated with T1D. T2D patients do not have the autoimmune antibodies exhibited in T1D and LADA.

HYBRID

T1D associated with T2D is increasingly being diagnosed in children and adolescents (6). This type of diabetes is termed *hybrid diabetes* or *double* or *type 1½ diabetes*. About 30% of children and adolescents diagnosed with T2D have autoimmune antibodies typical of persons with T1D. In addition, the youths were tested for C-peptide levels, a measure of β-islet cell function, and found to have lower levels than in normal youth. According to SEARCH for Diabetes in Youth (a study described in Chapter 1), the youths with hybrid diabetes tended to be older and obese and had evidence of insulin resistance.

IDIOPATHIC DIABETES (T1B)

Some types of diabetes have no known causations. They do not exhibit autoimmune antibodies, and the need for insulin is episodic. African or Asian ancestry increases the tendency of those who are diagnosed as T1D to fall into this category. For them, ketoacidosis (see Complications, Chapter 8) occasionally occurs and there are various degrees of insulinopenia. This form of diabetes is strongly inherited (7).

SECONDARY

Various conditions, illnesses, chemicals, and genetic defects can result in diabetes (1). These causes account for about 1–5% of total diabetics. Outlined in Table 6.1 (with one or more examples), we will deal with them in a more comprehensive manner in the following section.

Summary Box 6.1

- Diabetes is an array of conditions that have in common, hyperglycemia.
- T1D is an autoimmune disease in which the immune system attacks and destroys β-islet cells.
- T1D is characterized by the presence of specific β-islet cell autoimmune antibodies (GADA and ICA), resulting in insulinopenia and hyperglycemia.
- LADA is a subset of T1D in which the β-islet cells are slowly destroyed over time.
- T2D is a polygenic disease with insulin resistance as the principal characteristic of the disease.
- T2D displays both hyperinsulinemia and hyperglycemia.
- Obesity and hereditary play a prominent role in T2D, although thin people also contract T2D.
- Hybrid diabetes occurs in children and adolescents. It is T1D associated with T2D in the same individual. Usually, T2D is accompanied by obesity.
- Idiopathic diabetic patients do not demonstrate autoimmune antibodies and insulin is not always necessary.

GENETIC DEFECTS OF β-ISLET FUNCTION

MODY

Maturity-onset diabetes of the young (MODY) refers to autosomal dominant inherited[2] diabetes. The defect is described as being monogenic as it affects only one gene. MODY is usually treated by diet or medication, but occasionally insulin is required. Its onset is before 25 years. Because of its inherited nature, the condition can be traced through family history. Impaired β-islet cell function and insulin resistance are characteristics of MODY. In some cases, β-islet cell failure occurs late in the disease. This category constitutes 5% of total diabetics.

At present, there are eight identified types of MODY. They are described as follows:

> *MODY 1* Hepatocyte nuclear factor 4α (HNF-4α); a transcription factor that regulates the functioning of the pancreas and regulates insulin secretion, glucose 6-phosphatase, GLUT2, liver pyruvate kinase, glyceraldehyde

---

[2]Autosomal dominant inherited means that there is a defect in one gene and that you need to receive the gene from only one parent to cause the disorder even though the gene from the other parent is normal.

phosphate dehydrogenase, aldolase, and uncoupling protein is a rare form of MODY. Although insulin is not required initially, it does progress to an insulin requirement in its later stages.

MODY 2 MODY 2 is caused by any of several mutations in the glucokinase gene (a hexokinase located in the liver, pancreas, GI tract, and brain). Blood glucose levels are usually about 99 mg/dl (5.5 mmol/l) to 144 mg/dl (8.0 mmol/l). No symptoms or complications are encountered and no treatment is necessary. This gene defect represents 10–65% of MODY cases.

MODY 3 HNF-1α, a transcription factor that regulates GLUT2 and liver pyruvate kinase, represents 20–75% of MODY cases. Insulinopenia worsens as the patient possessing this defect ages. Insulin therefore becomes required later in life. Complications such as nephrology, retinopathy, and coronary heart disease often occur.

MODY 4 Insulin promoter factor-1 (IPF1) is extremely rare. MODY 4 results in the pancreas failing to develop. Little information is available as to its characteristics.

MODY 5 Heptocyte nuclear factor factor 1 homeobox β (HNF-1β), a transcription factor that controls the differentiation and growth of the embryonic pancreas, is a rare form of MODY. Kidney (renal) cysts, pancreatic atrophy, and uterine abnormalities often found in the fetus are common. Insulin is required.

MODY 6 Neurogenic differentiation 1 (NeuroD1) is an extremely rare MODY. Little information is available as to its characteristics.

MODY 7 Kruppel-like factor 11 (KLF11), a transcription factor that activates an insulin promoter, is a recently discovered MODY. Characteristics of this MODY are not yet known.

MODY 8 Carboxyl-ester lipase gene (CEL) is associated with pancreatic exocrine and β-cell dysfunction. It is a rare form of MODY.

Permanent Neonatal Diabetes Mellitus (KCNJ11). KCNJ11 is a newly discovered type of neonatal diabetes.

OTHER GENETIC DEFECTS OF THE β-CELL

Point mutations in mitochondrial DNA are associated with diabetes.
Inability to convert proinsulin to insulin.
Impaired insulin receptor binding of mutated insulin.

Genetic Defects in Insulin Action

- Women who are virilized and have enlarged, cystic ovaries. Originally termed *type A insulin resistance.*

- Leprechaunism[3] and Rabson-Mendenhall[4] syndrome have insulin receptor gene mutations, resulting in defects in insulin receptor function and insulin resistance.
- Insulin postreceptor signal transduction pathway[5] defects occur in lipotropic diabetes.

DISEASES OF THE EXOCRINE PANCREAS

- Trauma to the pancreas
- Pancreatitis
- Infection
- Pancreatectomy
- Pancreatic carcinoma
- Cystic fibrosis
- Hemochromatosis[6]
- Fibrocalculous pancreatopathy.[7]

ENDOCRINOPATHIES

Hormones that are in excess suppress insulin to abnormally low levels:
- Growth hormone (acromegaly), cortisol (Cushing's syndrome), glucagon (glucagonoma), and epinephrine (pheochromocytoma) cause diabetes.
- Somatostatinoma and aldosteronoma are cancers that inhibit insulin secretion because of excessive secretion of somatostatin and aldosterone.

DRUG OR CHEMICALLY CAUSED DIABETES

Certain drugs cause defectives in insulin secretion:
- Toxins, such as a rat poison called *Vacor*, destroy β-cells.
- Pentamidine, an antimicrobial medication, also destroys β-cells.
- Nicotinic acid and glucocorticoids interfere in insulin action.

---

[3] A genetic disorder that results in elfish and smaller than normal features.
[4] A genetic disorder that affects the insulin receptor site.
[5] Insulin postreceptor signal transduction pathway refers to the pathway after the insulin attaches to the insulin receptor.
[6] High quantities of iron build up in the body.
[7] Fibrocalculous pancreatic diabetes (FCPD) is a unique form of diabetes. It appears secondary to chronic calcific nonalcoholic pancreatitis and is observed mostly in the developing countries of the tropical world. The classical triad consists of abdominal pain, steatorrhea, and diabetes.

- Many hormones affect insulin action.
- The interferons produce islet cell antibodies, which result in insulin deficiency.

The list of drugs, hormones, and toxins is quite numerous and no attempt to list them will be attempted in this textbook.

Summary Box 6.2

- MODY refers to a mutation in one gene that causes diabetes.
- MODY 2 (glucokinase) and 3 (a transcription factor that regulates GLUT2 and pyruvate kinase) are the most prevalent.
- Any diseases or conditions that affect the pancreas cause diabetes.
- Cancers or conditions that increase hormones such as growth hormone, cortisol, glucagon, epinephrine, somatostatin, and aldosterone cause diabetes.
- Many drugs such as nicotinic acid and glucocorticoids that interfere with insulin action cause diabetes.

INFECTIONS

Infections have been found to cause diabetes. Viruses have been implicated in activating the autoimmune system to produce antibodies to portions of the β-cell that contains epitopes[8] to regions of the viral coat. As you may recognize, this effect leads to T1D. The viruses implicated are

- Coxsackievirus B
- Cytomegalovirus
- Adenovirous
- Mumps virus
- Congenital rubella (measles).

UNCOMMON FORMS OF IMMUNE-MEDIATED DISEASES

- An autoimmune condition of the central nervous system that causes stiffness of the axial muscles (muscles attached to head, neck, vertebral column, thorax, abdomen, and pelvis) with spasms called *stiff-man syndrome*.
- Anti-insulin receptor antibodies are antibodies that bind to insulin receptor sites, thus making it impossible for insulin to bind to them. The antibodies, although blocking insulin molecules, mimic insulin, thus keeping the insulin signaling pathway on and creating hypoglycemia.

---

[8] Amino acid sequences that are recognized by an antibody or T-cell.

- Anti-insulin receptor antibodies are also found occasionally in erythematosus and other autoimmune diseases.

OTHER GENETIC SYNDROMES SOMETIMES ASSOCIATED WITH DIABETES

Many diseases that are genetically derived are associated with diabetes.
- Extra chromosome
  - Down syndrome
  - Klinefelter syndrome
- Lacking a chromosome (specifically an X-chromosome)
  - Turner syndrome

Wolfram's syndrome is an autosomal recessive condition in which one of the disorders exhibited is insulin deficiency and absence of β-cells.

Summary Box 6.3

- Mumps and rubella (measles) viruses can result in the autoimmunity that destroys β-islet cells and cause T1D.
- Other viruses that have been reported to cause diabetes are coxsackievirus B, cytomegalovirus, and adenovirus.
- Anti-insulin receptor antibodies bind to insulin receptor sites keeping them in a constant state of being activated.
- Genetically derived diseases in which there is either a missing chromosome or added chromosome cause diabetes.

PREDIABETES

IGT and IFG are categories of hyperglycemia that do not reach the criteria for diabetes but nonetheless are above established normal levels. Individuals who have glucose levels that place them in either the IGT and/or categories often progress to T2D. As of 2003, 54 million (26%) Americans were classified IFG and in 1994, 32 million (15.8%) Americans were classified as IGT. The ADA has recommended an oral drug called *metformin* as well as lifestyle modification for a person classified as IFG and/or IGT (8).

The World Health Organization (WHO) defines IGT as a glucose value of 140–199 mg/dl (7.8–11.0 mmol/l) 2 h after an oral glucose tolerance test (OGTT) using 1.75 g of glucose per kilogram for adults to a maximum of 75 g for children. The OGTT is discussed in detail in Chapter 7. The ADA defines IFG as a blood glucose level at 100–125 mg/dl (5.5–6.9 mmol/l) after an 8-h fast. The normal glucose level was originally 110–125 mg/dl (6.1–6.9 mmol/l) but was lowered in 2003 in order to increase sensitivity and specificity of the test[9] (9). To place the above-mentioned data into perspective, the normal fasting glucose level after an 8-h fast is 70–99 mg/dl (3.9–5.5 mmol/l) and the 2-h glucose level after an OGTT is normally close to 155 mg/dl (8.6 mmol/dl) with >199 mg/dl (11.0 mmol/l) being considered diabetic.

Approximately 24.3% of individuals diagnosed as IFG by the old criteria (110–125 mg/dl) progressed to T2D within a mean of 29.0 months, while 8.1% of those diagnosed as IFG by the new criteria (100–125 mg/dl) progressed to T2D within a mean of 41.4 months (9). There are several reports using different years of follow up that had results that varied from 2.7% to as high as 33% for development of T2D. We can therefore conclude that there are a significant number of cases of IFG that progress to T2D. The length of time that a patient has spent classified as IFG affects their likelihood of progressing to T2D.

The National Institutes of Health-sponsored Diabetes Prevention Program studied over 3800 subjects with IGT (10). The study demonstrated an 11% per year progression to T2D. When treated with metformin, the rate was reduced to 7.8% per year. When physical activity was introduced and weight loss accomplished the progression to T2D, the rate was reduced to 4.8% per year. There are several studies using subjects with both IGT and IFG. They all indicate a significant increase in patients who progress to T2D over those who were categorized as IFG alone. Patients with IFG and/or IGT are at a two times greater risk for cardiovascular disease than those with normal glucose levels.

GESTATIONAL DIABETES MELLITUS (GDM)

Gestational diabetes mellitus (GDM) is probably the most controversial of the classes of diabetes we have discussed. Even its definition is debatable. The definition for GDM that appears in numerous articles is: GDM is "glucose intolerance that is first detected during pregnancy." (11). This definition leaves open the possibility that GDM is present in the woman previous to her pregnancy and is discovered during pregnancy. Priscilla White, a pioneer in the treatment of GDM, recognized this and developed what is now called the *White Classification Scheme*, which distinguishes between GDM and diabetes that existed before pregnancy. She determined two types of GDM: type A1 which is abnormal OGTT with normal fasting blood glucose levels and type A2 which is abnormal OGTT with increased fasting blood glucose levels. We will discuss the diagnosis of GDM in further detail in the next chapter.

Another source of contention is the degree of the hyperglycemia necessary to be classified as GDM (12). It should be noted that the GDM classification does not address whether the patient after pregnancy progresses to T2D or whether the diabetic symptoms cease. The last contentious issue we shall mention here

[9]Sensitivity and specificity will be discussed in Chapter 7.

is lack of definite diagnostic criteria for GDM. This aspect of GDM will also be covered in Chapter 7.

The Agency for Healthcare Research and Quality (AHRQ) reports that in 2008, 250,000 women who gave birth in hospitals had preexisting diabetes or developed GDM during pregnancy. This is 6.4% of the 4.2 million women who gave birth in hospitals in the United States in 2008 (13). Other reports indicate that diabetes is detected in 3–10% of pregnancies.

In normal pregnancies, insulin secretion increases up to fourfold in response to the demands of the fetus. The necessity for extra insulin secretion is partially caused by an increase in hormones such as prolactin, human placental lactogen (also called *human chorionic somatomammotropin*), cortisol, and progesterone that counteract insulin. In addition, there is the extra weight that a women gains during pregnancy and the increased nutrition necessitated by the fetus. GDM therefore is a subtype of T2D that is exhibited when the woman's pancreatic reserve becomes exhausted.

The symptoms of GDM are covert. They include increased thirst and increased urination. These symptoms are attributed to the pregnancy and not to hyperglycemia. The diagnosis of GDM will be discussed in Chapter 7. The maternal and fetal complications of GDM will be discussed in detail in Chapter 8. To summarize them here, maternal complications include increased probability that GDM will appear in subsequent pregnancies, likelihood that GDM will progress to T2D immediately after giving birth or later in life, and higher probability of Cesarean-section; fetus/neonatal complications include macrosomia/large for gestational age (greater than normal weight during gestation and at birth), shoulder dystocia (refers to the neonate's shoulder becoming lodged behind the mother's pubic bone during birth), and greater frequency that the neonate will develop T2D later in life.

STATISTICAL RISK CLASSES

Previous abnormality of glucose tolerance (PrevAGT) and Potential abnormality of glucose tolerance (PotAGT) are not clinical classes like those previously discussed. They are statistical risk classes. Individuals who have demonstrated one or more of the following criteria are classified as PrevAGT:

- Gestational diabetes but after delivery have normal blood glucose levels
- T1D individuals who gone into remission
- Individuals who showed hyperglycemia after trauma; however, when the trauma subsided, glucose levels returned to normal
- Formerly obese individuals who previously had T2D. Normal glucose levels returned after the individual slimmed down to normal weight.

Individuals who have exhibited one or more of the following criteria are classified as PotAGT:

- Prediabetes
- Individuals who have blood islet cell antibodies
- Identical twin of a diabetic (T1D or T2D)
- Sibling of a T1D
- Offspring of a T1D
- Relative of a T2D (usually first degree)
- Macrosomic at birth
- Certain racial or ethnic groups (such as Native Americans) (see Chapter 1).

The statistical risk classes are often used by physicians to counsel nondiabetic individuals who fall into either of the two classes as to the threat of future diabetes and what steps to take to avoid it.

Summary Box 6.4

- IGT includes persons who have a fasting blood glucose level of 140–199 mg/dl (7.8–11.0 mmol/l) 2 h after an OGTT.
- IFT includes persons with fasting blood glucose at 100–125 mg/dl (5.5–6.9 mmol/l).
- IGT and IFT glucose levels are just above normal.
- Individuals who fall into these categories often progress to T2D.
- GDM is defined as diabetes that suddenly appears during pregnancy with no previous history of the disease or diabetes previously diagnosed that continues during pregnancy.
- Maternal complications include Cesarean-section and progression to T2D.
- Fetus/neonatal complications include macrosomia, shoulder dystocia, and greater probability of T2D in later life.
- Statistical risk classes include individuals who are not presently diabetic but have a greater risk of diabetes due to such things as a relative (twin, sibling, parents) who have a history of diabetes, being previously obese, and macrosomic at birth, etc.

METABOLIC SYNDROME[10]

Metabolic syndrome is a cluster of metabolic risk factors found in individuals who are at risk for cardiovascular disease (heart attack and stroke) (14). The cluster includes insulin resistance, obesity (especially extra weight in the

---

[10]Strictly speaking, metabolic syndrome is not a class of diabetes and does not appear in Table 6.1. It is a cluster of metabolic disorders that includes glucose intolerance and therefore it seems appropriate that it is discussed in this chapter.

abdominal region), atherogenic dyslipidemia,[11] hypertension, elevated fibrinogen or plasminogen activator inhibitor-1 (PAI-1),[12] and elevated C-reactive protein (CRP) (15). Gerald Reave, an endocrinologist at Stanford University School of Medicine, was the first to allude to this cluster as a syndrome initially calling it *syndrome X* (16). Syndrome X was renamed metabolic syndrome in 2001 by the Expert Panel on Detection, Evaluation, and Treatment of High Blood Cholesterol in Adults: Adult Treatment Panel III (ATPIII). The specific criteria for the diagnosis of metabolic syndrome are (17) as follows:

- Abdominal obesity
  - For men—waist circumference >40 in. (102 cm)
  - For women—waist circumference >35 in. (88 cm)
- Fasting glucose ≥110 to <126 mg/dl (≥6.1 to ≤7.0 mmol/l)
- Blood pressure ≥130/85 mmHg
- Triacylglycerols ≥150 mg/dl (≥1.7 mmol/l)
- HDL-Cholesterol
  - For men <40 mg/dl (<1.0 mmol/l)
  - For women <50 mg/dl (<1.3 mmol/l).

If three or more of these five criteria are met, metabolic syndrome is present.

In recent years, the presence of the prothrombotic state (elevated fibrinogen and/or PAI-1) and proinflammatory state as indicated by elevated CRP have added to the criteria. Normal fibrinogen is 1.5–2.77 g/l, normal CRP is less than 10 mg/dl, and normal PAI-1 is 3.5–7.2 ng/ml. Insulin resistance is implicated in the development of most of these conditions, namely, glucose intolerance, hypertension, hypertriacylglycerolemia, hypercoagulability (caused by elevated fibrinogen and/or PAI-1), and vascular inflammation (suggested by elevated CRP) (18).

Researchers using the above-mentioned bulleted criteria and data gathered from 1988–1994 by the Third National Health and Nutrition Examination Survey (NHANESIII), determined the male and female combined prevalence of metabolic syndrome in the US population as 6.7% in the age group 20–29 years, 13.3% in the age group 30–39 years, 23.5% in the age group 40–59 years, 43.5% in the age group 60–69, and 42.0% in the age group ≥70 years. Males and females had similar prevalence. The Mexican American prevalence was highest and Caucasian and African American prevalence lowest (19). These results suggest that metabolic syndrome was highly prevalent in the US population in the late 1980s to the middle 1990s and is likely to be even higher in the present population.

---

[11] Atherogenic dyslipidemia refers to increased blood concentrations of small low density lipoprotein (LDL, the "bad" lipoprotein), decreased high density lipoprotein (HDL, the "good" lipoprotein), and increased triacylglycerols (triglycerides).
[12] Elevations in this inhibitor suggest that an inflammatory process is occurring.

Although the aforementioned categories of diabetes are defined and very specific, an individual placed into one category maybe defined in a different category in a few years. Clearly, we are looking at a continuum of conditions and not a distinct disorder. For example, an individual may have occasional hyperglycemia and therefore be judged as normal. Later, the individual progresses to prediabetes (IFG/IGT) and still later progresses to T2D. We are looking at a continuous spectrum of conditions and not a distinct disease.

Summary Box 6.5

- Metabolic syndrome consists of a cluster of risk factors; insulin resistance, obesity, atherogenic dyslipidemia, hypertension, elevated fibrinogen or PAI-1, and CRP.
- Individuals who meet three of these criteria are considered to have metabolic syndrome and are at risk for heart attack and stroke.
- An individual who is placed in one of the classes of diabetes described in this chapter may progress from that category to another during his/her lifetime.

GLOSSARY

Atherogenic dyslipidemia   Refers to increased blood concentrations of small low density lipoprotein (LDL, the "bad" lipoprotein), decreased high density lipoprotein (HDL, the "good" lipoprotein), and increased triacylglycerols (triglycerides).

Autosomal dominant inherited   A defect in one gene that is received from only one parent to cause the disorder even though the gene from the other parent is normal.

Chronic disease or condition   A condition or disease that lasts at least 3 months. In the case of diabetes mellitus, it usually lasts a lifetime.

Epitopes   Amino acid sequences that are recognized by an antibody or T-cell.

Fibrocalculous pancreatic diabetes   A unique form of diabetes. It appears secondary to chronic calcific nonalcoholic pancreatitis and is observed mostly in the developing countries of the tropical world. The classical triad consists of abdominal pain, steatorrhea, and diabetes.

Hemochromatosis   High quantities of iron build up in the body.

Insulin postreceptor signal transduction pathway   Refers to the pathway after the insulin attaches to the insulin receptor.

Leprechaunism   A genetic disorder that results in elfish and smaller than normal features.

Plasminogen activator inhibitor-1   Elevations in this inhibitor suggest that an inflammatory process is occurring.

Rabson–Mendenhall syndrome A genetic disorder which affects the insulin receptor site.

REFERENCES

1. American Diabetes Association. Diagnosis and classification of diabetes mellitus. Diabetes Care 2011;34(Suppl 1):S62. DOI: 10.2337/dcl11-S062.

2. American Diabetes Association. Standards of Medicare Care in Diabetes—2011. Diabetes Care 2011;34(Suppl 1):S11.

3. Palmer PP, Hampe CS, Chiu H, et al. Is latent autoimmune diabetes in adults distinct from type 1 diabetes or just type 1 diabetes at an older age? Diabetes 2005;54(Suppl 2):S62.

4. Stenstrom G, Gottsater A, Bakhtadze E, et al. Latent autoimmune diabetes in adults: definition, prevalence, beta-cell function, and treatment. Diabetes 2005;54:S68.

5. Lundgren VM, Isomaa B, Lyssenko V, et al. GAD antibody positively predicts type 2 diabetes in an adult population. Diabetes 2010;59:416.

6. Pozzilli P, Guglielmi C. Double diabetes: A mixture of type 1 and type 2 diabetes in youth. In: Cappa M, Maghnie M, Loche S, Bottazzo GF, editors. Volume 14. *Endocrine Involvement in Developmental Syndromes*. 2009. p 151. DOI: 10.1159/000207484.

7. Diabetes Association A. Diagnosis and classification of diabetes mellitus. Diabetes Care 2011;34(Suppl 1):S62. DOI: 10.2337/dc11-S062.

8. Karve A, Hayward RA. Prevalence, diagnosis, and treatment of impaired fasting glucose and impaired glucose tolerance in nondiabetic U.S. adults. Diabetes Care 2010;33:2355. DOI: 10.2337/dc09-1957.

9. Nichols GA, Hillier TA, Brown JB. Progression from newly acquired impaired fasting glucose to type 2 diabetes. Diabetes Care 2007;30:228. DOI: 10.2337/dc06-1392.

10. Ratner RE. An update on the diabetes prevention program. Endocr Pract 2006;12(Suppl 1):20.

11. International Association of Diabetes and Pregnancy Study Groups Consensus Panel. Internal Association of Diabetes and Pregnancy Study Groups recommendations on the diagnosis and classification of hyperglycemia in pregnancy. Diabetes Care 2010;33(3):676. DOI: 10.2337/dc09-1848.

12. Ryan EA. Diagnosing gestational diabetes. Diabetologia 2011;54:480. DOI: 10.1007/s00125-010-2005-4.

13. Agency for Healthcare Research and Quality. *One in 16 Women Hospitalized for Childbirth Has Diabetes*. Rockford (MD): AHRQ; 2010. Available at http://www.ahrq.gov/news/nn/nn121510.htm. Accessed December 2010.

14. Qiao Q, Gao W, Zhang L, et al. Metabolic syndrome and cardiovascular disease. Ann Clin Biochem 2007;44:232.

15. Dods R. Diabetes. In: Pesce A, Kaplan L, editors. *Clinical Chemistry: Theory, Analysis, Correlation*. 5th ed. St. Louis: Mosby; 2010. p 729.

16. Reaven GM. Banting lecture 1988. Role of insulin resistance in human disease. Diabetes 1988;37:1595. DOI: 10.2337/diabetes.37.12.1595.

17. Reaven GM. The metabolic syndrome: requiescat in pace. Clin Chem 2005;51:931.

18. Gogia A, Agarwal PK. Metabolic syndrome. Indian J Med Sci 2006;60:72.

19. Ford ES, Giles WH, Dietz WH. Prevalence of the metabolic syndrome among US adults: findings from the third national and nutrition examination survey. JAMA 2002;287:356. DOI: 10.1001/jama.287.3.356.

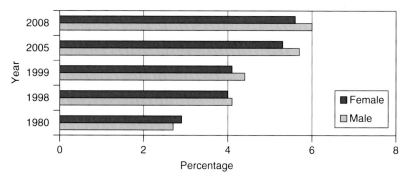

Figure 1.1 Age-adjusted percentage of civilian, noninstitutionalized persons with diagnosed diabetes by sex for selected years.

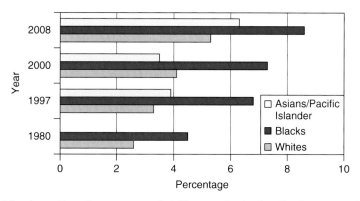

Figure 1.2 Age-adjusted percentage of civilian, noninstitutionalized persons with diagnosed diabetes by race: whites, blacks, and Asians/Pacific Islanders for selected years.

Understanding Diabetes: A Biochemical Perspective, First Edition. Richard F. Dods.
© 2013 John Wiley & Sons, Inc. Published 2013 by John Wiley & Sons, Inc.

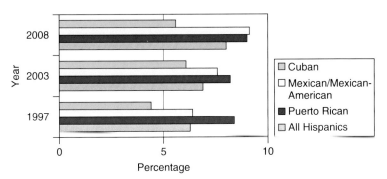

Figure 1.3 Age-adjusted percentage of civilian, noninstitutionalized persons with diagnosed diabetes among Hispanics: Puerto Ricans, Mexicans/Mexican-Americans, and Cubans for selected years.

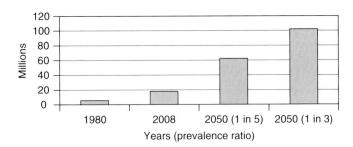

Figure 1.4 Number (in millions) of projected cases of diabetes for 2050 assuming prevalence of one in five and one in three compared to those reported in 1980 and 2008.

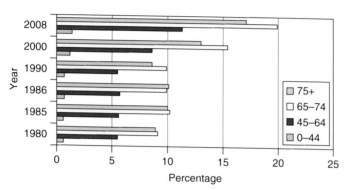

Figure 1.5 Percentage of civilian, noninstitutionalized persons with diagnosed diabetes by age (0–44, 45–64, 65–74, 75+) for selected years.

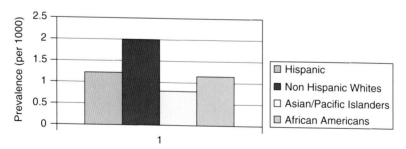

Figure 1.6 Diabetes prevalence in African Americans, Asian and Pacific Islanders, non-Hispanic whites, and Hispanics under the age of 20 years. The data for gender, age intervals, and type of diabetes were combined. This data will be expanded upon in Chapter 5.

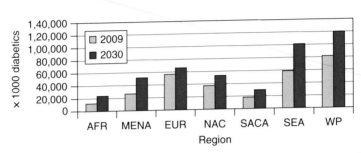

Figure 1.7 Estimated numbers of diabetics for 2030 contrasted with 2009 data for seven regions of the world; AFR, African Region; MENA, Middle East and North African Region; EUR, European Region; NAC, North America and Caribbean Region; SACA, South and Central American Region; SEA, South-East Asian Region; WP, Western Pacific Region. Data is compiled from the "IDF Diabetes Atlas," 4th ed., November 2009. The WP Region was recalculated to represent the data from the more recent estimates from Reference 14.

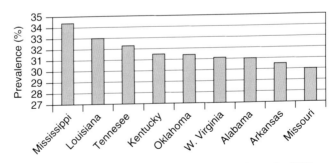

Figure 1.8 States with ≥30% obesity prevalence in 2009.

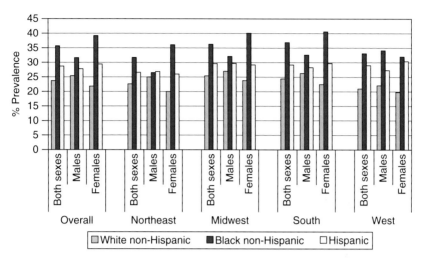

Figure 1.9 Obesity prevalence according to ethnicity, gender, and region. Data analyzed by the CDC from the Behavioral Risk Factor Surveillance System. Data collected by a random-dialed telephone survey of the US civilian noninstitutionalized ≥ 18 years. Surveys conducted in states, Washington, DC, and three territories. Pregnant women and those ≥ 500 lb or a height ≥ 7 ft were excluded. Surveys were conducted in 2006–2009. The data was age-adjusted to the US 2000 standard population. The prevalence relative standard error was less than 30%.

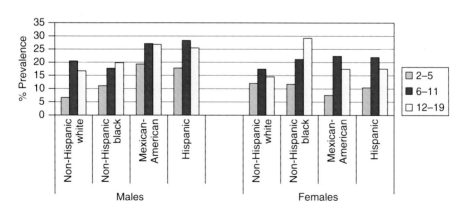

Figure 1.10 Obesity prevalence in US youths aged 2–19 by gender, age, and ethnicity in 2007–2008. Obesity defined as BMI at the 95th percentile or higher. Data are from Reference 22.

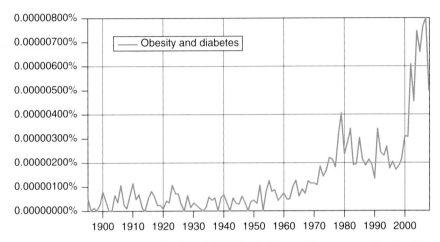

Figure 1.11 This figure shows the Googlelabs Books Ngram Viewer return for the phrase "obesity and diabetes" from the year 1895 to 2008.

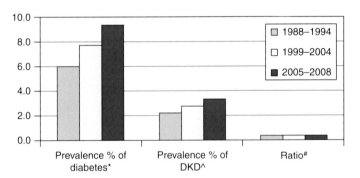

*% of US population with diabetes
^ % of diabetes diagnosed with DKD
# DKD population / Diabetes population

Figure 8.1 Prevalence (%) of diabetes and DKD and their ratios to each other. Prevalence (%) for diabetes and DKD was adjusted for age, sex, and race/ethnicity.

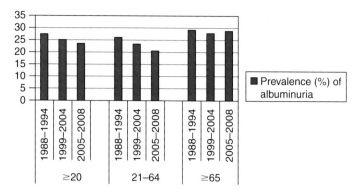

Figure 8.2 Prevalence (%) of albuminuria according to age.

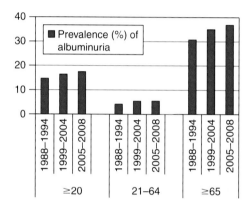

Figure 8.3 Prevalence (%) of impaired glomerular filtration rate.

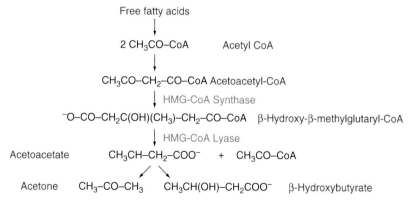

Figure 8.5 Synthesis of the ketone bodies.

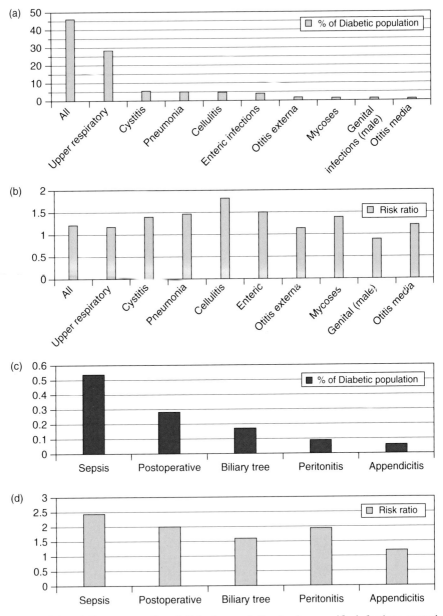

Figure 8.6 (a) Percentage of total diabetic population having specific infections treated as outpatients, (b) RR ratios for infections for diabetics treated as outpatients, (c) percentage of diabetic population having specific infections necessitating hospitalization, and (d) RR ratios for infections necessitating hospitalization.

1. The human Insulin gene is isolated.

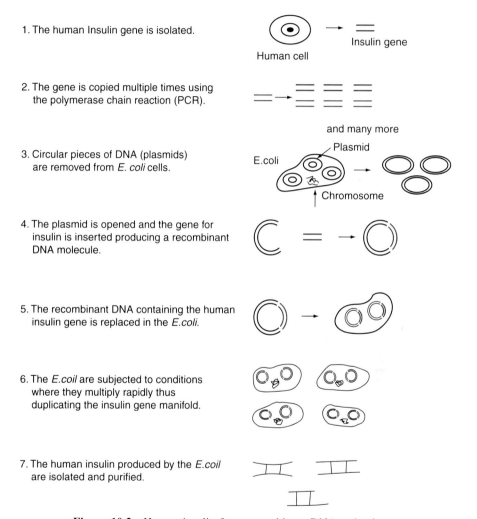

2. The gene is copied multiple times using the polymerase chain reaction (PCR).

3. Circular pieces of DNA (plasmids) are removed from E. coli cells.

4. The plasmid is opened and the gene for insulin is inserted producing a recombinant DNA molecule.

5. The recombinant DNA containing the human insulin gene is replaced in the E.coli.

6. The E.coil are subjected to conditions where they multiply rapidly thus duplicating the insulin gene manifold.

7. The human insulin produced by the E.coil are isolated and purified.

Figure 10.2 Human insulin from recombinant DNA technology.

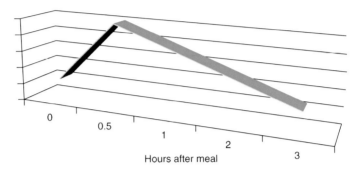

Figure 10.3 Insulin concentration in blood after a meal.

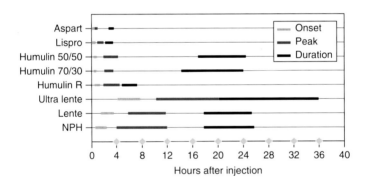

Onset - Length of time before insulin begins to reduce blood glucose
(as a range) (individual variability)

Peak - Length of time when insulin is most effective (as a range) (individual variability)

Duration - Length of time that insulin acts (as a range) (individual variability)

• Humulin 50/50 - consists of 50%, NPH + 50% recombinant DNA human insulin

• Humulin 70/30 - consists of 70%, NPH + 30% recombinant DNA human insulin

Figure 10.4 Onset, peak, and duration of various types of injectable insulin.

CHAPTER 7

DIAGNOSIS OF DIABETES MELLITUS

PART 1: ESTABLISHING A NORMAL RANGE

While the individual man is an insoluble puzzle, in the aggregate he becomes a mathematical certainty. You can, for example, never foretell what any one man will be up to, but you can say with precision what an average number will be up to. Individuals vary, but percentages remain constant. So says the statistician.

—Sir Arthur Ignatius Conan Doyle (1859–1930), author of the Sherlock Holmes novels.

Before we discuss assay methods for the quantitation of glucose and the detection of diabetes, we will take the mystery out of the "normal range." How is the normal range determined? What is the meaning of a test that is just outside of the normal range? How do we determine the efficacy of a test? These and other questions are answered in this part of the Chapter.

THE CONCEPT OF NORMAL AND ABNORMAL POPULATIONS

In Figure 7.1a–c, the Gaussian distribution curve[1] of the total population is represented by the bold line —. Healthy persons in that population have a Gaussian

[1]A Gaussian curve refers to a bell-shaped frequency distribution curve. There is no reason to assume a specific test will represent people in a Gaussian manner but many tests do fall into Gaussian distributions.

Understanding Diabetes: A Biochemical Perspective, First Edition. Richard F. Dods.
© 2013 John Wiley & Sons, Inc. Published 2013 by John Wiley & Sons, Inc.

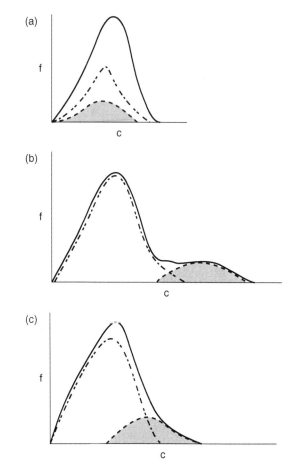

Figure 7.1 (a–c) Patterns for three different markers separating abnormal from normal populations.

distribution curve denoted by the line — · —; and those diagnosed as having a specific illness in that population have a Gaussian curve that is represented by the line ------ (and the shaded area). The y-axis is the frequency and the x-axis the laboratory value of the corresponding assay. The distribution in Figure 7.1a indicates that the disease state that is represented in this example does not affect the laboratory test studied. In Figure 7.1b, there is a significant increase in the laboratory test being studied in diseased individuals to the extent that it is effectively separated from the healthy population. This laboratory test would be very effective in determining the healthy from the ill population. In Figure 7.1c, there is a considerable overlap in the healthy and ill populations. This laboratory test would not be effective in discerning ill from healthy individuals.

If the disease condition were diabetes, Figure 7.1a could represent serum sodium levels; Figure 7.1b, could represent fasting blood glucose levels, and

Figure 7.1c could represent serum triacylglycerol (triglyceride) levels. Uncomplicated diabetes mellitus does not affect serum sodium levels, fasting glucose levels in overt diabetes are usually elevated, and triacylglycerol levels are often but not always elevated. One may conclude that the greater the overlap of the normal and abnormal populations the less useful is the assay in diagnosing the disease.

THE PROBABILITY FACTOR IN DIAGNOSING DISEASE

Figure 7.2 depicts typical bell-shaped distribution curves representing two populations; one population which has the disease being tested for (abnormal), the second population which does not have that condition (normal). In addition, we assume that the higher the result of the laboratory test the greater the probability that the individual has that specific condition. For a given laboratory test, a line erected perpendicularly from a specific test result (x-axis) to where it cuts the normal and abnormal Gaussian curves represents the probability that the disease occurs. For illustrative purposes, laboratory test a has a 100% probability of representing a normal result. Result b is in the overlap area, the ratio $d/(b + d) \times 100$ represents the percentage of the total population exhibiting the value and being truly abnormal. In this illustration, a patient with such a result is probably free of the specific disease by odds of about 4–1. For laboratory test value e the odds are 1–2 (50%) that the patient is free of the specific disease. In the case of f, the likelihood of disease is $f/(f + h)$ or about 25% of the total population having this result is normal and in I, the likelihood of the disease state is 100%.

PROBABILITY OF DISEASE AND PREVALENCE

The prevalence of a disease is defined as its frequency at a given moment in a representative population. Inherent in this definition is the frequency or incidence of the disease, that is, number of new cases of the disease per unit of population during a given period of time (usually a year). A number of factors affect the incidence, including the duration of the disease. More individuals in the population will manifest the disease if the disease is generally of long duration. Further, incidence of disease will be higher during epidemics and in selected populations. An increase in the physician's capability of detecting and defining the disease will increase its known incidence. For example, before Legionnaires' disease was defined, it had no incidence and therefore zero prevalence. The incidence of the disease times its duration equals prevalence.

In Figure 7.3, the prevalence of the disease is represented by the height of the abnormal curve or, more exactly, the area it encompasses. Figure 7.3 demonstrates how an increase in prevalence affects the probability that the assay being used for detection is accurate in health and disease. As the prevalence of the disease increases, the probability that the result a is normal decreases and the likelihood that result b is abnormal increases.

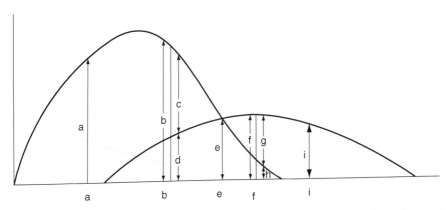

Figure 7.2   Distribution curves for a laboratory test that separates normal from abnormal populations.

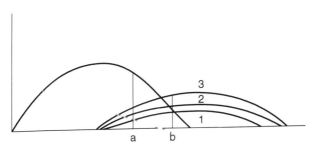

Figure 7.3   Effect of prevalence on accuracy.

THE NORMAL RANGE

The normal range is usually defined as the region of the Gaussian distribution for normal subjects that encompass 95% of the area under the curve (Fig. 7.4). In the language of statistics, 95% of the population will be included if one uses two standard deviations (2SDs) either side of the mean as boundaries. One standard deviation includes 80% of the population and three standard deviations, 98%. The inclusion of a normal (2SD) in Figure 7.5 allows the following conclusions in regard to disease prevalence.

1. Result (*a*) (which is in the normal range), would incorrectly identify abnormal individuals as normal to an increasing degree as the prevalence of the disease increases.
2. Result (*b*) which is in the abnormal range, would be more correct in identifying an abnormal individual as the prevalence of the disease increases.

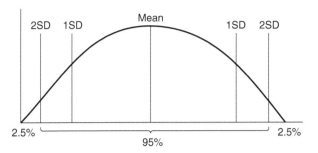

Figure 7.4   The normal range.

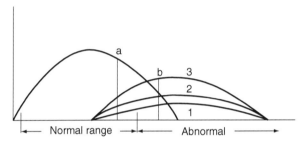

Figure 7.5   Sensitivity and specificity.

Summary Box 7.1

- A test that has its normal and abnormal populations completely overlap is a test that cannot be used to distinguish normal populations from abnormal.
- A laboratory test that has some overlap between normal and abnormal populations is a good test, but there will be false normals and false abnormals.
- The prevalence of a disease is its frequency at a given time in a representative population.
- Incidence is the number of new cases of the disease per unit of population per unit period of time.
- The normal range for a laboratory test is defined as the region of the Gaussian distribution for normal subjects that encompass 95% of the area under the curve (two standard deviations either side of the mean).

ASSAY SENSITIVITY AND SPECIFICITY

The above observations (1, 2) may be studied in a quantitative manner. This is done by defining the following terms:

The sensitivity of a laboratory test. This represents the incidence of abnormal results in patients known to have the disease. Sensitivity may be represented as the ratio of the number of patients correctly diagnosed as diseased to the total population of abnormal results. In terms of the graphic representation in Figure 7.5, sensitivity refers to the overlap of the normal state with the abnormal state in the abnormal range (i.e., outside the normal range as defined in the previous section).

The specificity of a laboratory test. This represents the incidence of normal results in patients known to be free of the disease being studied. Specificity may be represented as the ratio of the number of patients correctly diagnosed as healthy to the total number of patients in the normal range. In terms of the graphic representation, Figure 7.5 refers to the overlap of the normal state with the abnormal state in the normal range.

The predictive value of a normal result. This represents the incidence that the assay correctly defines an individual as normal from a population of normal and diseased individuals.

The predictive value of an abnormal result. This is the incidence that the assay correctly defines an individual as abnormal from a population of normal and diseased individuals.

These definitions are represented in equations as follows:

$$\text{Sensitivity} = \frac{TA}{TA + FN} \times 100$$

$$\text{Specificity} = \frac{TN}{TN + FA} \times 100$$

$$\text{Predictive value of an abnormal result} = \frac{TA}{TA + FA} \times 100$$

$$\text{Predictive value of a normal result} = \frac{TN}{TN + FN} \times 100$$

$$\text{Prevalence} = \text{Incidence} \times \text{Disease duration}$$

$$\text{Incidence} = \text{Occurrence of disease per unit population per unit time where per unit time is usually one year. Incidence is expressed in percent.}$$

Here,

TA(true abnormal) = correctly identified abnormal

TN(true normal) = correctly identified normal

FA(false abnormal) = normal, incorrectly identified as abnormal by assay

FN(false normal) = abnormal, incorrectly identified as normal by assay.

RELATIONSHIPS AMONG SENSITIVITY, SPECIFICITY, PREVALENCE, PREDICTABILITY, AND NORMAL RANGE

Rather than to strictly adhering to the mathematical definition of normal range, one may wish to adjust the normal range in order to maximize sensitivity or specificity.

Figures 7.6a–h show the effects of increasing the upper limit of the normal range on TN, FN, TA, and FA populations. For sake of clarity, Figure parts b, d, f, and h isolate the specific areas of the population. When the upper limit of normal is increased, more of the abnormal population is included in the normal range and less of the normal population is included in the abnormal range; TN and FN increase, TA and FA decrease. In fact, this example has FA equal to zero. Sensitivity is affected in the following manner:

$$\text{Sensitivity} = \frac{\text{TA} \downarrow}{\text{TA} \downarrow + \text{FN} \uparrow}$$

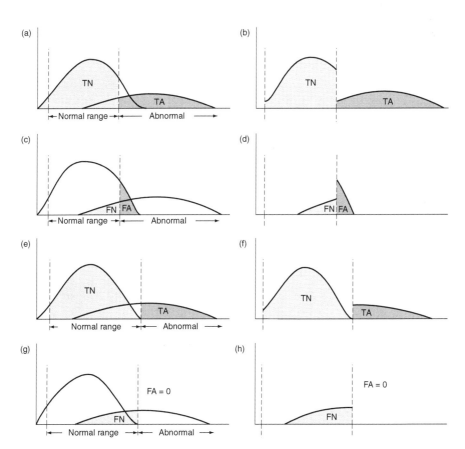

Figure 7.6   Effects of changing the upper limit of normal.

The overall effect is to decrease the denominator less than the numerator resulting in a decrease in sensitivity.

$$\text{Specificity} = \frac{\text{TN} \uparrow}{\text{TN} \uparrow + \text{FA} \downarrow}$$

The overall effect on specificity is to increase the numerator to a greater extent than the denominator, thus resulting in an increase in specificity. Thus shifting the normal range to higher values results in a decrease in sensitivity and an increase in specificity. Let us next examine the effect on predictability.

$$\text{Predictability of a normal result} = \frac{\text{TN} \uparrow}{\text{TN} \uparrow + \text{FN} \uparrow}$$

Predictability of a normal result decreases.

$$\text{Predictability of an abnormal result} = \frac{\text{TA} \downarrow}{\text{TA} \downarrow + \text{FA} \downarrow}$$

Predictability of an abnormal result increases.

Shifting the normal range to higher values results in a decrease in the predictability of a normal result and an increase in the predictability of an abnormal result. The reverse effects occur upon lowering the normal range.

EXERCISE

1. Serum albumin levels are decreased in certain liver diseases, for example, cirrhosis. Draw a concentration versus frequency graph assuming prevalence and overlap similar to Figure 7.4a.
2. The normal range for serum albumin is 3.5–5.5 g/dl. What effect on TN, TA, FN, FA, sensitivity, specificity, and predictability would occur if the normal range was changed to 4.0–6.5 g/dl?

Summary Box 7.2

- Sensitivity refers to the incidence of abnormal results in patients known to have the disease.
- Specificity refers to the incidence of normal results in patients known to be free of the disease.
- The predictive value of a normal result is the incidence that the laboratory test correctly defines a patient as normal.
- The predictive value of an abnormal result is the incidence that the laboratory test defines as abnormal.
- A change in the normal range affects the sensitivity, specificity, predictability of a normal result, and predictability of an abnormal result.

HOW DOES ONE CHOOSE A NORMAL RANGE?

When dealing with a life-threatening disease which is treatable, FA results can be tolerated. We wish to reduce the number of TA that avoid detection (i.e., reduce FN). The upper range of normal is therefore decreased. This action increases sensitivity at the expense of specificity. The predictability of a normal result is increased.

When dealing with a disease which is untreatable or FA results would lead to undue expense or worry for the patient, the normal range should be adjusted so as to decrease sensitivity and increase specificity. This is done by increasing the upper limit of normal. Predictability of an abnormal result is increased.

The prevalence of a disease as defined above adds a third variable to the situation. As may be seen in Figure 7.3, an increase in the prevalence of a disease increases the overlap of normal and abnormal populations. Thus an increase in prevalence results in a decrease in specificity and an increase in sensitivity. An alternative way of stating this is that an increase in prevalence results in a decrease in the predictive value of a normal result and an increase in the predictive value of an abnormal result.

The prevalence of a disease varies greatly according to the population that is studied. The prevalence of heart conditions in the American population is approximately 50 per 1000 persons. If the population studied is defined as American males, 65 years and over, the prevalence increases to nearly 200 per 1000. The prevalence of individuals with heart disease entering the intensive care unit of a city hospital may be as high as 500 per 1000 admissions or higher. If the sensitivity of the assay is maintained at 80% and a specificity of 95%, the predictability of an abnormal result would be 46% in the general population, 80% in males 65 years old and greater, and 94% among admissions to the intensive care unit. On the other hand, the predictability of a normal result would be 99%, 95%, and 83%, respectively. Thus, as the general population is narrowed to a segment more likely to exhibit the disease, the predictability of the assay improves. This is the reason that most screening tests fail. They become uneconomical as false abnormals are followed up.

TRUTHFULNESS (EFFICIENCY)

The truthfulness of an assay for a given disease state is the ratio of the number of times the test correctly defines a patient as normal or abnormal to the total population. The term truthfulness is the same as efficiency as used by other textbooks. It combines in one number the predictive value of a normal result with the predictive value of an abnormal result.

$$\text{Truthfulness} = \frac{\text{TA} + \text{TN}}{\text{TA} + \text{TN} + \text{FA} + \text{FN}} \times 100$$

NON-GAUSSIAN DISTRIBUTION

The preceding presentation assumes a normal distribution of results for a given disease, that is, it assumes a bell-shaped symmetric curve. However, results need not be Gaussian, they can be non-Gaussian. An example of a non-Gaussian laboratory test is that of hepatitis B-surface antigen.

The antigen is measurable only in the early phase of the disease. This results in the absence of a normal range. As can be seen in Figure 7.7, there is an ambiguous zone caused by the inaccuracy of the testing procedure near zero titer of antigen. Since not all hepatitis cases will exhibit measurable titers, the amplitude of the hepatitis B population is reduced relative to the total population diagnosed as ill due to the hepatitis B-viral antigen.

THE EFFECT OF REPRODUCIBILITY ON SENSITIVITY AND SPECIFICITY

In assessing a methodology for the determination of a parameter considerable weight is usually given to the economics of the method and its efficiency in regard to time. The reproducibility and accuracy of the test are frequently given little scrutiny. Terminology such as "reproducibility sufficient for clinical significance" often appears in the literature. Certainly, results that are grossly abnormal or well within the normal range will have the same diagnostic value whether they are assayed by a method having high precision or somewhat poorer precision. However, the effect of reduced reproducibility is to increase the overlap of the normal and abnormal populations. This effect is shown in Figure 7.8a and b. The broadness of the curves represents the reproducibility. In Figure 7.8a, the normal and abnormal only overlap slightly. When the standard deviation (reproducibility) is increased (the curve is broadened) a significant overlap becomes evident (Fig. 7.8b). The predictive value of normal and abnormal is reduced. The specificity and sensitivity are reduced.

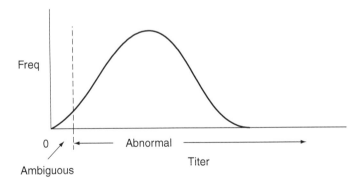

Figure 7.7 Non-Gaussian distribution.

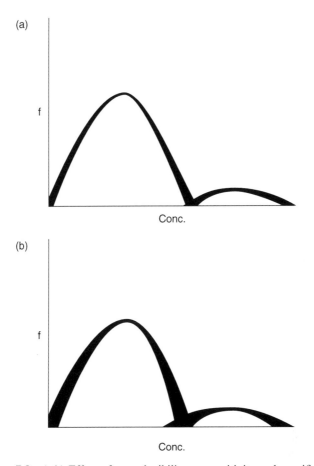

Figure 7.8   (a,b) Effect of reproducibility on sensitivity and specificity.

SEVERITY OF DISEASE AND ASSAY RESULTS

One of the reasons for a false normal result is that it is possible to have a several fold increase in a parameter and still remain in the normal range. For example; a total lactate dehydrogenase (LD) (normal range, 76–202 IU/l[2]) on day of admission of 80 IU/l is significantly changed if it increases to 185 on day 2. The increase is even more significant if an increase to 200 were to be found on day 3. This second elevation would tend to rule out normal day-to-day fluctuations in this blood component. Biological variations in parameters from day to day and even hour to hour are the rule and not the exception. The biological variation may be relatively small, as it is for blood pH ($\pm$0.01–0.03), or wide, as found for many enzyme activities. These variations are quite separate entities from

[2]IU/l refers to an enzyme activity and stands for International units per liter.

variations produced from the standard deviation of the methodology used for the assay. Although the above LD example remains in the normal range the increase from 80 IU/l to 200 IU/l is significant and is abnormal. Analysis by LD isozyme[3] electrophoresis may indicate a definite elevation in LD-5, thus suggesting liver damage. The severity of the damage to the liver might be equivalent to that of a patient with an LD of 800 IU/l. For the most part, the severity of disease does not correlate well with the elevation (or decrease) of a given parameter. The reasons for this lack of correlation are beyond the limits of this text. Suffice it to say that factors such as availability of a blood supply to the affected region, the half-life of the component in the blood, and the presence of naturally occurring inhibitors play a role in reducing the correlation between the severity of the disease and the degree to which a particular parameter will be abnormal.

PARALLEL AND SERIES MULTIPARAMETER TESTING

Parallel multiparameter testing is the simultaneous testing of two or more biological parameters. It is frequently referred to as *multiphasic, panel, battery,* or *combination testing*. The occasion rarely arises where a physician will order only a single test on a patient upon their initial visit to the office. Usually, a battery of as many as 24 tests will be ordered. The tests chosen in this initial battery is usually dependent upon a thorough history and physical examination. Although it is not usually thought of in this light, a physical examination and history taking is a form of parallel multiparameter testing. In this manner, the physician narrows the diagnosis to a few possible disease entities. Physical examinations lack the objectivity and quantitative properties of laboratory testing. However, they are utilized by the physician in a manner similar to laboratory testing.

In parallel multiparameter testing, the greater the number of tests in the panel, the more probable that a false abnormal will occur. In this type of testing, a single abnormal result in the panel defines the patient as abnormal. Thus, for truly healthy individuals, the greater the number of tests performed, the greater the likelihood that they will be considered abnormal. Conversely, a truly abnormal will be less likely to escape detection. Once parallel multiphasic testing has narrowed the diagnosis to a few possible disease entities, the physician will order additional tests pertinent to the illnesses still under consideration. For example, a battery shows an elevated alanine aminotransferase (ALT) indicating possible liver damage. The physician will order a second battery of tests known to be elevated in liver disease, for example, γ-glutamyltransferase (GGT), and/or LD-5. The ordering of tests in this manner is termed *series multiparameter testing*.

Some of the complexity of parallel and series multiparameter testing is reduced by expert systems that utilize computerized algorithms to diagnose the patient (1).

---

[3]There are five LD isozymes separable by electrophoresis. LD-5 is predominantly found in liver.

Summary Box 7.3

- The upper range of normal can be reduced for a test that deals with a life-threatening disease that is treatable. This action increases sensitivity, decreases specificity, and increases the predictability of a normal result.
- The upper range of normal can be increased for a test that deals with a disease that is untreatable. This action decreases sensitivity, increases specificity, and increases the predictability of an abnormal result.
- As the general population for a specific disease is narrowed to a segment more likely to have the disease, the predictability of the laboratory test increases.
- The truthfulness of an assay for a given disease state is the ratio of the number of times the test correctly defines a patient as normal or abnormal to the total population.
- Non-Gaussian distributions represent tests where the result is a yes or no such as hepatitis B.
- As an assay has decreased reproducibility, the predictive value of normal and abnormal is reduced.
- A given assay that has significant increases in values but stays in the normal range suggests that it is a false normal.
- Severity of a disease is often independent of its elevation of an assay specific for the disease.
- Parallel testing is simultaneous testing. Series testing is consecutive testing.
- In parallel testing, a single abnormal test labels the patient as abnormal. In series testing, an abnormal result in the initial test leads to the second test. If the second test is abnormal, the patient is considered abnormal.

EXERCISE

1. Compare the sensitivity and specificity of parallel multiparameter testing to a single assay.
2. Compare the predictability of an abnormal result of series multiparameter testing to parallel multiparameter testing.

One aspect of testing not taken into consideration by the preceding is the extent to which the result is abnormal. As demonstrated earlier, the more abnormal the result, the more likely it is truly abnormal.

In order to compare series and parallel multiparameter testing in regard to sensitivity and specificity, two panels of testing for cardiovascular disease will be utilized. The first panel consists of four tests aspartate aminotransferase (AST), lactate dehydrogenase (LD), creatine phosphokinase (CK), and electrocardiogram

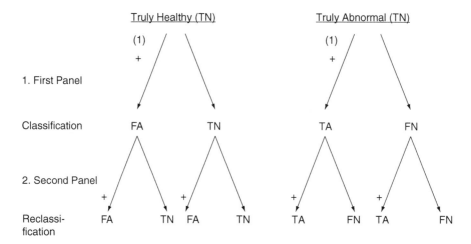

(1) – The + sign above the arrow indicates a single abnormal result in the panel.

Figure 7.9 Multiparameter testing.

(EKG or ECG),[4] If the patient has a myocardial infarct (MI)(heart attack in the vernacular), then only one of the above assays needs to be elevated in order to classify the individual as abnormal. On the other hand, only one false abnormal need occur in order to incorrectly classify a healthy individual as abnormal. In contrast, four false normals would be needed in order to label a truly abnormal individual incorrectly as normal.

In the initial battery, a truly healthy person may be labeled TN or FA and a truly abnormal, TA or FN (refer to Fig. 7.9). This may be symbolized by

$$TN \longrightarrow FA$$
$$TA \longrightarrow FN$$

In the second panel of testing, which may include CK and LD-5 in the example, there are several crossover possibilities. A truly healthy patient may be reclassified as TN after classification as FA in the first panel,

$$FA \rightarrow TN$$

or reclassified as FA after classification of TN,

$$TN \rightarrow FA$$

---

[4]The author realizes that there are more sensitive biomarkers that are commonly used to detect MI such as CK-MB, myoglobin and troponins but for illustration I will use the four tests listed above.

EXAMPLE **215**

A true abnormal may be classified FN after initial classification as TA,

$$TA \rightarrow FN$$

or TA after incorrect classification as FN,

$$FN \rightarrow TA$$

TN $\rightarrow$ FA is more probable than FA $\rightarrow$ TN and

FN $\rightarrow$ TA occurs more frequently than TA $\rightarrow$ FN

Thus, TN and FN decrease and FA and TA increase in series testing. Sensitivity increases because

$$\frac{TA \uparrow}{TA \uparrow + FN \downarrow}$$

and specificity decreases,

$$\frac{TA \downarrow}{TA \downarrow + FA \uparrow}$$

relative to one panel, parallel multiparameter testing.

This complex topic is summarized as follows (2):

Parallel multiparameter testing is simultaneous testing. A single abnormal test labels the patient as abnormal. If all tests in the panel are in the normal range the patient is normal. Sensitivity is therefore increased at the cost of specificity.

EXAMPLE

Test	Sensitivity, %	Specificity, %
CK	88.7	60.2
AST	85.6	63.5
LD	90.5	85.5
CK-MB	90.5	83.3
All 4 markers in parallel	96.6	31.0

CK-MB is creatine kinase-MB isozyme.

Series multiparameter testing is consecutive testing. If the first test is abnormal, then the second test is administered. If the second test is abnormal, the patient is labeled abnormal. If the first test is normal the patient is normal, and no further testing is done. Specificity is increased at the cost of sensitivity.

EXAMPLE

Test	Sensitivity, %	Specificity, %
CK	88.7	60.2
AST	85.6	63.5
LD	84.5	85.5
AST and LD in series	84.1	90.9
CK and AST in series	85.0	82.6
CK and LD in series	86.2	89.1
CK, AST, and LD in series	75.2	94.3

Readers who wish to learn more about how laboratory tests are evaluated with regard to their value in recognizing a healthy person from one with disease are referred to the groundbreaking book in this area: RS Galen and SR Gambino's seminal book "Beyond Normality: The Predictive Value and Efficiency of Medical Diagnoses" New York: John Wiley & Sons; 1975.

REFERENCES

1. Hobart ED Jr, Dods RF. MUMPS in a medical decision support system. MUG Quart 1987;16:29.
2. Dods RF. Section A, *Introduction to Clinical Chemistry, Clinical Chemistry*, American Chemical Society, Washington DC, 1990, p. 18. An audiocassette course.

PART 2: MODERN TECHNIQUES FOR THE QUANTITATION OF GLUCOSE

METHODS OF HISTORICAL INTEREST

The advances made in the study of diabetes paralleled the advances made in the assay of glucose. Otto Folin stated as follows in his paper presenting an adaptation of his micro method for quantification of blood glucose levels in 1929 (1):

Our efforts to enlarge the range of the method so as more nearly to meet the needs of diabetic clinics have been more successful than was at first thought possible. We have in fact secured a much wider dependable range than is obtainable by any other colorimetric method for blood sugar and, at the same time; we have made the method distinctly easier than it was before.

Thus, diabetes gave impetus to the improvement of glucose assay with respect to accuracy, reproducibility, and size of sample. In Chapter 2, we followed the advance of the assay of glucose in biological fluids from the crude observation of yeast fermentation to the use of methods that were dependent on the reduction of cupric to cuprous ion in alkaline solution. The early copper reduction methods were qualitative rather than quantitative and differed from one another in regard to the anion associated with the copper and the additives keeping the copper ion in solution.

The copper reduction methods were made more sensitive and specific by the utilization of chemical compounds which were reduced to colored complexes by the cuprous ion. Thus phosphomolybdate and arsenomolybdate are reduced by the cuprous ion to molybdenum blue. The discovery and advancements in the spectrophotometer permitted scientists to quantitate glucose levels using the spectrum of molybdenum blue. Significant quantities of protein found in blood samples interfered with the assay for glucose. Two methods became widely used for the pretreatment of the serum specimen; the Folin–Wu procedure, which uses tungstic acid and the Somogyi–Nelson procedure, which precipitates the proteins with a barium sulfate–zinc hydroxide solution. The latter method became the more popular method because the resultant solution is near neutral, pH 7.4 and naturally occurring interfering substances such as uric acid and creatinine are precipitated with the barium sulfate–zinc hydroxide.

2,9-Dimethyl,1,10-phenanthrolene hydrochloride (neocuproine) was widely utilized in place of phospho or arseno molybdic acid. Its sensitivity[1] is nearly 30 times greater than that of phosphomolybdic acid. However the trade-off is a greatly reduced specificity;[2] glucose assays were frequently falsely elevated by drugs interacting with the neocuproine and adding to the color reaction.

Another popular method for the quantitation of glucose, which was given up long ago, is based on the reduction of ferricyanide to ferrocyanide by glucose in alkaline solution. The ferri compound has a yellow color and the ferro compound is colorless. In contrast to copper reduction methods the ferricyanide reaction results in a decrease in absorption (at 420 nm). This method called the *Prussian blue reaction* is the one referred to by Folin, quoted at the start of this section. The glucose levels assayed by the ferricyanide method are 7–10% higher than those obtained from the modern enzyme methods. The principal interferences are the same as in the copper reduction methods; namely creatinine and uric acid.

The reduction of picric acid to picramic acid also has been utilized to measure glucose. However creatinine rapidly reduced picrate and thus the method has only historical importance.

o-Toluidine produces a Schiff base[3] when reacted with glucose. The assay developed by Eric Hultman (2) in 1959 is conducted at 100 °C and results in the production of glycosylamine through the intermediate Schiff base. Maximum absorbance of the complex occurs at 630–635 nm. Excellent correlations between the *o*-toluidine method and enzymic methods have been reported. There are many modifications of the *o*-toluidine method in current usage. Optimization of the

---

[1] In this context sensitivity refers to the lowest concentration the assay can accurately measure.
[2] In this context specificity refers to number of compounds that interfere in the assay. They can interfere by lowering or increasing the result.
[3] A Schiff base forms when an amine reacts with an aldehyde. The compound formed has a bridge between the two molecules, $-NH = C-$.

method by the use of 50% or 65% acetic acid or additives such as borate or benzyl alcohol have been described. The replacement of acetic acid entirely by glycolic acid and propylene glycol has been reported.

In summary the copper reduction methods with their poor specificity led to other methods such as the ferricyanide and picric acid approaches. These had their own specificity problems. But these methods, historically led to major discoveries in the field of diabetes. The advent of the use of o-toluidine gave greater accuracy to the assay of glucose and this is used in some laboratories today. But enzymology was the field that led to the most accurate and precise measurements of glucose in biological fluids and thus we enter into the modern age of the assay of glucose.

MODERN-DAY METHODS OF MEASURING GLUCOSE

Glucose Oxidase/Peroxidase/Chromogen

Müller (3) first reported in 1928 the presence of an enzyme isolated from extracts of *Aspergillus niger* and *Penicillium glaucum* that acted on glucose to produce D gluconic acid and hydrogen peroxide. Little did Muller know that he had launched the ultimate specific test for glucose—an enzymic assay for glucose. Glucose oxidase was purified under the name Penicillin B (later changed to Penicillin A (4)) by a research team from St. Louis University School of Medicine. Another team (5) from London School of Hygiene and Tropical Medicine also purified the same enzyme under the name Notatin (from *Penicillium notatum*). Glucose oxidase, Notatin, and Penicillin B (or A) were the same enzyme.

Glucose exists in aqueous solutions such as blood, urine, and cerebrospinal fluid as α-D-glucose ($\sim$36%) and β-D-glucose ($\sim$64%). These two substances are called *anomers*. Glucose oxidase acts specifically on the β-anomer of glucose. However as the β-anomer reacts the equilibrium between α-anomer and β-anomer shifts toward the β-anomer. In addition most preparations of glucose contain a trace of mutarotase which accelerates the conversion of α-D-glucose to β-D-glucose. The overall reaction of glucose oxidase on glucose is given by

$$\beta\text{-}D\text{-Glucose} + H_2O + O_2 \rightarrow D\text{-Gluconic acid} + H_2O_2$$

This reaction, by itself, poses some problems in regard to quantitation. D-gluconic acid does not have properties that allow simple measurement. Initially, the consumption of oxygen was measured. But the most practical method for quantitation and the one which became popular was the colorimetric procedure, which couples peroxidase and an oxygen accepting chromogen.

$$H_2O_2 + \text{Chromogen} \rightarrow H_2O + \text{Oxidized chromogen}$$

The oxidized chromogen absorbs light at a specific wavelength in the visible. The reduced form of the chromogen either does not absorb or absorbs at a wavelength far from the oxidized chromogen. Some of the chromogens appear below:

Chromogen	Wavelength (nm) Read
o-Tolidine	500
o-Dianisidine	540
DL-Adrenaline	500
Diethylaniline 4-aminoantipyrene	553
4-Aminoantipyrine(4-aminophenazone)	505

A more complex means of measuring the glucose oxidase reaction utilizes two compounds which undergo a condensation reaction in the presence of H_2O_2 and peroxidase. Examples for this appear below:

1,5 Dimethyl-2-phenyl 4-aminopyrazolone +

$$2,4,6\text{-Tribromophenol} \xrightarrow{H_2O_2} \text{Bromoquinonimine}$$

3-Methyl-2-benzothiazolineone +

$$N, N\text{-Dimethylhydrazone (MBTH)} \xrightarrow{H_2O_2} \text{Complex dye}$$

The "Achilles heel" of the reaction is the nonspecificity of the peroxidase enzyme. Bilirubin, uric acid, hemoglobin, ascorbic acid, reduced glutathione, and many drugs react with H_2O_2 in the presence of peroxidase, thus contributing a negative error to the assay. Deproteinization utilizing the Somogyi–Nelson method, which also precipitates bilirubin, uric acid, and reduced glutathione, significantly reduces these interferences. Removal of protein is mandatory since protein inhibits glucose oxidase. Hemolysis increases the negative error by releasing hemoglobin and reduced glutathione. The glucose oxidase preparation has to be catalase-free. Catalase is a potent enzyme that uses one molecule of H_2O_2 as an electron acceptor. It causes a negative bias by destroying the peroxide produced by the glucose oxidase.

By the late 1960s the enzymic procedure utilizing glucose oxidase/peroxidase/ chromogen (especially o-tolidine and o-dianisidine) and the chemical method utilizing o-toluidine were the methods of choice for the analysis of glucose in biological fluids. In addition, a semi quantitative method for the identification of glucose in urine had been developed by Miles–Ames Research Laboratory. The method, termed *Clinitest*, involved dipping a strip of paper impregnated with Benedict's copper reduction reaction in the specimen of urine. With the advent of glucose oxidase (GOD) methodology, Free (6) developed a strip imbedded with glucose oxidase, peroxidase, and o-tolidine. A blue color signified a positive test for glucose. The intensity of the color of the strip is proportional to the glucose level. The strip was called *Clinistix*. A common interference for Clinistix is ascorbic acid. Five-10 mg/dl of ascorbic acid reduces the sensitivity of the test and 5–100 mg/dl causes false negatives in samples that contain glucose in the range of 100 mg/dl.

Hexokinase/NADP⁺

Es wird eine enzymatische Methode beschrieben, die eine spezifsche Bestimmung von Glucose und Fructose im gleichen Versuchsansatz gestattet. Die gekoppelten enzymatischen Reaktionen benötigen die Enzyme Hexokinase, Glucose-6-Phosphatdehydrogenase und Phosphoglucoseisomerase, TPNH ist Meßgröße. Neben den beiden Hexosen sind auch deren 6-Phosphoräureester bestimmbar. Die hohen Umsatzzahlen der Boehringer-Enzympräparationen gestatten kurze Meßzeiten.

Mittels GOD-POD-Reaktion gefundene Blutglucosewerte werden durch Parallelbestimmungen mit der beschriebenen Methode bestätigt. (7)

This relatively short abstract by Schmidt in 1961 introduced the enzymic procedure that is the most prevalent methodology used today for the analysis of glucose in biological fluids. For those who do not understand German or do not wish to use a German–English dictionary (online or otherwise) a translation follows:

"An enzymic method is described which allows a specific determination of glucose and fructose in the same sample. The coupled enzymic reactions need the enzymes hexokinase, glucose 6-phosphate dehydrogenase and phosphoglucose isomerase. TPNH[4] is the indicator. In addition to the two hexoses; their 6-phosphates are also determinable. The high turnover of the Boehringer Enzyme products allow the short assay times. Blood glucose values determined by the GOD–POD method are confirmed by parallel determinations with the described method."

—Translation by Joachim Witzke

As may be construed from the above, the hexokinase methodology is really a two-step procedure, the phosphoglucose isomerase composing the initial procedure is to convert fructose to glucose. When used for the assay of glucose only, the first step is a specific phosphorylation of glucose at the 6-hydroxyl position.

$$\text{Glucose} + \text{ATP} \xrightarrow{\text{Hexokinase}} \text{Glucose 6-phosphate} + \text{ADP}$$

The second step is oxidation of the glucose 6-phosphate.

$$\text{Glucose 6-phosphate} + \text{NADP}^+ \xrightarrow{\text{Glucose 6-phosphate dehydrogenase}}$$

$$\text{6-Phosphogluconate} + \text{NADPH}$$

---

[4]TPNH refers to reduced triphosphopyridine nucleotide an obsolete name for reduced NADPH.

Since NADPH absorbs at 340 nm, the reaction is followed spectrophotometrically.

The purity of the enzymes is important since the hexokinase must be free of ATPase, 6-phosphogluconic dehydrogenase, and phosphoglucose isomerase. Glucose 6-phosphate dehydrogenase, derived from brewer's, baker's, or Torula yeast or from *Leuconostoc mesenteroides* must be sulfate free since it is a potent inhibitor of the enzyme. All preparations must be free of NADPH oxidase, phosphoglucomutase, and phosphoglucose isomerase.

Summary Box 7.1

- The Folin–Wu method using tungstic acid and the Somogyi–Nelson method using barium sulfate-zinc hydroxide were the principal procedures used to precipitate proteins from biological fluids.
- Copper reduction methods coupled to various dyes, ferricyanide, picric acid, *o*-toluidine assays are no longer used in clinical laboratories.
- The GOD/POD procedure coupled to chromogens became the method of choice.
- The "Achilles heel" of the GOD/POD method was the peroxidase enzyme.
- The hexokinase method composed of two steps; phosphorylation of glucose and oxidation of $NADP^+$ is presently the reference method for the assay of glucose in biological fluids.

EXERCISE

How does NADH oxidase, phosphoglucomutase, phosphoglucose isomerase, and ATPase interfere with the hexokinase method for the assay of glucose?

Figure 7.1 compares the effects of potential interferents on three methodologies for the assay of glucose. You may note that the hexokinase method is least affected by the substances listed. However both GOD and hexokinase methods are affected by hemolysis. The release of glucose 6-phosphate by lyzed erythrocytes interferes with the hexokinase assay, whereas hemoglobin and glutathione release interfere with the GOD method. If serum or plasma is used as the specimen deproteinization is not necessary.

GLYCATED HEMOGLOBIN

Glycated hemoglobin also called *glucosylated hemoglobin*, hemoglobin A1c, HbA_{1c}, and HbA1c, is measured to monitor average blood glucose levels over approximately 60–90 days; the erythrocyte life span is 120 days. Several methods

Substances	Concentration (mg/dl) at Which Interference is significant			Direction of Interference		
	GOD	Hexo	TOL	GOD	Hexo	TOL
Bilirubin	3	3–7	3–7	↓	↓	↑
Ascorbic acid	4	NE (25)	15–25	↓	↔	↑
Glutathione	40	—	—	↓	—	—
Creatinine	NE (13)	NE (12.5)	NE (12.5)	↔	↔	↔
Uric acid	20	NE (25)	NE (25)	↑	↔	↔
Hemolysis	NE (220)	NE (220)	—	↔	↓	—
EDTA	NE	1000[a]	NE	↔	↓	↔

[a]7ml BD Vacutainer tubes contain 10.5mg K_3 EDTA; well below inhibition levels for hexokinase.

NE( ) - No effect (concentration in mg/dl)

— - Information not readily available

Figure 7.1 Effects of various substances on three methods for the assay of glucose.

are in use including a minicolumn approach (8, 9), boronate affinity chromatography (10), enzyme immunoassay (11), and high pressure liquid chromatography (12). Previously, isoelectric focusing, agar gel electrophoresis, and a colorimetric procedure (the Fluckiger–Winterhalter method) after column chromatography have been used.

Glycosylation of proteins such as hemoglobin occurs through the reaction of glucose with the free amino groups of the N-terminal and internal lysine residues of the polypeptide chains composing hemoglobin. As illustrated in Figure 7.2, a Schiff base (see Footnote 3) is initially formed (step 1). The Schiff base subsequently undergoes rearrangement (Amadori rearrangement) to form a ketoamine (step 2). The intermediate Schiff base is also known as an *aldimine*.

The globin portion of hemoglobin is glucosylated[5] during the erythrocytes' exposure to the plasma. Since the reaction takes place spontaneously (no enzyme required), its rate of formation is dependent solely on the concentration of the glucose and the quantity of available amino groups. Glucose permeates freely into and out of such cells as the erythrocyte, lens, glomerulus, and nerve. These cells are contrasted with liver cells which control the movement of glucose by the mediation of insulin. A minor hemoglobin, HbA_1 is formed by the reaction of glucose with the globin of hemoglobin. Because of the enzymic independence of the reaction of glucose with the globin and the permeability of the erythrocyte to glucose, the concentration of HbA_1 is directly proportional to the average plasma glucose concentration during the 120-day life span of

---

[5]Glucosylated is a specific reference for the reaction of glucose with protein. Glycosylation is more general term that refers to other monosaccharides such as mannose, galactose, xylose, ribose, and fructose reacting with proteins. However these monosaccharides are rarely present in plasma in significant quantities.

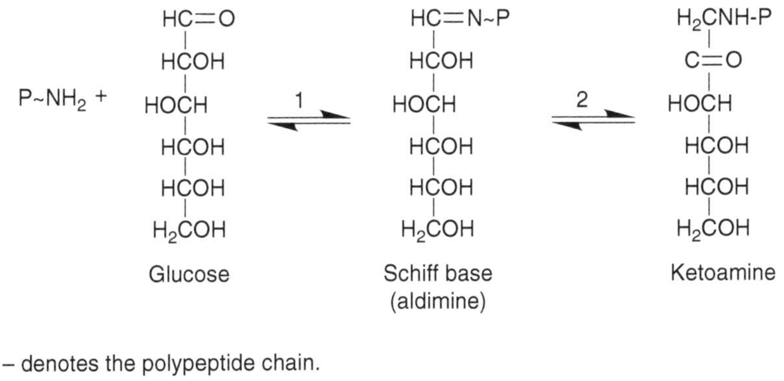

P – denotes the polypeptide chain.

Figure 7.2 Reaction of glucose aldehyde with the free amino groups of hemoglobin to form a Schiff base and then via an Amadori rearrangement a ketoamine (glycated hemoglobin).

the erythrocyte. A blood specimen is composed of erythrocyte populations of varying degrees and exposures to glucose. Thus an increase in plasma glucose levels just prior to blood sampling does not significantly affect HbA_1 levels. However, in long-term hyperglycemia, HbA_1 levels are significantly elevated proportionately to the average glucose concentration. A sudden increase in glucose increases HbA_1 levels gradually until a maximum is reached (i.e., until the maximum possible erythrocyte population has been exposed to the elevated glucose levels).

HbA_1 is separable into three fractions, namely; HbA_{1a}, HbA_{1b}, and HbA_{1c}. Although all three HbA_1 components are elevated in hyperglycemia, HbA_{1c} is affected to the greatest extent. In nondiabetic persons HbA_{1a} and HbA_{1b} together comprise 1–2% of the total hemoglobin and show an increase to 2–3% in diabetics. HbA_{1c} normally composes 3.3–3.5% of the total hemoglobin and increases to 10% and higher in diabetics.

Each fraction consists of two subfractions, the labile (the aldimine) fraction and the stabile (the ketoamine) fraction. The labile fraction reflects recent plasma levels and the stabile fraction represents long term blood glucose levels. The labile fraction is removed by incubation of washed erythrocytes in saline for 6 h at 37 °C.

SPECIMEN COLLECTION

Specimen collection and handling are crucial to good laboratory testing. There are three types of blood specimens commonly used in the clinical laboratory, whole blood, serum, and plasma. Whole blood is used as drawn and contains clotting proteins and formed elements, erythrocytes and leukocytes, white blood cells, and platelets. The plasma is the liquid over the formed elements after centrifuging whole blood, and serum is the liquid above the formed elements after permitting the blood to clot and then centrifuging. Therefore, the plasma contains the soluble clotting elements; the serum does not. Delays in the separation of the clot from serum and formed elements from plasma result in bacterial contamination, producing inaccurate results. In the absence of appropriate preservatives, uncentrifuged blood shows a rapid decrease in glucose levels that can be as much as 75% per hour at room temperature. The principal reason for this is glycolysis by erythrocytes and leukocytes. When clotting is permitted and followed by centrifugation, glucose levels in the serum remain stable for two days if the specimen is kept refrigerated at 4–8 °C. Bacterial contamination is the primarily cause for decreases in glucose levels after the serum has been separated from the clot.

The most effective approach to the glycolysis problem is rapid separation of the components of the blood from serum or plasma followed by refrigeration at 4 °C to guard against contamination. If the whole blood cannot be immediately processed, preservatives must be used. Fluoride or iodoacetate are common antiglycolytic reagents. Fluoride ion at 4–10 mg/ml of blood is effective in inhibiting glycolysis. This is accomplished by fluoride inhibition of the glycolytic enzyme, aldolase. Fluoride also acts as an anticoagulant by binding with calcium, an ion of major importance in the clotting process. However, the anticoagulation effect of fluoride is weak and standing for more than an hour or two will result in a clot. At concentrations higher than 10 mg/ml one runs the risk of inhibiting glucose oxidase. Fluoride should not be used as a preservative for glucose when enzyme assays are to be assayed from the same specimen.

When plasma or whole blood is used for glucose assay, anticoagulants such as oxalate, ethylenediaminetetraacetic acid (EDTA), or heparin is used. Oxalate (1–2 mg/ml), citrate (4 mg/ml), and EDTA (1–2 mg/ml) chelate calcium and thereby inhibit coagulation. Oxalate and fluoride are commonly used together to prevent glycolysis and coagulation. It should be noted that sodium, potassium, lithium, and ammonium salts of the anticoagulant are generally used and therefore assaying for that cation is eliminated for that specimen.

Summary Box 7.2

- Glycated hemoglobin is measured to determine average blood glucose levels during the previous 60–90 days.
- Glucose passes freely through the plasma membrane of erythrocytes and forms glycated hemoglobins via its formation of Schiff bases with free amino groups on the globin of residues of hemoglobin.
- There are three fractions of HbA_1, of which HbA_{1c}, is the principal fraction normally composing 3.3–3.5% of the total hemoglobin.
- Serum is the liquid found after whole blood has clotted and the clot centrifuged to the bottom of the blood drawing tube. The serum does not contain clotting proteins.

- Plasma is the liquid found after centrifuging whole blood to which an anticoagulant has been added. The formed elements, (white blood cells, erythrocytes, and platelets) are found at the bottom of the blood drawing tube and the clotting proteins remain in the liquid.
- Anticoagulants are oxalate, EDTA, heparin, and citrate.
- Fluoride and iodoacetate inhibit glycolysis.

EXERCISE

How many mEq/l does the anticoagulant sodium oxalate contribute to a sodium determination assuming the following:

(a) Sodium oxalate is present at 2 mg/ml.

(b) The molecular mass of sodium oxalate is 134; the atomic mass of sodium is 23.0?

Calculate the result in mg/ml and mEq/l of sodium.

How much does the anticoagulant K_3EDTA add to the results of plasma potassium assay assuming

(a) K_3EDTA concentration is 2 mg/ml of blood.

(b) The molecular weight of K_3EDTA is 406 and the atomic mass of potassium is 39.1?

Glucose values vary depending on where the blood is collected. For example, the glucose concentration of venous blood is lower than that of arterial blood by 10–20% due to its conversion to glycogen in muscle tissue. Capillary blood is approximately equal in glucose content to a simultaneously drawn arterial specimen. This is true because the blood pressure in the arteriole limbs of capillaries is considerably greater than the pressure in the venule limb and therefore the blood collected is primarily arteriole blood.

THE GOLD STANDARD

Biological testing can be compared to shooting at a bull's-eye. If several shots at the bull's-eye all go in the same location even though the shooter misses the center target, they have excellent reproducibility but poor accuracy. If the shooter occasionally hits the center but often misses the mark, those shots that hit the center are accurate, but they have poor reproducibility. If the shooter hits the center mark every time, they are both accurate and have excellent reproducibility—the person is quite a sharp shooter! This analogy applies to biological testing. We shall use glucose testing as an example. If one uses copper reduction methods to assay for glucose, one will not only be off target but also scattered in one's results. This is because this approach to testing for glucose is especially affected by interferents in the sample and the type and quantity of the interferents are different per sample. Each sample will produce reproducible results but comparing different samples having the same concentrations of glucose yields varying results; poor accuracy. If we use the GOD/peroxidase (POD) method the aim for the center might be better but samples that were not deproteinized will be scattered although they had the same glucose value. They would not be reproducible or accurate. Hexokinase would have more accurate results and better reproducibility because it is less affected by interfering substances. Finally, using a technique called *isotope-dilution mass spectroscopy* will hit the bull's-eye every time (reproducible) and is accurate every time. The isotope-dilution mass spectroscopy method will not be described in detail, but the author is certain that the method is both reproducible and accurate. The isolation-dilution mass spectroscopy method is the definitive method (or gold standard) for the assay for glucose. The GOD/POD method using deproteinization and the hexokinase method are neck in neck for being the *reference method*. o-Toluidine comes in a close third.

To summarize:

- The isotope-dilution mass spectroscopy method was right on target each time. However, this method is too complex and the instrumentation too expensive to be practicable in most laboratories. This would be the *definitive method or gold standard.*
- A methodology that gives the best value, has the least number of possible interferences, and is practical for application in the clinical laboratory is termed the *reference method.* "Practical" refers to ease of operation and the cost of the test.

INSTRUMENTATION

Parallel to the improvement in the methodologies for the assay of glucose was the development of analytical instruments to perform them. As the demand for glucose and other assays increased, the need for instruments that were capable of running large volumes of tests in the shortest time became a necessity.

In 1940 Beckman and colleagues at the National Technologies Laboratories built the first spectrophotometer from an amplifier from a pH meter, a glass prism, and a photocell (13). This instrument was later refined into the famous Beckman DU spectrophotometer. Arnold O. Beckman was born in 1900 in Cullom, Illinois, and earned his doctorate degree from California Institute of Technology. He founded National Technologies Laboratories while a member of the Cal Tech faculty. One of the designers of the Model DU was Howard Cary who later developed the Cary Instrument Company, which made some of the very best spectrophotometers.

Glucose samples were originally assayed separately but as the volume of glucose and many other tests increased there was a need for automated, large-throughput instruments. The Technicon autoanalyzer, using a technique termed *continuous flow analysis* was one such instrument invented in 1957 by Leonard Skeggs (1918–2002) and manufactured by the Technicon Corporation. Skeggs earned his doctorate from Case Western Reserve University in 1948. While working in a hospital laboratory he became aware of the length of time and unreliability of manual assays. He set up a workshop in his home and developed an automated chemistry analyzer, the prototype of the autoanalyzer. This autoanalyzer separated the samples after dilution by bubbles (segmented flow) that passed through tubing at rates of 2–3 ml/min. The samples were mixed with chemicals that were also passed along through tubing intercepting the sample in the appropriate sequence. The samples and chemicals were pushed through the tubing by a simple pump system. The sample or standard travels from the sampler, and then, in some cases, passes through a dialyzer where proteins are removed; it is next reacted with reagents in mixing coils. Depending on the methodology used the sample or standard may pass through a 37 °C heating bath. In most cases the sample terminates its path through the autoanalyzer by passing into a colorimeter. The colorimeter transmits its reading to a recorder.

The flow diagram for the determination of serum or urine glucose by a GOD method on an autoanalyzer I is shown in Figure 7.3 with commentary below (14). At the bottom right of the Figure a sampler with a 50 sample/h throughput is shown. The sample is diluted with saline and Tween (a surfactant) and separated from other samples by air bubbles. The sample is pumped into a dialyzer at 37 °C, and then is mixed with glucose oxidase/peroxidase and phenol and the

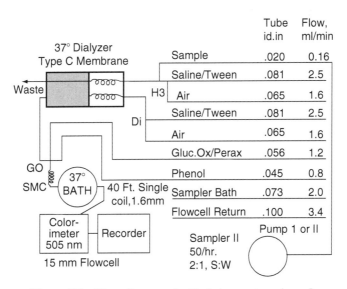

Figure 7.3 Flow diagram of a Technicon autoanalyzer I.

mixture next passes into a 37 °C bath. After the reaction, it is passed into the colorimeter where the sample is read at 505 nm. The colorimetric reading is electronically passed to the recorder where the deflection of the pen signifies the result. The diagram also indicates the internal diameter (id) of the tubing used which regulates the flow of the liquid; the larger id (in inches) the faster the flow (in ml/min).

The Technicon Sequential Multiple Analyzer (SMA) and the Technicon Sequential Multiple Analyzer, Computer (SMAC) were introduced in 1974. Now as many as 24 tests could be assayed from a single specimen. Some of the parameters were calculated (such as the BUN/creatinine ratio). Other instruments introduced soon after the Technicon instruments were the DuPont Automated Clinical Analyzer (ACA) in 1971. One problem with the Technicon autoanalyzer and SMA (and SMAC) was difficulty in assaying a stat.[6] The ACA did this easily. The ACA used packets containing sections for diluents and reagents and did away with the tubing that made Technicon instruments look like a virtual bowl of spaghetti that maximized the possibility for clogs. On the other hand, researchers appreciated the Technicon instruments for their adaption to new regimens for tests. The DuPont ACA also was more easily adaptable to small laboratories. Another high throughput instrument that invaded the scene was the American Monitor KDA. The KDA-automated chemical analyzer was brought on the market in 1977. Both the DuPont ACA and the American Monitor KDA reduced "carry over" from sample to sample. Carry over refers to the mixing of one specimen with the next or the mixing of the reacted specimen's color with the previous one, traveling behind it. Carry over was a major concern with the Technicon instruments.

Needless to say, these innovative instruments are no longer used; they have been replaced by more modern instruments. There are a multitude of high powered instruments being used today. A varied number of instruments have been designed and manufactured by such companies as Beckman, Roche, Siemens, and Johnson & Johnson. The new models utilize state-of-the-art photometric, luminescent, and nephelometric methodologies. Photometric refers to techniques that use several wavelengths in the visible. Luminescent methods, such as luminescent oxygen channeling immunochemistry (LOCI) utilizes sensitive photometers used to measure a very low light signal from a luminescent source. Nephelometric techniques measure the amount of light scattered by particles suspended in solution. To explain these techniques in detail would be to go far beyond the scope of this book. To learn more about these techniques, the reader may refer you to References (15, 16). Thus while in the past immunoassays were used exclusively for hormones, therapeutic drug monitoring (TDM), and drugs of abuse detection were measured on instruments other than those used for general tests such as glucose, protein, etc. Thus the new technology made available cardiac biomarkers, plasma

---

[6]A stat test refers to a test which must be conducted immediately (from the Latin "statim" which translates "immediately").

proteins, TDM panels, drugs of abuse, anemia, thyroid testing, electrolytes, and diabetes monitoring including HbA_{1c} all on one instrument. Modern-day instruments provide the following:

- Throughput; including instruments that analyze over 1500 total tests/h or better than 200 samples/h;
- Total number of analytes that can be tested; as many as 90–100 separate assays;
- Capacity to assay "stat" tests;
- Bar coding to identify the sample and determine which tests to perform on it;
- Approach to provide quality control;
- Error codes for insufficient sample volume, interferences such as hemolysis, bilirubinemia, and lipemia;
- Stability; no need for frequent standardization;
- Sample volume; microvolumes for each test;
- Reagent volume; microvolumes that are cost effective;
- Methods used; methods for each analyte should be a reference method;
- Upgradable to new tests;
- User-defined methods;
- Integration of all the results on a sample into one separate report;
- Little down time for quality control and malfunctions.

Summary Box 7.3

- The definitive or gold standard assay has absolute accuracy and reproducibility but may not be feasible for a clinical laboratory to routinely perform.
- The reference method gives the best value, has the least number of interferents and is practical for application in the clinical laboratory.
- Breakthroughs in the assay for glucose spurred inventors to discover instruments that would be rapid, necessitate the smallest volume of sample, and be able to accurately read the results of the assay.
- Thus each generation of instrument worked better and more efficiently than the past generation.

REFERENCES

1. Folin O, Malmos H. An improved form of Folin's micro method for blood sugar determinations. J Biol Chem 1929;83:115.

2. Hultman E. Rapid specific method for determination of aldosaccharides in body fluids. Nature 1959;183:108. DOI: 10.1038/183108a0.

3. Muller D. Studien uberein neus enzym glykoseoxydase I. Biochemische Zeitschrift 1928;199:136.

4. Van Bruggen JT, Reithel FJ, Gaby WL. Penicillin B: preparation, purification, and mode of action. J Biol Chem 1943;148:365.

5. Coulthard CE, Michaelis R, Short WF, et al. Notatin: an anti-bacterial glucose-aerodehydrogenase from *Penicillium notatum* Westling. Nature 1942;150:634. DOI: 10.1038/150634a0.

6. Free AH, Adams EC, Kercher ML, et al. Simple specific test for unrine glucose. Clin Chem 1956;3:163.

7. FH S. Die enzymatische Bestiimung von Glucose und Fructose nebeneinander. Klin Wochenschr 1961;39:1244. DOI: 10.1007/BF01506150.

8. Dods RF, Bolmey C. Glycosylated hemoglobin assay and oral glucose tolerance test compared for detection of diabetes mellitus. Clin Chem 1979;25:764.

9. Trivelli LA, Ranney HM, Lai NT. Hemoglobin components in patients with diabetes mellitus. N Engl J Med 1971;284:353.

10. Garlick RL, Mazer JS, Pi H, et al. Characterization of glycosylated hemoglobin. Relevance to monitoring of diabetic control and analysis of other proteins J Clin Invest 1983;71;1062. DOI: 10.1172/JCII 10856.

11. John GW, Gray MR, Bates DL, et al. Enzyme immunoassay-a new technique for estimating hemoglobin Ale. Clin Chem 1993;39:663.

12. Ellis G, Diamandia E, Glesbrecht EE, et al. An automated "high-pressure" liquid chromatographic assay for hemoglobin Ale. Clin Chem 1984;30:1746.

13. Simoni RD, Hill RL, Vaughn M, et al. A classic instrument: the Beckman DU spectrophotometer and its inventor, Arnold O. Beckman. J Biol Chem 2003; 278:79.

14. Lott JA, Turner K. Evaluation of Trinder's glucose oxidase method for measuring glucose in serum and urine. Clin Chem 1975;21:1754.

15. Pesce A. Spectral techniques, Chapter 2. In: Pesce A, Kaplan L, editors. *Clinical Chemistry: Theory, Analysis, Correlation*. 5th ed. St. Louis: Mosby/Elsevier; 2010. p 44.

16. Thompson SG. Principles for competitive-binding assays, Chapter 9. In: Pesce A, Kaplan L, editors. *Clinical Chemistry: Theory, Analysis, Correlation*. 5th ed. St. Louis: Mosby/Elsevier; 2010. p 180.

PART 3: SYMPTOMS AND TOOLS FOR THE DIAGNOSIS OF DIABETES MELLITUS

Correct is to recognize what diseases are and hence they come; which are long and which are short; which are mortal and which are not; which are in the process of changing into others; which are increasing and which are diminishing; which are major and which are minor; to treat the diseases that can be treated, but to recognize the ones that cannot be, and to know why they cannot be; by treating patients with the former, to give them the benefit of treatment as far as it is possible.

—Hippocrates, From "Hippocrates: Affections, Diseases 1. Diseases 2", translated by P. Potter, Loeb Classical Library No. 472, volume 113, p 113, 1988.

Having discussed the basis for the normal range, the truthfulness of a laboratory test and the best method for the assay of glucose in biological fluids, we will now use this knowledge for the diagnoses of diabetes mellitus.

THE SYMPTOMS OF DIABETES MELLITUS

A person who exhibits any or some of the following conditions may have some form of diabetes mellitus:

- Excessive thirst (polydipsia)
- Increased urinary output (polyuria)
- Frequent urinations
- Constant hunger (polyphagia)
- Unexplained sudden weight loss
- Tiredness much of the time
- Sudden changes in vision
- Tingling or numbness in the extremities
- Sores that do not or are slow to heal
- Infections
- Women with polycystic ovary syndrome[1]

INDIVIDUALS WHO SHOULD BE TESTED FOR DIABETES

Any person having one or more of the symptoms described above you should be tested for diabetes. When individuals visit their physician for a routine physical they often have blood drawn for a comprehensive screening panel, which includes fasting blood glucose (FBG). As of 2010, Medicare pays in full for a yearly "Wellness" visit. The American Diabetes Association and other organizations often have free or nominal charge clinics for determining the presence of diabetes. But unfortunately, many people do not have the insurance or funds for yearly checkups. It is difficult to imagine how large the total of prediabetes or overt diabetes that are missed and would be added to the figures presented in Chapter 1.

[1]Polycystic ovary syndrome is often found with insulin resistance and leads to T2D.

Diabetes testing should be considered in asymptomatic adults who have or are[2] the following (1):

- a BMI $\geq 25\,\text{kg/m}^2$ and have additional risk factors from the list that follows;
- morbid obesity;
- little or no physical activity;
- a sibling or offspring with diabetes;
- African American, Latino, Native American, Asian American, or Pacific Islander;
- given birth to an infant weighing greater than 9 lb;
- gestational diabetes;
- a resting blood pressure of $\geq 140/90\,\text{mmHg}$;
- HDL cholesterol levels of less than 35 mg/dl;
- triacylglycerol (triglyceride) levels of greater than 250 mg/dl (2.82 mmol/l);
- $HbA_{1c} \geq 5.7\%$;
- Been diagnosed as IGT or IFG previously;
- acanthosis nigricans.[3]

Diabetes testing should be considered in asymptomatic children who have or are the following (1):

- a BMI greater than 85th percentile for age and sex, a weight for height greater than 85th percentile or weight greater than 120% of ideal for height,

and any two of the following.

- first- or second-degree relative who has T2D;
- Native American, African American, Latino, Asian American, or Pacific Islander;
- insulin resistance or conditions associated with insulin resistance (acanthosis nigricans, hypertension, elevations in triglycerides, cholesterol, light density lipoprotein (LDL), and low HDL,[4] polycystic ovary syndrome, or small for gestational age birth weight).

Most of the above conditions are risk factors for T2D. T1D individuals usually are easier to diagnose. They have highly elevated blood glucose levels and exhibit many of the symptoms listed above soon after the onset of the disease. Those who are at risk for T1D have islet cell autoantibodies (ICAs) for the disease years before onset of the disease and thus screening could identify them.

---

[2] You may have come across some of these conditions in previous chapters of this book.

[3] Acanthosis nigricans is brown to black regions of the skin. The cause is insulin resistance, a characteristic of T2D.

[4] These conditions are collectively called dyslipemia.

However, screening large populations for autoantibodies is not recognized as cost effective and is only recommended for individuals who are at high risk for diabetes.

Having looked at the populations that should be tested for diabetes we will look next at the protocols used.

TOOLS FOR THE DIAGNOSIS OF DIABETES

Urinary Glucose

Urinary glucose levels are ineffective in detecting prediabetics or even certain cases of overt diabetes. Normal urinary volumes vary from about 1–2 l/24 h. Normal urinary glucose levels are 0–20 mg/dl (1.11 mmol/l) (2). The normal renal threshold for glucose is about 180 mg/dl (10 mmol/l) to 198 mg/dl (11 mmol/l) (3). In other words, blood glucose levels must exceed at least about 180–198 mg/dl before glucose appears in the urine in significant quantities. Patients with T2D have elevated and highly variable (between subjects) renal thresholds of glucose. T2D patients have been reported to have elevated renal thresholds [as high as 220 mg/dl (12.2 mmol/l)] (3). This effect defeats the efficacy of urinary glucose as a reliable test for the detection of diabetes.

Fasting Blood Glucose

FBG refers to testing for glucose in the blood after a fast. If glucose alone is being assayed then an 8-h fast is sufficient. If other tests are also assayed along with glucose such as cholesterol and triacylglycerols (triglycerides), a 12-h fast is required. During the fasting interval, no food or drinks other than water is permitted. Coffee and tea are usually prohibited. But if they are allowable, table sugar is not permitted. Sugarless gum is allowable. The normal range for FBG is usually 70 mg/dl (3.9 mmol/l) at the lower limit and is less than 100 mg/dl (5.5 mmol/l) at the upper limit. The upper limit[5] or cut point was reduced from 110 mg/dl (6.1 mmol/l) to 100 mg/dl (5.5 mmol/l) in 2003 by the International Expert Committee on the Diagnosis and Classification of Diabetes Mellitus (4). At the same time, the cut point for diabetes was lowered from ≥140 mg/dl (7.8 mmol/l) to ≥126 mg/dl (7.0 mmol/l).

It may be have been noticed that there is a gap between the upper limit of normal and the cut point for diabetes, that is, 100 mg/dl (5.5 mmol/l) and 126 mg/dl (7.0 mmol/l) respectively. It may be recalled from Chapter 6, individuals who fall into this range are diagnosed as having IFG, a prediabetic condition that frequently progresses to T2D.

---

[5]The upper or lower limits of a range are often called cut points.

Oral Glucose Tolerance Test

The OGTT involves a glucose load. The actual load originally was dependent on the patient's weight and height but has been changed to 75 g of glucose to make the test less cumbersome. The standard test conditions for the test follow (5):

- Minimum carbohydrate intake should be 150–200 g/day for 3 days prior to test.
- An 8–16 h fast
- Patient should be ambulatory
- Exercising strenuously for at least 8 h before the tested is not permitted
- Avoid testing when the patient has an illness
- Hormonal abnormalities such as thyroxine, growth hormone, cortisol, and catecholamines interference
- Medications including oral contraceptives, aspirin, nicotinic acid (from cigarettes, cigars, pipes, chewing tobacco, and patches), diuretics, and hypoglycemic agents must be avoided.
- Test between 7 AM and 12 noon.
- Seventy-five grams of glucose in about 300 ml of water is ingested within a 5-min time period. The glucose drink may be flavored with orange, lime, or cola (caffeine free).
- Blood samples are collected just before the ingestion of the glucose and at 2 h after ingestion. Previously, samples were taken before the ingestion of the glucose and at 1, 2, and 3 h after ingestion.
- Minimal risks are associated with this procedure. Adverse reactions are nausea, abdominal bloating, headache, and vomiting.
- This protocol is for individuals 12 years and older.
- Use this protocol for children less than 12 years with the exception that the glucose load is 1.75 g of glucose/kg body weight. Do not exceed a total of 75 g glucose.

Two-hour plasma glucose levels (2-h PG) $\geq$200 mg/dl (11.1 mmol/l) indicate a positive test for diabetes. Although considered the "gold standard" for diabetes testing, there is some criticism of the method. Many variables listed above for the test are difficult to control. The hardest variable to control is the individual's previous carbohydrate intake. Also age, presence of infection, time of testing, and physical activity influence results. Because of this, the reproducibility of the procedure is poor especially in borderline situations with the same individual giving different interpretations when retested. In general, the test tends to overdiagnose diabetes (6, 7).

Summary Box 7.1

- Urinary glucose is not used to diagnose diabetes because the renal threshold for glucose is reached when the blood glucose has reached 180–198 mg/dl (10–11 mmol/l).
- The upper limit for fasting blood glucose is 100 mg/dl (5.5 mmol/l).
- For fasting blood glucose the cut point for diabetes is $\geq$126 mg/dl) (7.0 mmol/l).
- For the OGTT 2-h PG between $\geq$140 (7.8 mmol/l) and 200 mg/dl (11.1 mmol/l) defines IGT.
- For OGTT $\geq$200 mg/l (11.1 mmol/l) defines diabetes.

HbA_{1c}

For many years the HbA_{1c} was used exclusively for monitoring the blood glucose levels of diabetic individuals over time. As you remember from Chapter 7, Part 2, HbA_1 is composed of at least three fractions, HbA_{1a}, HbA_{1b}, and HbA_{1c}. Of these HbA_{1c} has the greatest increase when blood glucose levels are above normal. Five methodologies have been reported for the measurement of HbA_{1c}: ion-exchange chromatography, isoelectric focusing, electrophoresis, high performance liquid chromatography (HPLC), and immunological approaches. Suggestions that HbA_{1c} could be utilized to diagnose diabetes came in the late 1970s (8–10). But one thing was lacking—no acceptable reference was available. The development of calibrators to allow standardization of HbA_{1c} methodology proved to be quite challenging. The development of a reference method was just as challenging. In 1997, researchers reported on procedures for the development of calibrators as well as two definitive methods (11). The calibration and standardization process that derived from this study is called the International Federation of Clinical Chemistry (IFCC) Reference Method although it seems more similar to a definitive or gold standard method.

Although a thorough description of the technology used for this standardization of the HbA_{1c} test is beyond the scope of this text, a summary of the methodology is summarized below:

- Calibrators
  - HbA_0[6] and HbA_{1c} were isolated from the blood of healthy, nondiabetic subjects using SP-Sepharose cation-exchange chromatography, affinity chromatography, and SP-Sepharose cation-exchange chromatography a second time. HbA_0 and HbA_{1c} are stored stabilized by KCN in a buffer.

---

[6]HbA_0 is a mixture of glycated and non-glycated hemoglobin but lacks HbA_{1a}, HbA_{1b}, and HbA_{1c}.

     ○ Calibrators in the range of $0-15\%$ HbA_{1c} relative to total Hb is composed from purified HbA_0 and HbA_{1c}.
- Reference[7] Procedures
  - Hemoglobin (Hb) is irreversibly glycated at the N terminus of one or both β chains. Therefore both reference procedures proceed as described in the following.
    - Blood specimens are washed, a hemolysate prepared and cell debris removed by centrifugation
    - An aliquot of the sample or calibrator is incubated with endoproteinase for 2 h at $37°C$ in order to cleave the N terminus of the β chain.
    - The samples or calibrators is assayed by reversed-phase HPLC[8] followed by electrospray ionization-mass spectrometry (ESI-MS)[9] (Method A) or
    - HPLC followed by capillary electrophoresis[10] (Method B)[11]

    IFCC accepted the above procedures for the calibration and standardization of the HbA_{1c} test. Methods A and B became the gold standards for the HbA_{1c} test. Methods commonly used in clinical laboratories for the measurement of HbA_{1c} such as ion-exchange chromatography, electrophoresis, affinity chromatography, and immunoassay now had the tools for standardization.

    The IFCC Reference Measurement System for HbA_1 established a network of 14 reference laboratories in the United States, Japan, and Sweden (12). This network of laboratories is called the IFCC Network of Reference Laboratories. Intercomparison studies using the calibrators described above have been reported by the IFCC (13). The studies showed excellent reproducibility and the calibrators were shown as stable for at least 2 years. HbA_{1c} is reported as the ratio of glycated to nonglycated fragments times 100 to provide a percentage. The results of the IFCC Network of Reference Laboratories demonstrated that the new calibration and standardization results were $1.5-2.0\%$ points lower than those that were current at that time.

    As everyday monitoring and therapy of diabetics has been based on glucose levels and new HbA_{1c} results are lower than unstandardized results, HbA_{1c} values have been translated into estimated average glucose values (eAG) (14). The cut point for the diagnosis of diabetes is 6.5%, which equals 140 mg/dl (7.8 mmol/l) eAG. This turns out to be closer to the old FBG cut point.

---

[7]These procedures are referred to by their authors as reference methods. Nonetheless, according to the definitions in Chapter 7, Part 2 they are also definitive methods.

[8]Reversed-phase HPLC separates the glycated and nonglycated N-terminal peptides of the β-chains from the peptide mixture.

[9]ESI-MS allows quantification of glycated and nonglycated fragments.

[10]CE allows quantification of glycated and nonglycated fragments by ultraviolet spectroscopy.

[11]For an explanation of HPLC, ESI-MS, and CE the reader is referred to p 89 and 113, Chapter 4, Lehrer M, and p 144, Chapter 6, Brewer JM. In: Kaplan L, Pesce A, editors. *Clinical Chemistry: Theory, Analysis, Correlation*. St. Louis: Mosby/Elsevier; 2010.

Advances in the standardization of the HbA_{1c} test has made this test one that is used globally for the detection of diabetes. It has the advantage of being insensitive to most of the factors that affect the results of the OGTT and FBG test. There is no need for fasting or any prior need for control of carbohydrate intake. The time of day the test is conducted makes no difference in results. HbA_{1c} levels are quite stable to mode of collection and storage. Depending on storage conditions, glucose levels significantly decrease after collection. This does not happen with HbA_{1c}. The only interferences reported are shortened erythrocyte survival time and the presence of fetal hemoglobin.

Summary Box 7.2

- The development of calibrators and definitive methods for HbA_{1c} has resulted in the test's use as a diagnostic tool for diabetes.
- The preparation needed for the fasting blood glucose and OGTT is not needed for the HbA_{1c}.

CUT POINTS FOR THE DIAGNOSIS OF DIABETES

At this point in the discussion, one may wonder how the cut points (limits) for diabetes were decided. Originally, the limits (lower and upper) for IFG and 2-h PG were as explained in Chapter 7, Part 2, by statistical analysis. The upper cut point for FBG was determined as the 95th percentile, the lower as the 5th percentile of the Gaussian curve of a large population of healthy individuals. As stated above, the upper limit was 110 mg/dl (6.1 mmol/l), the lower, 70 mg/dl. In the same way the 2-h PG gave a number of ≥140 mg/dl (7.8 mmol/l). Arbitrarily, the cut point for diabetes for FBG was decided as ≥140 mg/dl (7.8 mmol/l). The 2-h PG for diabetes was ≥140 mg/dl (7.8 mmol/l). FBG blood values falling between 110 mg/dl (6.1 mmol/l) and 140 mg/dl (7.8 mmol/l) and 2-h PG between 140 mg/dl (7.8 mmol/dl) and 200 mg/dl (11.1 mmol/l) were designated as prediabetes, IFG and IGT, and respectively.

A better way of determining the cut points is to link them to some early complication of diabetes. This was done in three studies in which retinopathy (an early complication of diabetes to be more fully discussed in Chapter 8) was detected by digital fundus photography and direct opthalmoscopy[12] and correlated with FBG, 2-h PG, and HbA_{1c}. The three populations were Pima Indians, Egyptians, and 40–70 year participants NHANES III. Figure 7.1 shows the results of these three studies (15). The cut point when the prevalence of retinopathy begins to increase linearly for FBG is approximately ≥126 mg/dl (7.0 mmol/l); for 2-h PG, this is ≥200 mg/dl (11.1 mmol/l); and for HbA_{1c} 6.5%.

---

[12]Digital fundus photography and direct ophthalmoscopy visualize small hemorrhages, arterial occlusions, and small clots called dots on the interior surface of the eyeball.

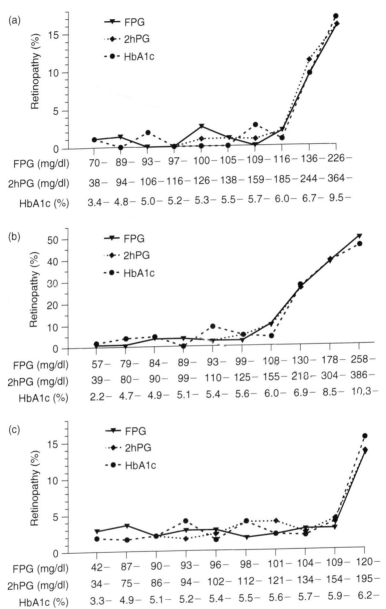

Figure 7.1 Prevalence of retinopathy by deciles of the distribution of FBG, 2-h PG and HbA₁c in three populations: (a) Pima Indians, (b) Egyptians, and (c) 40- to 74-year-old participants in NHANES III.

The upper cut point for FBG was lowered to ≥ 100 mg/dl (5.5 mmol/l) in order to include a greater percentage of the individuals who are diagnosed with IGT (prediabetes) to coincide with those diagnosed with IFT. Even with this change, not all of those diagnosed with IFT will also be diagnosed with IGT or vice versa.

DIAGNOSIS OF DIABETES USING FBG, 2-H PG, OR HBA$_{1C}$

The cut points for the diagnosis of diabetes for FBG, 2-h PG, and HbA$_{1c}$ are summarized in Table 7.1. The criteria for the diagnosis of diabetes using a single test are shown in Figure 7.2a. The criteria for the diagnosis of diabetes using two tests are shown in Figure 7.2b.

Summary Box 7.3

- The best approach for determining the upper limit of an assay is to link it to a physiological state of the disease it is directed.
- The upper limit for FBG is 100 mg/dl (5.5 mmol/l). The cutoff point for diabetes is 126 mg/dl (7.0 mmol/l). The values between 100 and 126 mg/dl (5.5 and 7.0 mmol/l) define IFT.
- The upper limit for OGTT 2-h PG of 140 mg/dl (7.8 mmol/l) defines IGT.
- The OGTT 2-h PG cut point of 200 mg/dl (11.1 mmol/l) defines diabetes.
- The upper limit for HbA$_{1c}$ is 6.5%.

DIAGNOSIS OF GESTATIONAL DIABETES MELLITUS

At present the diagnosis of GDM is in a state of flux. The International Association of Diabetes and Pregnancy Study Groups Consensus Panel (IADPSG) made recommendations for the diagnosis of GDM (16) based on a study of 25,505 pregnant women at 15 centers in nine countries. The Hyperglycemia and Adverse Pregnancy Outcome (HAPO) study (17), tested the population with a 75 g OGTT test at 24–32 weeks of gestation. Venipuncture occurred before the ingestion of the 75-g drink and at 2 h after the drink. Birth weight above the 90th

TABLE 7.1   FBG, 2-h PG and HbA$_{1c}$ Cut Points for the Diagnosis of Diabetes

Test	Criteria for the Diagnoses of Diabetes
FBG	≥ 126 mg/dl (7.0 mmol/l)
2-h PG	≥ 200 mg/dl (11.1 mmol/l)
HbA$_{1c}$	$\geq 6.5\%$

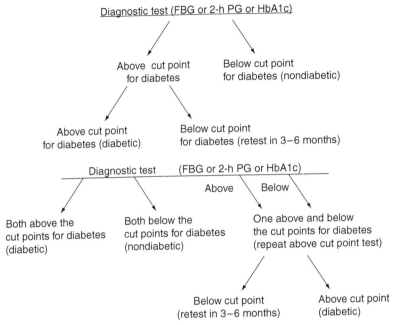

Figure 7.2 (a) Criteria for the diagnosis of diabetes using (a) a single test and (b) two tests.

percentile for gestational age (macrosomia), Cesarean section, neonatal hypoglycemia, cord-blood serum C-peptide level above the 90th percentile, delivery before 37 weeks of gestation, shoulder dystocia, birth injury, intensive neonatal care, hyperbilirubinemia, and preeclampsia were noted. The criteria for diabetes was FBG of greater than 105 mg/dl (5.8 mmol/l), 2-h PG of greater than 200 mg/dl (11.1 mmol/l), or any random PG of greater than 160 mg/l (8.9 mmol/l.

Maternal hyperglycemia leads to fetal hyperglycemia (called the *Pedersen hypothesis* (18)). The data from this study showed that increasing FBG, 1-h PG, and 2-h PG (from OGTT testing) was strongly associated with increasing numbers of neonates whose birth weight and cord serum C peptide were above the 90th percentile. In addition, there were associations between increasing FBG and Cesarean section, neonatal hypoglycemia, premature delivery, shoulder dystocia, birth injury, intensive neonatal care, hyperbilirubinemia, and preeclampsia.

C-peptide levels are in a ratio of 1/1 with insulin secreted from the pancreas. Thus increases in cord-blood C peptide levels suggest increases in fetal insulin levels. In turn increases in fetal insulin levels explain the excessive growth in fetuses of mothers that have GDM. Unlike retinopathy in T2D, none of the aforementioned complications had a point where they increased dramatically; rather, associations of the complications with the blood tests were on a continuum even in ranges considered normal.

The new recommendations for the diagnosis of GDM issued by the IADPSG and adopted by the American Diabetes Association (1) are as follows:

- Diabetes testing should be considered in women undiagnosed for T2D and having the risk factors listed above for asymptomatic adults. Testing should be conducted during their first prenatal visit.
- At 24–28 weeks of gestation, screen for GDM using 75 g, 2-h PG, using the cut points
  - FBG $\geq$ 92 mg/dl (5.1 mmol/l)
  - 1-h PG $\geq$ 180 mg/dl (10.0 mmol/l)
  - 2-h PG $\geq$ 153 mg/dl (8.5 mmol/l)
- GDM is diagnosed when any of the above criteria are exceeded.
- Women with persistent GDM should be screened 6–12 weeks postpartum.
- Women with a history of GDM should be screened every 3 years.

Ryan (19) has suggested that glucose is a weak predictor of macrosomia. In the HAPO study, the majority of macrosomic infants were born to women not exhibiting GDM. The new criteria would have to diagnose and treat 1702 additional cases of GDM in order to avoid 140 macrosomic neonates, 21 shoulder dystocias, and 16 cases of birth injury. Ryan states that a single 75-g OGTT lacks the reproducibility needed to diagnose for GDM. He also states maternal weight or weight gain during pregnancy is a more important factor in determining whether macrosomia and birth injury will occur during the pregnancy.

Summary Box 7.4

- The HAPO study showed that increasing values for FBG, 1-h PG, and 2-h PG was associated with neonate abnormalities such as Cesarean-section, shoulder dystocia, delivery prior to 37 weeks, neonatal hypoglycemia, hyper-bilirubinemia, and preeclampsia.
- The Pedersen hypothesis states that maternal hyperglycemia leads to fetal hyperglycemia.
- Thus, increased cord blood C-peptide suggests increased fetal insulin levels. Increased fetal insulin explains overweight infants.
- New recommendations for the diagnosis of GDM by the ADA have been criticized by Dr. Ryan as overdiagnosing GDM.

AUTOIMMUNE ANTIBODIES AS PREDICTORS FOR T1D AND LADA

As explained in Chapter 6, autoimmune antibodies play a significant role in T1D and LADA. At present, the four autoantibodies[13] that have diagnostic value

[13] Each autoantibody has an antigen that it is directed towards.

are ICAs, glutamic acid decarboxylase 65 antibodies (GADA65[14]), IAA, and insulinoma-antigen 2 antibody (protein tyrosine phosphatase antibody[15]) (IA-2A). Other autoantibodies have been found but either they have not been investigated to the degree that the four above have been or assays for them are not sensitive enough to permit accurate and reproducible results.

GADA is directed at GAD, which catalyzes the conversion of glutamic acid to γ-aminobutyric acid and IA-2A is directed to protein tyrosine phosphatase. ICA is assayed using indirect fluorescence microscopy. Immunofluorescence uses two antibodies. Patient sera containing ICA is incubated with a section of pancreas that has been sliced in a cryostat. A cryostat is an instrument that permits the thin slicing of frozen tissue samples. The β-islet cells present in the pancreas section act as the antigen. Thus the pancreatic β-islet cells bind any ICA present in the sera. Next, a fluorescein-labeled secondary antibody directed toward ICA is added. The pancreatic slice is applied to a microscope slide and the slide observed under a fluorescent microscope. If the β-cell islets fluoresce ICA is present (20).

GADA and IA-2A are assayed using immunoprecipitation. For GADA, the patient sample was incubated with biotinylated GAD and ^{35}S-methionine-labeled GAD in streptavidin-coated microtiter plates. We shall digress for a moment. As shown in Figure 7.3, antibodies (immunoglobulins) are tetramers composed of two types of polypeptide chains: the heavy chain (H-chain) and the light chain (L-chain) joined together by a disulfide bond. The H- and L-chains have amino acid regions called constant regions that determine the immunoglobulin class (IgG, IgA, IgM, IgD, and IgE). H- and L-chains also have regions called variable regions. These regions vary according to the immunoglobulin and the antigen it is directed towards. The immunoglobulin has a fork or Y appearance. At the tips of the prongs of the fork are hypervariable regions responsible for the recognition and binding of antigen. One may say that the antigen is speared by the prongs of the fork. Now we go back to the assay.

One set of prongs of the GADA binds biotin-GAD, the other ^{35}S-methionine-labeled GAD. The biotin-GAD becomes attached to the streptavidin, which coats the well. All other antibodies, unreacted substances, and so on are washed away and the ^{35}S is counted in a γ-counter (21).

IA-2A is assayed using a fluorimetric immunoprecipitation approach. The method is essentially the same used for GADA described above except that glutathione S-transferase (GST) was used instead of ^{35}S-methionine. The patient sample is incubated with biotinylated IA-2 and GST-IA-2A. Next, an anti-immunoglobulin is added to GST. This antibody has europium (Eu^{+3}) attached. On one set of prongs of the IA-2A is IA-2-biotin, which in turn is bound via biotin to streptavidin, which coats the well. The second set of prongs has IA-2-GST bound to Eu^{+3} antibody. The fluorescence is measured in a fluorometer (22).

---

[14]The 65 derives from the antigen's molar mass in daltons, 65,000 Da.

[15]IA-2A is directed towards a class of enzymes called protein tyrosine phosphatases.

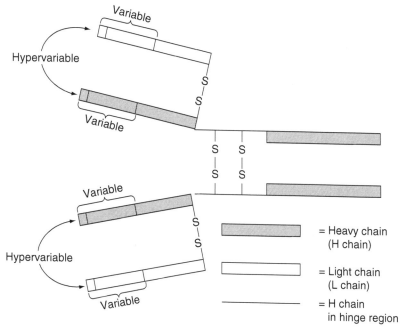

Figure 7.3 Immunoglobulin structure: variable regions.

Testing for IAA should not be conducted after the administration of insulin has begun. The development of antibodies to insulin often occurs soon after a regimen of injectable insulin has been started. IAA is assayed by a simple radioactive method. Patient serum is incubated with ^{125}I-labeled insulin. The IAA bound to the labeled insulin is precipitated with polyethylene glycol; the precipitate is centrifuged and the washed precipitate is counted in a γ-counter (23).

The pathophysiology for the appearance of islet cell antibodies is not known at present. It is possible that β-islet cell death is programmed into the DNA of those who develop T1D. An alternative theory is that an environmental agent, such as a virus, has a sequence of amino acids in the viral coat that is recognized by the immune system. The same sequence of amino acids is also present on the plasma membrane of the β-islet cell and is recognized by the antibodies produced by the immune system. This relationship is often called molecular mimicry. The antibodies not only bind to the causative agent but also to the β-islet cell. Thus with the immune system activated in this manner both the causative agent and the β-islet cell are destroyed. β-Islet cell proteins, glutamic acid decarboxylase, protein tyrosine phosphatases, and insulin are released from the destroyed or damaged β-islet cells. Antibodies are produced for each of these antigens.

Numerous studies on the prevalence of each autoantibody and the extent that they predict the occurrence of T1D have been reported. In a study (24) of three autoantibodies, IAA, GADA, and ICA512bdcAA[16] in 882 offspring or siblings of one or more T1D parents, 98% had one or more of the autoantibodies and 80% had two or more. Those with two or more of the autoantibodies had a risk of T1D within three years of 39%, and within 5 years, 68%. Those who had three autoantibodies had a risk of 100% within 5 years. A project aimed at determining whether autoantibodies appear sequentially or simultaneously before the onset of T1D suggests that the former is true (25). Although the number of autoantibodies that is elicited by an individual is highly predictive for the occurrence of T1D and there is evidence that autoantibodies appear in a sequential manner, the determination of the specific pattern has proven elusive.

A recent study (26) using data from the Diabetes Prevention Trial-Type 1 (DPT-1) of 29,035 individuals who were relatives of patients with T1D showed a prevalence of 3.8% for ICA, 2.0% for ICA512, 4.4% for GAD65, and 2.3% for IAA. The report also notes that the determination of more than one of the autoantibodies studied in an individual increased the risk for T1D. IAA was not considered a predictor of T1D. Also IAA did not add to the risk of T1D even if it was present with one or two of other autoantibodies. Another study from the same research group using data from the DPT-1 concludes that IA-2A increases and GADA decreases prior to progression to the diagnosis of T1D (27).

It may recalled from Chapter 6 that the presence of autoimmune antibodies is not only found in T1D patients but also LADA patients. It may also be recalled that LADA is a very slowly progressing disease. Patients are not initially insulin requiring, do not elicit autoantibodies, and are frequently typed as T2D. Since the destruction of β-islets is slow in LADA patients, they do not show symptoms such as polyuria, polydipsia, and weight loss. Although LADA is confined primarily to adults, there have been some cases found in children. These cases have been termed *latent autoimmune diabetes in the young*, LADY, and latent autoimmune diabetes in children, LADC. The United Kingdom Prospective Diabetics Study (28) measured GADA and ICA in 3672 patients age 25–65 years in 1997 and found that the proportion of patients exhibiting the autoantibodies, GADA and/or ICA decreased with increasing age. Ninety-four percent of the patients with ICA and 84% exhibiting GADA required insulin injections by 6 years compared with 14% of those who did not exhibit either of the autoantibodies. Another study (29) showed that the presence of GADA clustered in Finnish families with T1D or LADA. GADA was the principal predictor of diabetes. Other studies from Tasmania, New Zealand, and Sweden appearing around the same time as the English and Finnish studies concurred.

The appearance of GADA does not occur in the early stages of LADA. Therefore GADA does not distinguish LADA from T2D early in the disease. But

---

[16]One of the ICAs is ICA512. A more complete name for ICA512 is ICA512/IA-2A, an insulinoma-associated antigen-2. IA-2A has been identified as a type of tyrosine phosphatase-like protein. An isoform of ICA512/IA-2AA is ICA512bdc. ICA512bdc/IA-2AA is an antibody directed at amino acid residues 256–979 (the region that spans the interior section of IA-2). ICA512/IA-2A plays a role in pancreatic β-cell proliferation (32).

LADA is diagnosed by C-peptide measurements. As stated earlier, C-peptide secretion is a measure of insulin production. Early in T2D, C-peptide levels are elevated or high normal, while in LADA, they are low or low normal.

GADA and IA-2A have been reported (30) to differentiate between T1D and MODY (refer to Chapter 6). GAD and/or IA-2A were present in less than 1% patients with MODY and present in 82% of patients diagnosed with T1D.

In Chapter 5, we noted that β-cell volume and mass are decreased in T2D. An inflammatory process was postulated for islet cell death in T2D. The possibility that T2D could be an autoimmune disease was advanced by a recent publication (31). The investigators hypothesized that the death of adipocytes triggers an autoimmune response. The cell death is postulated to be caused by a surplus of adipocytes that outstrip the blood supply. The blood-deprived cells die. The cell death activates the immune system to respond to the resulting cell debris. A report describing a mechanism of this type was first reported in 2005. It describes T2D as an autoimmune-inflammatory disorder (32).

Summary Box 7.5

- ICAs, GADA65, IAA, and IA-2A are the principal autoimmune antibodies that are assayable.
- These autoimmune antibodies are assayed by indirect fluorescence microscopy, immunoprecipitation, and radioimmunoassay.
- The greater the number of autoimmune antibodies, the greater the risk of T1D.
- As LADA is a slowly progressing disease, in the initial stages, it is often mistaken as T2D.
- MODY is defined by its absence of autoimmune antibodies.
- Recent research suggests that T2D is also an autoimmune disease.

GLOSSARY

Acanthosis nigricans   Brown to black regions of the skin. The cause is insulin resistance, a characteristic of T2D.

Capillary electrophoresis   Separates charged substances based on their mass-to-charge ratio in a narrow diameter capillary. Allows quantification of glycated and nonglycated fragments by ultraviolet spectroscopy.

Digital fundus photography and direct ophthalmoscopy   Visualizes small hemorrhages, arterial occlusions, and small clots called *dots* on the interior surface of the eyeball.

Electrospray ionization-mass spectroscopy   Liquid containing substance(s) of interest are dispersed into a fine aerosol, ionized, and sprayed into a mass spectrometer for analysis. Allows quantification of glycated and nonglycated fragments.

Glucosylated A specific reference for the reaction of glucose with protein.

Glycosylation A more general term than glucosylated that refers to other monosaccharides such as mannose, galactose, xylose, ribose, and fructose reacting with proteins. However, these monosaccharides are rarely present in plasma in significant quantities.

HbA$_0$ A mixture of glycated and nonglycated hemoglobin but lacks HbA$_{1a}$, HbA$_{1b}$, and HbA$_{1c}$.

Reversed phase HPLC High performance liquid chromatography (HPLC) that utilizes a nonpolar stationary phase and thereby elutes polar compounds first while nonpolar compounds lagged behind. Separates the glycated and nonglycated N-terminal peptides of the β-chains from the peptide mixture.

Schiff base Forms when an amine reacts with an aldehyde. The compound formed has a bridge between the two molecules, $-NH=C-$.

Stat test Refers to a test that must be conducted immediately (from the Latin "statim," which translates "immediately").

TPNH Refers to reduced triphosphopyridine nucleotide an obsolete name for reduced nicotinamide adenine dinucleotide phosphate, NADPH.

REFERENCES

1. American Diabetes Association. Standards of medical care in diabetes—2011. Diabetes Care 2011;34:S11. DOI: 10.2337/dc11-S011.

2. Peterson JI, Young DS. Evaluation of the hexokinase/glucose-6-phosphate dehydrogenase method of determination of glucose in urine. Anal Biochem 1968;23:301. DOI: 10.1016/0003-2697(68)90361-8.

3. Rave K, Nosek L, Posner J, et al. Renal glucose excretion as a function of blood glucose concentration in subjects with type 2 diabetes-results of a Hyperglycaemic Glucose Clamp Study. Nephrol Dial Transpl 2006;21:2166. DOI: 10.1093/ndt/gfl175.

4. The Expert Committee on the Diagnosis of Diabetes Mellitus. Follow-up report on the diagnosis of diabetes mellitus. Diabetes Care 2003;26:3160.

5. Statistics Committe on Statistics of the American Diabetes Association. Standardization of the oral glucose tolerance test: Report of the committee on statistics of the American Diabetes Association 1968. Diabetes 1969;18:299.

6. Siperstein MD. The glucose tolerance test: a pitfall in hte diagnosis of diabetes mellitus. Adv Intern Med 1975;20:297.

7. Sherwin RS. Limitations of the oral glucose tolerance test in diagnosis of early diabetes. Primary Care 1977;4:255.

8. Dods RF, Bolmey C. Glycosylated hemoglobin assay and oral glucose tolerance test compared for detection of diabetes mellitus. Clin Chem 1979;25:764.

9. Koenig RJ, Peterson CM, Kilo C. Hemoglobin A1c as an indicator of the degree of glucose intolerance in diabetes. Diabetes 1976;25:230.

10. Santiago JV, Davis JE, Fisher F. Hemoglobin A1c levels in a diabetes detection program. J Clin Endocrinol Metab 1978;47:578.

11. Kobold U, Jeppsson J-O, Dulffer T. Candidate reference methods for hemoglobin A1c based on peptide mapping. Clin Chem 1997;43:1944.

12. Hoelzel W, Weykamp C, Jeppson J-O, et al. IFCC reference system for measurement of Hemoglobin A1c in human blood and the national standardization schemes in the United States, Japan, and Sweden: A method-comparison study. Clin Chem 2004;50:166. DOI: 10.1373/clinchem.2003.024802.

13. Weykamp C, John WG, Mosca A, et al. The IFCC reference measurement system for HbA1c: a 6-year progress report. Clin Chem 2008;54:240. DOI: 10.1373/clinchem .2007.097402.

14. Nathan DM, Kuenen J, Borg R, et al. Translating thre A1c assay into estimated average glucose values. Diabetes Care 2008;31:1473. DOI: 10.2337/dc08-0545.

15. The Expert Committee on the Diagnosis and Classification of Diabetes Mellitus. Report of the Expert Committee on the diagnosis and classification of diabetes mellitus. Diabetes Care 2003;26(Suppl 1):S5.

16. International Association of diabetes and Pregnancy Study Groups Consensus Panel. International Association of Diabetes and Pregnancy Study groups recommendations on the diagnosis and classification of hyperglycemia in pregnancy. Diabetes Care 2010;33:676. DOI: 10.2337/dc09-1848.

17. The HAPO Study Cooperative Research Group. Hyperglycemia and adverse pregnancy outcomes. N Engl J Med 2008;358:1991.

18. Pedersen J, Bojsen-Moller B, Poulsen H. Blood sugar in newborn infants of diabetic mothers. Acta Endocrinol 1954;15:33.

19. Ryan EA. Diagnosing gestational diabetes. Diabetologia 2011;54:480. DOI: 10.1007 /s00125-010-2005-4.

20. Winter WE, Schatz DA. Autoimmune markers in diabetes. Clin Chem 2011;57:168.

21. Hillman M, Torn C, Landin-Olsson M. Determination of glutamic acid decarboxylase antibodies (GADA) IgG subclasses-comparison of three immunoprecipitation assays (IPAs). Clin Exp Immunol 2007;150:68. DOI: 10.1111/j.1365-2249.2007.03473.x.

22. Westerlund-Karlsson A, Suonpaa K, Ankelo M, et al. Detection of autoantibodies to protein tyrosine phosphatase-like protein !A-2 with a novel time-resolved fluorimetric assay. Clin Chem 2003;49:916.

23. Palmer JP, Asplin CM, Clemons P, et al. Insulin antibodies in insulin-dependent diabetics before insulin treatment. Science 1983;222:1337.

24. Vedrge CF, Gianani R, Kawasaki E, et al. Prediction of type 1 diabetes in first-degree relatives using a combination of insulin, GAD, and ICA512bdc/IA-2 autoantibodies. Diabetes 1996;45:926. DOI: 10.2337/diabetes.45.7.926.

25. Yu L, Rewers M, Gianani R, et al. Antiislet autoantibodies usually develo sequentially rather than simultaneously. J Clin Endocrinol Metab 1996;81:4264. DOI: 10.1210 /jc81.12.4264.

26. Orban T, Sosenko JM, Cuthbertson D, et al. Pancreatic islet autoantibodies as predictors of type 1 diabetes in the diabetes prevention trial-type 1. Diabetes Care 2009;32:2269. DOI: 10.2337/dc09-0934.

27. Sosenko JM, Skyler JS, Palmer JP, et al. A longitudinal study of GAD65 and ICA512 autoantibodies during the progression to type 1 diabetes in Diabetes Prevention Trial-Type 1 (DPT-1) participants. Diabetes Care 2011;34:2435. DOI: 10.2337/dc11-0981.

28. Turner R, Stratton I, Horton V, et al. UKPDS 25: autoantibodies to islet cytoplasm and glutamic acid decarboxyulase for prediction of insulin requirement in type 2 diabetes. Lancet 1997;350:1288. DOI: 1016/50140-6736(97)03062-6.

29. Lundgren VM, Isomaa B, Lyssenko V, et al. GAD antibody positively predicts type 2 diabetes in an adult population. Diabetes 2010;59:416. DOI: 10.2337/db09-0747.

30. McDonald TJ, Colclough K, Brown R, et al. Islet autoantibodies can discriminate maturity-onset diabetes of the young (MODY) from type 1 diabetes. Diabetic Med 2011;28:1028. DOI: 10.1111/j.1464-5491.2011.03287.x.

31. Winer DA, Winer S, Shen L, et al. B cells promote insulin resistance through modulation of T cells and production of pathogenic IgG antibodies. Nat Med 2011;17:610. DOI: 10.1038/nm.2353.

32. De Souza CT, Araujo EP, Bordin S, et al. Consumption of a fat-rich diet activates a proinflammatory response and induces insulin resistance in the hypothalamus. Endocrinology 2005;146:4192.

CHAPTER 8

COMPLICATIONS OF DIABETES MELLITUS AND THEIR PATHOPHYSIOLOGY

Ibn Sina (Avicenna) (refer to Chapter 2) was the earliest scientist to describe complications arising from diabetes. In his 14-volume The Canon of Medicine *(known as the* Law of Medicine, *"Al-Qanun fi al-Tibb" in Arabic; as Law "Qanun" in Persian; and Canon of Medicine "Canon Medicinae" in Latin) he described "the collapse of sexual functions" and "neurological impairment of the bladder" as diabetic complications. Ibn Sina also described diabetic gangrene (1)*

In this chapter we shall discuss the complications of diabetes and their pathophysiology. It is not the hyperglycemia that makes diabetes mellitus a devastating disease; it is the complications that arise from diabetes that make it such. The following complications make diabetes a destructive disease to those who are suffering from it.

THE COMPLICATIONS OF DIABETES MELLITUS

Retinopathy and Other Eye Complications

As you may recall, Chapter 7, Part 3 discusses the sharp increase in the prevalence of retinopathy that begins at fasting blood glucose (FBG) levels of 126 mg/dl (7.0 mmol/l). This level defines the cut point between impaired glucose tolerance and overt diabetes.

Understanding Diabetes: A Biochemical Perspective, First Edition. Richard F. Dods.
© 2013 John Wiley & Sons, Inc. Published 2013 by John Wiley & Sons, Inc.

TABLE 8.1 Stages and Classes of Diabetic Retinopathy (DR)

Nonproliferative Retinopathy	
Mild	Small areas of swelling occur in capillaries of the retina (microaneurysms).
Moderate	Complete blockages occur in some capillaries.
Severe	Many blockages occur depriving parts of the retina for blood. Some of these microaneurysms can burst forming small, round hemorrhages (<100 µm) called *dot and blot hemorrhages*.
Proliferative Retinopathy	
Proliferative	The blood deprived areas of the retina secrete substances that promote the proliferation of blood vessels in the retina (angiogenesis). The neovascularization produces new vessels that bypass the blocked vessels and feed blood into the blood starved regions (termed *collateral blood supply*). These new blood vessels are fragile and leak blood thereby limiting eyesight.
Macula edema	In advanced stages of proliferative retinopathy plasma leaks out of the blood vessels into the macula region (part of the retina where straight-ahead vision occurs) and promotes swelling of the macula thus blurring the image. Macula edema is the most common cause of blindness in diabetics.
Detached retina	In some cases the new blood vessels tug on the retina and eventually pull the retina off of the surface of the choroid (the layer that the retina is attached to) resulting in a detached retina.

There are four stages of DR that fall into two classifications summarized in Table 8.1. As you can see from Table 8.1, DR is a complication that is caused by damaged small blood vessels of the eye, the capillaries, and is therefore termed a *microvascular complication*. Other diabetic complications that are microvascular are nephropathy (kidney disease) and neuropathy (nerve damage). On the other hand a macrovascular complication is said to occur when large blood vessels such as arteries and veins are involved such as in myocardial infarction (heart attack) and cerebrovascular accident (stroke).

An estimated 40–45% of people with diabetes over the age of 40 years have retinopathy. Approximately 8.2% of those with retinopathy have progressed to the vision threatening stage. If left untreated, half of those with proliferative retinopathy (see Fig. 8.1) would eventually become legally blind (2, 3). The Joslin 50-year Medalist Study investigated 351 patients who had been diagnosed with T1D for ≥ 50 years. The study determined that a high proportion of those diagnosed with diabetes were free from proliferative diabetic retinopathy (DR) (42.6%) (4). Intensive metabolic control (i.e., glucose and glycated hemoglobin kept at normal or near normal levels) reduces the incidence and progression of DR. In addition, medications that inhibit the renin–angiotensin system also reduce the incidence and progression of DR. The PPAR-α activator, fenofibrate, reduces the progression of DR by close to 40% among patients that have non-proliferative retinopathy.

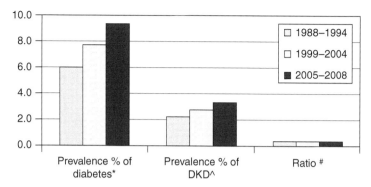

*% of US population with diabetes
^ % of diabetes diagnosed with DKD
# DKD population / Diabetes population

Figure 8.1 Prevalence (%) of diabetes and DKD and their ratios to each other. Prevalence (%) for diabetes and DKD was adjusted for age, sex, and race/ethnicity. (*See insert for color representation of the figure.*)

Another eye complication arising from diabetes is glaucoma. Glaucoma occurs when there is a buildup of pressure from the accumulation of fluid (aqueous humor) inside the eye. The accumulation of fluid is due to the slowing or blockage of fluid draining from the eye through channels called the *anterior chamber angle*. The pressure (called the *intraocular pressure*, IOP) can eventually reach a level where the optic nerve is damaged. Table 8.2 lists and defines the five principal classes of glaucoma. Two of these classes of glaucoma are primarily associated with diabetes. The use of vascular endothelial growth factor (VEGF)-neutralizing antibodies,[1] bevacizumab and ranibizumab, increases vision in patients with macular edema.

The third eye problem related to diabetes is cataract formation. A cataract is a clouding of the lens, a condition that occurs most frequently and earlier in diabetics. Twenty percent of all cataract procedures such as lens transplants are conducted on diabetics. Diabetics are 2–5 times at greater risk for the cataract development than nondiabetics and they occur more frequently at a young age, the risk increasing to 15–25 times in diabetics younger than 40 years (5). The lens opacity was proportionate to the degree of hyperglycemia (6).

The *Morbidity and Mortality Weekly* (MM) published by the CDC places retinopathy, glaucoma, and cataracts in the same data category called visual impairment (VI) and blindness. They reported that a National Health Interview Survey (NHIS) states that between 1997 and 2010 the age-adjusted percentage of adults diagnosed with diabetes who reported VI declined significantly from 23.7% to 16.7%. During this period age-adjusted prevalence declined among adult men, women, whites, Hispanics, and those aged ≥45 years. The actual population of

---

[1]VEGF is a protein that causes new blood vessel production (angiogenesis).

TABLE 8.2   Classes of Glaucoma

• Chronic (open-angled)	Most common
	Asymptomatic
	Cause unknown
	Inherited
	Afro-Americans at high risk
	Slowly progresses
	Associated with diabetes
• Neovascular	Rare
	New blood vessels derived from diabetic retinopathy proliferate on the iris cutting off fluid flow from the channels, increasing intraocular pressure and damaging the optic nerve.
	Presently no good means of treatment
• Acute (angle-closure) Also termed *acute narrow angle*	Occurs suddenly
	Sudden severe pain in one eye
	Cloudy vision with halo effect
	Rarely caused by diabetes
	Occurs in less than 10% of all glaucoma's
• Congenital	Unrelated to diabetes
• Secondary	Caused by drugs, trauma to eye but not diabetes

VI adults increased from 2.7 to 3.9 million. It should be noted that this was a self-reported study. It surveyed only civilian, noninstitutionalized respondents. Undiagnosed diabetics were not approximated and added to the data. Severity of VI was not assessed. The data were age-adjusted according to the 2000 total population. The number of patients with diabetes who visited optometrists and ophthalmologists remained constant through the period studied at 63%. This low percentage may have underrepresented the VI group. However the decrease may also be due to the large increase in new cases of diabetes during the 1990s that did not have a chance to develop VI. On the optimistic side the decreased frequency of VI may be due to better control of blood glucose. It is likely that all three of these statements are in operation.

With the current trends (refer to Chapter 1) for diabetes, the projection for DR is 16 million adults aged $\geq$40 years by 2050. VI DR would be projected as 3.4 million. The projection for individuals with cataracts is forecast as 9.9 million and those with glaucoma as 1.4 million (7).

Neuropathy and Related Conditions

Neuropathy is present in 60–70% of diabetic individuals. It is caused by reduced nerve conduction that manifests itself as impaired sensation, pain in the extremities, slowed gastric emptying, erectile dysfunction (as noted by Ibn Sina), decreases in blood pressure upon standing or sitting, urination problems (also noted by Ibn Sina), and weakness. The Joslin 50-Year Medalist Study described

TABLE 8.3   Classes of Diabetic Neuropathy

• Peripheral (distal)	Most Common
	In extremities: numbness, tingling, burning, pain, sensitivity to touch
	Loss of balance and coordination
• Autonomic	Control of blood pressure upon changes in position
	Delay in gastric emptying
	Erectile dysfunction
	Bladder function
	Unawareness of symptoms of hypoglycemia (<70 mg/dl, 3.9 mmol/l)
	Excessive sweating at night
• Proximal	More common in T2D
	Pain in hips, buttocks, thighs
	Weakness in legs
• Focal	Affects groups of nerves
	Muscle weakness and pain
	Eye problems: pain, inability to focus, double vision
	Back pain, pain in chest, stomach, pelvis, chest, abdomen, foot
	Bell's palsy

earlier found that 60.6% showed symptoms of neuropathy; 39.4% did not. The highest risk of neuropathy is for those who had diabetes for 25 years or longer. The classes of neuropathy and their symptoms are tabulated in Table 8.3 (8).The nerve conduction velocity test is used to determine the velocity that electrical signals move along a nerve. Electrodes are placed on the skin near the nerves being tested. An electrical impulse stimulates the nerve and the time that the impulse takes to travel to a receiver electrode is recorded as the impulse time. The impulse time is reduced for diabetics who have neuropathic complications.

Gastric Emptying (Gastroparesis)   Gastric emptying (time interval for food to leave the stomach and enter the duodenum) is delayed in diabetics. Delayed gastric emptying (gastroparesis) is found in 30–50% of diabetics. The mechanism underlying the delay is controversial but clearly concerns hyperglycemia and nerve damage (9, 10).

Sexual Complications   Damaged autonomic nerves as well as damaged blood vessels caused by diabetes can contribute to sexual dysfunction. The prevalence of erectile dysfunction from T2D diabetes is approximately 50% or about 10 million Americans. In well-controlled diabetics, it is about 30%[2] (11). Among women with T1D, 18–27% and among those with T2D, 42% are reported to have sexual dysfunctions. A sexual dysfunction is manifested by decreased vaginal secretion, painful intercourse, reduced desire, and decreased sexual response.

[2]Data from the 1994 Massachusetts Male Aging Study.

Urologic Complications Damage to nerves that control bladder function occurs in 50% of men and women with diabetes. The symptoms include overactive bladder and leakage of urine. In addition, nerve damage can lead to urine retention. Urine retention is the back up of urine in the bladder due to nerve damage that leads to the inability to empty the bladder when the bladder becomes full. Diabetics have recurrent urinary tract infections that may lead to cystitis (bladder infection) and pyelonephritis (kidney inflammation) (12).

Summary Box 8.1

- The fasting blood glucose levels where there is a sharp increase in the prevalence of retinopathy defines the limit between impaired glucose tolerance and overt diabetes. This value is 126 mg/dl.
- DR is due to damaged small blood vessels of the eye.
- Glaucoma occurs when there is pressure from accumulated fluid inside the eye.
- Cataract formation is a clouding of the lens.
- Neuropathy is nerve conduction that is slowed in diabetes and causes impaired sensation, extreme pain, slowed gastric emptying (gastroparesis), erectile dysfunction, decreased blood pressure on sitting or standing, urination problems, and weakness.

Nephropathy, Diabetic Kidney Disease (DKD), and End-Stage Renal Disease

Nephrology refers to kidney damage involving the capillaries surrounding the filtering apparatus of the nephron (the glomerulus). Diabetic nephropathy or diabetes kidney disease (DKD) as it is sometimes called is insidious and if left untreated progresses to end-stage renal disease (ESRD), which is life-threatening. Nearly 40% of diabetics develop nephropathy. Fifty percent of ESRD patients had nephropathy previous to developing ESRD. Biological markers for DKD are albuminuria and impaired glomerular filtration rates (GFRs). Urinary albumin is expressed as the fraction of albumin to creatinine and its cut point is 30 mg albumin/g creatinine. GFR is measured by using radioisotopes (such as ^{125}I-iothalamate). Its cut point is 60 ml/min/1.73 m^2. Usually DKD is diagnosed when either or both markers are abnormal.

An older method for the measurement of GFR is the creatinine clearance test. Creatinine is formed from creatine. Phosphocreatine is a high energy compound found in muscle tissues. As shown below, phosphocreatine provides energy for muscle contraction by supplying ATP. Creatine forms a five membered ring thereby producing creatinine.

$$Phosphocreatine + ADP \rightarrow creatine + ATP$$

$$NH_2$$
$$|$$
$$HN=C-N-(CH_3)CH_2COOH \longrightarrow$$

(structure: Creatinine)

Creatine   Creatinine

After this brief digression we will look more closely at creatinine clearance. Creatinine enters the blood and is cleared by glomerular filtration in the kidney. An insignificant amount of creatinine is reabsorbed by the tubules. Thus

$$\text{Creatinine clearance} = \frac{UV}{P} \ (\text{ml/min})$$

U = creatinine (urinary) expressed as mg/dl
P = creatinine (plasma) expressed as mg/dl
V = volume of urine in ml/min collected over 24 h.

Creatinine clearance approximates GFR. It usually overexpresses the value because there is a small amount of tubular secretion of the creatinine. The normal value in males is 85–125 ml/min and in females is 75–115 ml/min. Serum creatinine also is proportional to GFR and it alone has been used for assessment of glomerular filtration. The normal range for serum creatinine concentrations is 0.6–1.2 mg/dl.

Recently serum Cystatin C has been used as a biological marker for kidney function. Cystatin C is determined by immunoassay. Cystatin C accumulates in the blood as GFR decreases. The normal range for Cystatin C in women is 0.55–1.18 mg/l and 0.60–1.11 mg/l for men.

Utilizing the albumin excretion rate (AER), serum creatinine, and creatinine clearance test, the Diabetes Control and Complications Trial (DCCT) Research Group assessed the development and progression of DKD (13). They studied two groups of subjects, 726 and 715 subjects with T1D ages 13–39 years involved in the DCCT. At the start of the study the populations did not show any complications from their diabetes, had normal creatinine clearance and serum creatinine, and had normal blood pressure levels. The first group (726), called the *primary prevention cohort*, in addition to the above criteria, had T1D for 1–5 years, serum C-peptide less than 0.5 pmol/ml, no retinopathy, and AER of less than 28 μg/min. The second group (715), termed the *secondary intervention cohort*, had T1D 1–5 years, C-peptide less than 0.2 pmol/l, minimal to moderate retinopathy, and AER of less than 139 μg/min. The average follow-up for both populations was 6-and-half years with a range of 3–9 years. The participants in this study were divided between those receiving therapy designed to keep blood glucose levels at 81–101 mg/dl (4.5–6.0 mmol/l) (intensive therapy) and those receiving therapy designed to keep blood glucose below 180 mg/dl (10 mmol/l) (conventional therapy). The primary prevention group was divided between 378 subjects undergoing conventional treatment and 348 undergoing intensive treatment. The secondary intervention group was divided as 352 conventionally treated subjects and 363 intensively treated subjects.

The researchers determined that intensive therapy resulted in reduced development of microalbuminuria[3] and no significant changes in creatinine clearance, and concluded that intensive therapy prevents or delays nephropathy. A follow-up study also demonstrated the benefits of intensive treatment. Named the Epidemiology of Diabetes Interventions and Complications (EDIC) Study (14), it utilized serum creatinine, AER, blood pressure,[4] and HbA$_{1c}$. EDIC subjects were from the DCCT study described earlier and were studied for 8 more years. The conclusions from the study were that intensive treatment of T1D diabetics protects the patient from neuropathy and, if present, its progression.

A recent study found the prevalence of DKD to have increased from 2.2% in 1988–1994 to 2.8% in 1999–2004. In 2005–2008, it increased further to 3.3%. The data came from the NHANES. As you can see from Figure 8.1, the increases in DKD were proportional to the increased prevalence of diabetes. These increases occurred despite increased use by diabetics of hypoglycemic medications and renin-angiotensin-aldosterone inhibitors (angiotensin-converting enzyme inhibitors and angiotensin-receptor blockers) (blood pressure lowering drugs) (15). In this study albuminuria is defined as the ratio of albumin to creatinine and impaired glomerular filtration rate is defined by the Chronic Kidney Disease Epidemiology Collaboration formula (16). The cut point for albuminuria was ≥30 mg/g and that of impaired GFR was less than 60 ml/min/1.73 m^2. DKD (diabetic nephropathy) was defined as diabetes concomitant with either albuminuria or impaired GFR or both. Figures 8.2 and 8.3 show the prevalence of albuminuria and impaired GFR. Figure 8.2 shows a decrease in the prevalence of albuminuria in each time period for the entire population studied. When the

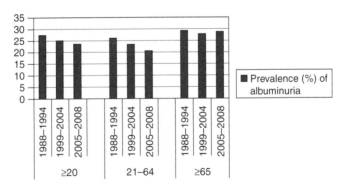

Figure 8.2 Prevalence (%) of albuminuria according to age. (*See insert for color representation of the figure.*)

---

[3]Albuminuria is defined as urinary albumin concentrations of ≥208 μg/min. Microalbuminuria is defined as urinary albumin concentrations of ≥28 μg/min.

[4]Why is blood pressure included? Hypertension damages the blood vessels surrounding the glomerulus and leads to kidney disease.

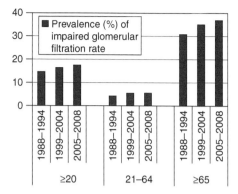

Figure 8.3   Prevalence (%) of impaired glomerular filtration rate. (*See insert for color representation of the figure.*)

data are broken down into two age brackets, the 21–64 year bracket shows a decline but the ≥65 does not. Figure 8.3 shows the prevalence of impaired GFR. Impaired GFR shows a consistent increase in prevalence for the total population as it proceeds to more recent years. When the data are broken down into 21–64 and ≥65 years brackets, the increase in prevalence persists in both categories as we go toward more recent years. These data are unadjusted for age, sex, and race/ethnicity. When adjusted for the listed demographic factors the differences become insignificant. Below normal GFR defines DKD. What is apparent from these data is that DKD can exist even without albuminuria.

As stated earlier, nephropathy or DKD often progresses to ESRD. ESRD requires dialysis or kidney transplantation; otherwise death ensues. In 2007 as many as 110,000 persons in the United States and Puerto Rico were newly diagnosed with ESRD. Forty-four percent of the new cases had been previously diagnosed with diabetes. Data from the US Renal Data System (USRDS) and the BRFSS indicate that since 1996, the age-adjusted incidence of diabetes caused ESRD had decreased by approximately 35%. However despite this optimistic statistic the total number of newly diagnosed diabetics increased to the extent that the increase in ESRD was offset. The decrease in the ratio of diabetic caused ESRD to the total diabetic population may reflect earlier and more potent treatment of kidney disease, more powerful drugs becoming widely prescribed such as the angiotensin-converting enzyme inhibitors and angiotensin-receptor blockers that delay and block progression of kidney disease, and better control of hyperglycemia (17).

Summary Box 8.2

- Damaged capillaries surrounding the glomerulus cause nephrology.
- Nephrology or DKD may proceed to end-stage renal disease which is life-threatening.

- Markers for DKD are albuminuria and abnormal GFRs.
- Creatinine clearance test is an older method for determining kidney function.
- Researchers of the Diabetes Control and Complications Trial Research Group determined that intensive therapy (which brings fasting blood values to near normal) prevents or delays nephropathy.
- Another study by the National Health and Nutrition Examination Study group showed that the prevalence of DKD continued to increase in proportion to the increase in the prevalence of diabetes despite the use of hypoglycemic and blood pressure lowering drugs.
- The incidence of end-stage renal disease has decreased due to more potent drugs that delay and block the progression of DKD.

Cardiovascular Disease (CVD), Hypertension, Coronary Heart Disease or Coronary Artery Disease (CHD), Cerebrovascular Accident (CVA), Pathophysiology of CVD: Endothelial Dysfunction

Cardiovascular Disease (CVD)   Cardiovascular disease (CVD) is a class of disorders that involves the circulatory system and includes hypertension, coronary heart disease (CHD), and cerebrovascular accident (CVA) (stroke).

A study published in 2007 used data from the Framingham Heart Study to determine the prevalence of CVD in diabetics. The study (18) used subjects aged 45–64 in two time periods, 1952–1974 and 1975–1998. The earlier population numbered 4418 nondiabetic subjects and 181 diabetics; the latter population numbered 4590 nondiabetic subjects and 351 diabetics. The subjects were primarily white and were not ethnically diverse. CVD events such as myocardial infarction (fatal and nonfatal), CHD death, CVA, transient ischemic attack, angina, coronary insufficiency, claudication, congestive heart failure, and recognized myocardial infarction were recorded. The conclusion was that despite a marked reduction in CVD morbidity and mortality in nondiabetic subjects that of diabetic subjects had significantly increased.

According to recent data released by the American Heart Association from 1998 to 2008 (19), the mortality from CVD declined 30.6%. Yet as of 2008 CVD accounted for one of every three deaths in the United States. CHD accounted for one of every six deaths, CVA one of every eighteen deaths, and heart failure accounted for one in nine deaths. The Framingham study data show that HR risk of CVD is increased 2.5 for women and 2.4 for men having diabetes. Additionally, death when CVD is diagnosed is increased for diabetics at the rate of 2.2 for women and 1.7 for men[5] (19).

---

[5]This is a hazard ratio (HR). HR is defined as the ratio of an event occurring in an experimental population divided by the event occurring in a control population. For example, persons with CVD and diabetes die at a rate per unit time of 2.2 persons per year, whereas those who have neither condition die at a rate of 1.1 persons per year. The HR is 2.

Hypertension (Atherosclerosis)   Hypertension (high blood pressure) often is a precursor to CHD and CVA. Hypertension defines the force exerted against the walls of your arteries as the blood is pumped by the heart to your body. Hypertension progresses to higher levels as atherosclerosis advances in the arteries. Atherosclerosis, commonly known as *hardening of the arteries*, and often called *arteriosclerosis*, develops from the formation of athermatous plaques on the walls of the arteries. The pathophysiology of atheroma formation will be discussed more thoroughly later in this chapter. Hypertension is found to coexist in 40–80% of diabetic individuals. The risk of cardiovascular events in diabetics is increased by coexistent hypertension (20).

CHD   CHD is the narrowing of the coronary (heart) blood vessels caused by atherosclerosis. It shares the same mechanism of causation as does hypertension, and therefore is frequently associated with hypertension. According to the National Institutes of Health it accounted for the death of 405,309 Americans in 2008. As the blood supply to the heart is decreased CHD manifests itself as chest pain, and tightness in the area of the heart. Known as *angina pectoris* the pain may spread to the jaw, throat, teeth, and shoulders and arms. This is an early manifestation of CHD. If untreated, angina may progress to complete blockage of a coronary blood vessel: a myocardial infarction (heart attack). Longitudinal data from the Framingham Study suggest that the RR for CHD in diabetics is about 1.38 and the risk for death is 1.86 higher for each 10-year period that the diabetes is present[6] (19).

CVA   There are two classes of CVA. Ischemic stroke is when a blood vessel in the brain has been blocked, usually a clot (thrombus and embolus). Hemorrhagic stroke occurs when a blood vessel ruptures and blood enters the brain. Hemorrhagic stroke is usually caused by hypertension. Ischemic stroke is the more common form of CVA. When the blood is blocked from reaching parts of the brain, the cells die (necrosis) and cannot ever be replaced. Although brain cells in other regions may take over the original function of the dead cells the function is not completely restored. If blood is quickly restored to the region damaged cells can repair themselves and restore some of the function. But if blood is not restored in a timely fashion that region of the brain permanently loses it function.

Diabetes significantly increases the chances of ischemic stroke. The RR has been reported to be in the range of 1.8–10.0. The Greater Cincinnati–Northern Kentucky Stroke Study (21) estimated stroke incidence and fatality in Caucasians and African Americans. The study found that diabetes was the principal risk factor for ischemic stroke. Diabetes risk of stroke was strongest in the younger than 65-year age group in whites and younger than 55-year age group in African

---

[6]This is a relative risk (RR) ratio. RR is the probability of developing a disease after being exposed to it for longer time periods. For example, RR is the probability that a person with diabetes will die of CHD the longer he has the diabetes. As contrasted with HR the RR ratios are cumulative over time and the HR represents instantaneous risk.

Americans. The presence of diabetes alone or with hypertension increased the risk of ischemic stroke. Twenty-five percent of ischemic strokes can be related to diabetes alone or with hypertension.

The bottom line of this survey was that diabetic subjects who had ischemic strokes were more likely to be African American, more likely to be hypertensive, younger, previously have had a heart attack and have high cholesterol.

Summary Box 8.3

- CVD includes hypertension, CHD, and CVA.
- The CVD hazard ratio is increased nearly two and a half fold for those with diabetes and the HR for mortality from CVD is increased 2.2 for women and 1.7 for men.
- Hypertension is a precursor to CHD and CVA.
- Forty to eighty percent of diabetics have hypertension.
- CHD is narrowing of the coronary blood vessels due to athersclerosis.
- A symptom of CHD is angina pectoris; chest pain spreading to the jaw, throat, teeth, shoulders, and arms.
- The RR for CHD is increased in diabetics about 1.38 and the risk of death is 1.86.
- There are two types of CVA: ischemic stroke when a blood vessel is blocked by a clot and a hemorrhagic stroke when a vessel bursts.
- RR for ischemic stroke is 1.8–10.0 in diabetics.

The Pathophysiology of CVD: Endothelium Dysfunction

To understand the role that diabetes plays in CHD, CVA, and hypertension we first must understand the biochemistry of the inner layer of blood vessels called the *endothelium* that serves as the interface between the blood and the blood vessel wall. First thought merely to maintain the permeability of the blood vessel wall, it is now recognized as the mediator of a great number of vascular functions.

As you may see from Figure 8.4 there are four layers encompassing the wall of a blood vessel. From outer layer to inner layer, first there is a protective layer, called the *tunica adventitia*, consisting of connective tissue including elastic fibers and in arteries small capillaries; next there is a layer consisting of smooth muscle called the *media* that has embedded nerves that control the diameter of the blood vessel; adjacent to the media resides an elastic layer called the *intima* that contracts and expands according to the heart beat, and finally the innermost layer, the epithelium, composed of a monolayer of endothelial cells. The functions of this last layer will be described later. The four layers composing the blood vessel wall surround the lumen, which transports in a plasma milieu the formed elements of the blood: leukocytes, erythrocytes, platelets, and various dissolved proteins such as albumin, immunoglobulins, enzymes, coagulation proteins, complement,

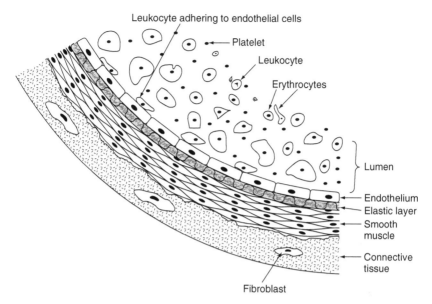

Figure 8.4   The endothelium.

and other dissolved elements such as electrolytes, nutrients, vitamins, lipids, and hormones.

As may be seen in Table 8.4 there are three principal functions of the endothelium (reviewed in Reference 22). First, it maintains vascular tone and permeability. At least three substances produced by endothelial cells are involved; endothelin-1 (ET-1), which causes vasoconstriction and nitric oxide (NO) and prostacyclin, which cause vasodilation. Second, it regulates the "stickiness" of leukocyte to leukocyte, platelet to platelet, and the "stickiness" of leukocyte and platelet to blood cell walls. NO and prostacyclin prevent adhesion, and secretin increases this tendency. And third, it regulates the coagulation cascade and fibrinolysis. Heparan sulfate (HS) and glycoaminoglycans inhibit coagulation. Tissue factor (TF) activates coagulation. Tissue-type plasminogen activates the fibrinolytic pathway, and plasminogen activator inhibitor-1 inhibits fibrinolysis. Thrombin activatable fibrinolysis inhibitor, a compound born of the complex from the binding of thrombin to thrombomodulin, inhibits fibrinolysis.

As long as there is no trauma, wound, microorganism attack, or other insults or damage to the blood vessel, coagulation and fibrinolysis remain in equilibrium. But if damage does occur to the blood vessel the endothelium releases substances that swing the balance over toward coagulation, constriction, and inflammation.

Injuries to the endothelial layer of capillaries results in the small blood clots that are the earliest signs of retinopathy. The glomerulus is composed of a network of capillaries. The endothelial cells of the glomerulus contain numerous pores with a diameter slightly smaller than the diameter of the protein, albumin. Thus molecules equal or greater than albumin in diameter are effectively blocked from

TABLE 8.4   Functions of the Endothelium

Vascular hemodynamics

- The endothelium (E) regulates the degree of contraction of the smooth muscle.
- The E determines the permeability of the vessel wall to nutrients and hormones, etc.
- Release of endothelin-1 (ET-1) from granules in endothelial cells (ECs) constricts smooth muscle.
- EC production of prostacyclin results in vasodilation.
- The EC produces NO (formerly called *endothelium-derived relaxing factor*) which functions as a vasodilator.

Adhesion and aggregation of leukocytes and platelets to each other and to walls of blood vessels

- EC production of NO regulates leukocyte adhesion to the inner wall of the blood vessel and aggregation (leukocyte–leukocyte) and thus curbs inflammation.
- NO also inhibits leukocyte rolling.*
- Selectin secreted by the EC makes endothelium cell membranes sticky to leukocytes. Selectin also promotes leukocyte tight adhesion to the cell wall and eventual migration into the smooth muscle layer.
- NO counters the vascular leukocyte adhesion that is produced by selectin.
- E inhibits platelet adhesion and aggregation by producing prostacyclin and NO.
- The release of NO and prostacyclin prevents the platelets from adhering to the EC membrane.

The balance between coagulation and fibrinolysis

- The E limits the activation of the coagulation cascade.
- Heparan sulfate (HS) and glycoaminoglycans (GAGs) promote the activity of antithrombin III (AT-III). Thrombin is an integral component of the coagulation cascade. Both HS and GAGs are located in the E. Coagulation is a very elegant and beautiful pathway and substances found in the E serve to regulate it.
- But if there is a Ying there must be a Yang. TF activates factors X and IX of the coagulation pathway. TF is found in the E and appears in the blood only after damage has occurred to blood vessels.
- The E releases tissue-type plasminogen (t-PA) and plasminogen activator inhibitor-1 (PAI-1) and thus regulates fibrinolysis.
- T-PA binds to PA receptors found on EC. t-PA binding to the receptor site promotes fibrinolytic activity.
- Thrombomodulin binds to thrombin to form a compound called *thrombin activatable fibrinolysis inhibitor* (TAFI) that affects fibrin so that t-PA binding sites on the fibrin are lost. Fibrinolysis is thus impeded.

---

*Leukocyte rolling describes the momentary adherence to endothelium cells and release and rolling on until the next endothelium cell captures the leukocyte. A leukocyte that is rolling travels at a velocity less than that of a leukocyte freely suspended in the plasma of the lumen.

passing through into the developing urine. Injuries to the endothelial cells lead to albuminuria and abnormal GFR, the earliest symptoms of nephropathy. These conditions largely isolated to small blood vessels are called *microvascular disease* (or *microangiopathy*). For large blood vessels the term used is *macrovascular disease* (or macroangiopathy). In most cases of CHD and CVA, both micro- and macrovascular diseases are involved.

Shear Stress   The force acting on the artery wall is composed of two components. One is directed perpendicular to the vessel wall and is the blood pressure. The second force is parallel to the vessel wall and is called the *shear force*. The shear force can be considered for use in computational models as the friction of the flowing blood against the cell wall and is expressed in Newton per meter squared or dyne per centimeter squared. The shear force places a stress on the surfaces of the endothelial cells. This stress is exerted on the surfaces of the endothelial cells and transmitted throughout the cells. Increased shear stress occurs concomitantly with increased blood pressure. Other structures that may change with increased shear stress are attachment sites to the surface of the cell, which may consist of receptor sites for enzymes, hormones, immunoglobulins, etc. A comprehensive review of mechanical forces derived from blood flow and their effects on endothelial cells appears in Reference 23.

An indication that the endothelium has been damaged is suggested when large numbers of monocytes are found adhering to the endothelium. Endothelial adherence of monocytes has been found in experiments on diabetic animals. In other experiments animals bred to have both hypertension and elevated cholesterol levels showed a significant increase of atherogenic lesions found in the abdominal aorta.

More direct experiments showed that the endothelium was involved in the atherosclerosis process. As stated earlier NO is a vasodilator. It is synthesized by endothelial nitric oxide synthase, which is activated by acetylcholine. When atherosclerosis narrowed, segments of coronary arteries were exposed to acetylcholine, a potent vasodilator; they constricted while normal segments dilated. The conclusion was that the narrowed sections lacked the synthase activity that normal sections possessed.

Experiments (24) on mice suggest that a lipid, palmitic acid attaches to nitric oxide synthase, allowing the enzyme to attach to the endothelial cell membrane. Decreases in the lipid by eliminating the enzyme that produces it, fatty acid synthase, result in decreased nitric oxide synthase activity. These mice, which were endothelial fatty acid synthase inactivated, also manifested increased membrane permeability, increased endothelial inflammatory markers, increased leukocyte migration, and were susceptible to endotoxin-induced death. These effects were reversed by NO. Thus the investigators concluded that the availability of NO identifies a system that may contribute to diabetic vascular disease.

Atherosclerosis has been linked with hyperlipidemia and abnormal lipoproteins for many years. As evidence of this association macrophages overloaded with lipid and therefore called *foam cells* have been identified as composing an atheroma. Foam cells can infiltrate the arterial wall where they form a yellow–white streak called a *fatty-like streak*. The fatty-like streak is the initial stage of an atheroma. A mature atheroma is a bulge in an artery wall. The atheroma contains foam cells, cholesterol, lipoproteins especially low-density lipoproteins (LDLs), fatty acids, monocytes, and other debris. In the early stages of atheroma formation, there is little danger but upon their growth and depending on their position in the circulatory system, a thrombus (a blood clot) can form leading to the blockage of the vessel or any other vessel, which the contents of the thrombus may be transported to, resulting in myocardial infarct (heart attack) or stroke. When a thrombus breaks loose from the blood cell wall and is relocated to another part of the circulatory system its name changes to embolus.

In recent years it has been demonstrated that atherosclerotic plaques develop at very specific regions of the circulatory system. These regions are at positions in the blood flow where low and oscillatory endothelial shear stress (ESS) occurs. Two definitions! Low ESS occurs when ESS is in one direction but has a varying magnitude thus averaging a low value, less than $10-12$ dyne/cm^2. Oscillatory ESS occurs when ESS is in both directions and the resulting magnitude is near 0. In contrast in straight sections of the artery ESS can be as high as 70 dyne/cm^2.

Low ESS causes the endothelial cells to change their structure and function thus promoting

- The formation of atheroma
- Development of athersclerosis
- Impaired production of NO
- Inflammation
- Development of thrombi (thrombogenecity)
- Arterial enlargement (called *expansive remodeling*)

For those interested in this topic and wishing to read further about it, the readers are referred to Reference 25.

Summary Box 8.4

- A blood vessel has four layers: connective tissue including elastic fibers, smooth muscle with embedded nerves, elastic layer, and a monolayer of endothelial cells.
- The endothelium maintains vascular tone and permeability, regulates the "stickiness" of leukocytes to leukocyte, platelet to platelet and leukocyte and platelet to blood cell walls, and regulates coagulation and fibrinolysis.
- The earliest signs of retinopathy and neuropathy result from injuries to the endothelial cells.
- The shear stress is made up of two forces: blood pressure directed perpendicular to the blood vessel and a force parallel to the blood vessel. The latter is called the *shear force.*
- NO produced by the endothelium is a vasodilator.
- Disruptions in the production of NO contribute to diabetic vascular disease.
- An atheroma is a bulge in the arterial wall.
- Foam cells, cholesterol, LDL, fatty acids, and monocytes are components of an atheroma, which can form a blood clot.

- Regions in blood vessels where endothelial shear stress (ESS) occurs produce atherosclerotic plaques.
- Low ESS produces formation of atheromas, athersclerosis, impaired production of NO, inflammation, production of thrombi, arterial enlargement.

Diabetic Ketoacidosis (DKA)

The reader may wish to refer to Chapters 3 and 4 for this section.

The ketone bodies, acetoacetate, β-hydroxybutyrate,[7] and acetone are normally found in low concentrations in the circulation. They are derived from a minor pathway shown in Figure 8.5, which becomes more prominent in uncontrolled T1D because of the scarcity of insulin. In T1D patients, insulinopenia or insulin resistance causes glucose starvation in cells (predominantly muscle and adipose cells), which are dependent on insulin for the transport of glucose. These cells respond to insulinopenia by mobilizing free fatty acids from triacylglycerols (triglycerides). The catabolism of free fatty acid produces acetyl CoA. Acetyl CoA enters both the tricarboxylic acid pathway and the pathway for the synthesis of ketones. In addition, without the suppression of insulin glucagon levels increase. Increased glucagon results in increased lipolysis and is an additional source of acetyl CoA.

Ketone bodies serve as a source of energy for the brain, kidneys, and skeletal muscle. Increased production of ketone bodies results in ketonemia and ketonuria. Acetoacetate and β-hydroxybutyrate are excreted by the kidneys with the concomitant loss of their counter ions, sodium, and potassium. During this process the hydrogen ion concentration increases resulting in an acidosis (blood

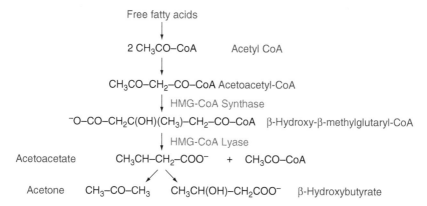

Figure 8.5 Synthesis of the ketone bodies. (*See insert for color representation of the figure.*)

[7]Although included among ketones, β-hydroxybutyrate is technically a carboxylic acid or fatty acid.

pH < 7.35). In diabetics a low pH coupled with the higher than normal levels of ketones is termed *diabetic ketoacidosis* (DKA). The definition of DKA is blood glucose greater than 250 mg/dl, blood bicarbonate less than 15 mEq/l, pH less than 7.35, ketonemia, and increased anion gap.[8]

Although ketoacidosis is predominantly found in T1D patients, DKA has been reported in some T2D patients (26). DKA accounts for 8–29% of all hospital diabetic admissions. Uncontrolled DKA can lead to coma and death. Coma is caused by acute cerebral edema, which occurs in about 1% of DKA patients. Mortality rates are 1–10% of DKA cases. DKA is treated with insulin. β-Hydroxybutyrate converts to acetoacetate as the pH moves closer to normal. The Centers for Disease Control and Prevention (27) determined that in 2005 the number of hospital discharges with DKA as the first-listed diagnosis was 120,000; whereas in 1985 the number was 84,000.

Hyperglycemic Hyperosmolar Non-Ketotic Syndrome

Hyperglycemic Hyperosmolar Non-Ketotic Coma (HHNC) Syndrome has been reported with increasing frequency. It is characterized by

- Hyperglycemia (>600 mg/dl)
- Normal ketone levels in blood and urine
- Some degree of acidosis but usually not as severe as DKA
- Hyperosmolarity (350 mOsm/kg H_2O)—normal is 289–308; indicates dehydration.
- Lethargy or coma
- Occurs primarily in T2D patients
- Brought on by illness, stroke, heart attack, surgery, or stressful events
- Potentially fatal.

Hospitalizations for HHNC Syndrome are less than 1% for diabetics. Approximately 30% of those hospitalized for HHNC Syndrome do not have a history of diabetes. It is hypothesized that delayed recognition of diabetic symptoms may lead to the dehydration, which is characteristic of HHNC Syndrome (28).

Hypoglycemia

A common complication of treated diabetes is hypoglycemia. Hypoglycemia occurs more frequently in patients who are being treated intensively. You may recall that intensive treatment is defined as that treatment, which keeps blood glucose levels close to the normal range. Hypoglycemia is defined as a blood glucose of less than 50 mg/dl (2.8 mmol/l). A study (29) encompassing 11,140 T2D subjects, ≥55 years, from 20 countries, with 5 years follow-up determined that 2.1%

---

[8]The anion gap is $[Na^+] + [K^+] - ([Cl^-] + [HCO_3^-])$

had at least one severe episode of hypoglycemia. The study divided the participants into two classes. One had treatment that gave intensive control of glucose levels. The second used treatment that gave standard control of glucose levels. The group that had intensive treatment had nearly two times greater number of episodes of severe hypoglycemia than the standard treated group (2.7% vs 1.5%). Furthermore 52.0% of the intensively controlled group, in contrast with 37.3% of the standard group reported a minor hypoglycemic episode. The conclusion is that severe hypoglycemia is more frequent when control is tight. The severe hypoglycemic episodes occurred primarily in older individuals with longer-term diabetes, lower BMI, higher creatinine levels, two or more hypoglycemic drugs, and a history of smoking or microvascular disease. The higher creatinine levels may suggest that the kidneys did not clear the drug efficiently. Refer to the section on nephropathy for another study that contrasted intensive control of blood glucose and standard control.

Infections

Diabetics have an increased frequency for infection. A study completed in 2001 consisting of 513,749 diabetic residents of Ontario, Canada, matched with an equal number of nondiabetics showed the frequency for infection to be as represented in Figure 8.6 (30). Nearly 46% of diabetics in this study had one hospitalization or outpatient visit for infection in contrast to 38% nondiabetics. Most of the nonhospitalized infections were for upper respiratory conditions. Increased death from upper respiratory infections is four times more frequent in diabetics suffering from pneumonia or influenza than nondiabetics. Next in prevalence were cystitis, pneumonia, cellulitis, and enteric infection. Infections requiring hospitalization were sepsis, postoperative infection, and biliary tree infections. In all cases except one, male genital infections, the risk ratio indicated the infection occurred more frequently in diabetics then nondiabetics.

A study (31) of 8,675 nondiabetic subjects and 533 diabetic subjects showed that the risk of mortality derived from infection was enhanced by coexisting cardiovascular disease. During 12–16 years of follow-up, 2103 deaths occurred in which 301 were related to infection. When compared to the nondiabetic population the age-adjusted risk for infection related death in diabetic women was 2.4 and diabetic men 1.7. Furthermore, the risk ratio for the deaths from infection of diabetics with a history of congestive heart failure was 3.2.

Amputations generally involving gangrenous extremities were necessary in 65,700 diabetic patients. This number (32) represents approximately 60% of nontraumatic lower-limb amputations in the United States in 2006.

Diabetics have more skin infections than nondiabetics. It appears from the above statistics that diabetics have defects in their immune defense systems that increase their likelihood of having more frequent infections than nondiabetics.

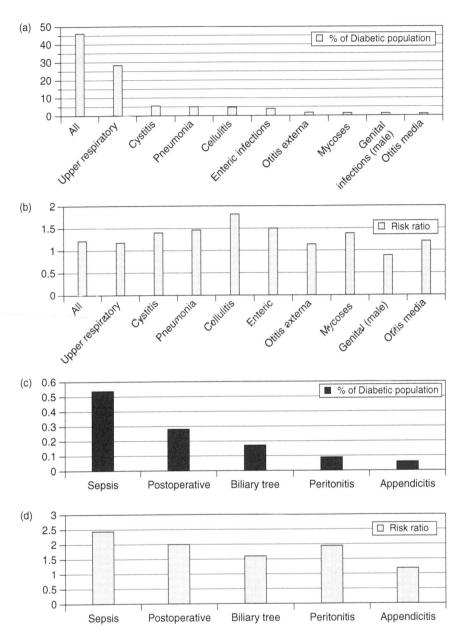

Figure 8.6 (a) Percentage of total diabetic population having specific infections treated as outpatients, (b) RR ratios for infections for diabetics treated as outpatients, (c) percentage of diabetic population having specific infections necessitating hospitalization, and (d) RR ratios for infections necessitating hospitalization. (*See insert for color representation of the figure.*)

Alzheimer's Disease or Alzheimer Disease (AD)

Alzheimer's Disease or Alzheimer (AD) has become a newly discovered complication of diabetes, but maybe not as new as suggested, because some Alzheimer patients were recognized as having disruptions in their glucose metabolism as early as 1988 if not earlier. In 1988 investigators speculated that glycolysis was disturbed at the pyruvate dehydrogenase level in early onset AD on the basis of a 44% decrease in brain metabolism of glucose with a fourfold increase in lactate production (33).

The primary sign of AD, a neurodegenerative condition, are dystrophic neuritis,[9] neurofibrillary tangles,[10] and neuritic amyloid plaques.[11] For the deposition of amyloid in beta cells and its subsequent effects, refer to Chapter 5. Recent studies suggest that diabetics have a 30–65% risk of developing AD. An investigation that relates insulin resistance to AD is derived from the Hisayama Study (34). The study utilized autopsies of 135 persons who had undergone a 2-h oral glucose tolerance test. On autopsy the presence of neuritic plaques and neurofibrillary tangles were looked for. The apolipoprotein E4 (ApoE4)[12] gene is associated with AD. In this study, AD was defined as the presence of neuritic amyloid plaques and/or neurofibrillary tangles. The study suggested that hyperinsulinemia and hyperglycemia were associated with the presence of neuritic plaques and the presence of the ApoE4 gene added to this relationship.

Another study (35) also utilizing data from the Hisayama Study was based on 1017 dementia free subjects. These subjects older than 60 years underwent a 2-h oral glucose tolerance test and were followed for 15 years. Of these, 232 developed AD, vascular dementia, and other forms of dementia; 150 of the subjects tested positive for diabetes; 49 (27%) of the subjects with diabetes developed dementia; 559 subjects remained free of diabetes. Of the diabetes-free population 115 (20.5%) developed dementia. The conclusion of the study was that diabetes was a risk factor for AD as well as vascular dementia[13] and other forms of dementia.

There have been several hypothesis developed for the association of AD with diabetes. One group of investigators is led by a neuroscientist, Suzanne M. de la Monte at Brown Medical School. She believes that AD is a manifestation of the impaired glucose metabolism present in diabetics that affects biochemical processes that involve memory and learning. She has demonstrated that the

---

[9]Dystrophic neurites are aggregated hyperphosphorylated tau proteins associated with amyloid plaques. Tau protein is a protein that normally stabilizes microtubules.

[10]Neurofibrillary tangles are also aggregated hyperphosphorylated tau proteins. They are found inside nerve cells. They consist of tau protein filaments wound about each other.

[11]Neuritic amyloid plaques are similar to neurofibrillary tangles but do not have the tangled appearance of them.

[12]ApoE4 is found in approximately 15% of the population. Individuals who inherit this gene from just one parent have a susceptibility for AD. The gene dictates the synthesis of ApoE4.

[13]Vascular dementia occurs after a stroke.

expression of insulin, insulin-like growth factor I and II, and their receptor genes are defective in the brains of AD; because of this, brain synthesis of insulin decreases. She believes that AD is a type of T2D and has labeled it as type 3 diabetes (36, 37).

The second group is led by Jeffrey M. Burns, a neurologist who directs the Alzheimer and Memory Center at the University of Kansas Medical Center. Burns and associates determined that glycogen synthase 3β, which phosphorylates tau protein, becomes exaggerated and produces the hyperphosphorylated tau proteins found in AD patients as neurofibrillary tangles and neuritic amyloid plaques. He determined that a decreased quantity of insulin in the brains of AD patients leads to brain atrophy and cognition dysfunction. Presently he is exploring the use of insulin nasal spray and its effects on AD. Insulin carried by nasal spray gets past the blood brain barrier and only enters the brain (38).

The third group is led by David R. Schubert, a neurobiologist at the Salk Institute for Biological Studies in La Jolla, California. Schubert and colleagues are studying the effects of hyperglycemia and glycation of proteins and its association with AD. Their studies showed that when endothelial cells were exposed to high levels of glucose and amyloid precursor protein there is an increased accumulation of glycated proteins. Furthermore, cognitive deficit in diabetic mutant mice was demonstrated when the mutant mice were made to produce amyloid precursor protein. Associated with the dull mice were increases in glycated proteins (39). Protein glycation will be discussed in detail later in this chapter.

The fourth group exploring this topic is that of neurobiologist, William L. Klein of the Department of Neurobiology and Physiology, Northwestern University in Evanston, Illinois. Klein and associates (40, 41) showed that signal transportation from insulin receptors located in neurons is inhibited by amyloid-β peptides (Aβ) organized in a polymeric package called an *amyloid β-oligomer*. Aβ are characteristic protein polymers found in AD and are commonly bound to neuronal synapses. They are also called *Aβ-derived diffusible ligands* (ADDLs). ADDLs caused a loss of neuronal surface insulin receptors especially on dendrites bound to ADDLs. The neuronal response to insulin was significantly reduced. This effect can be prevented by insulin, which blocks ADDL binding to synapses. Thus equilibrium is established between ADDL binding to the neuronal synapse and insulin binding to synaptical insulin receptor sites. Klein hypothesizes that with aging diabetes insulin signaling is decreased and AD becomes more likely.

Research on the association of diabetes with AD is only in its infancy; but it is an intriguing connection between diabetes and AD.

Diabetes and Cancer

Diabetics and obese persons have a higher risk of getting cancer. In fact T2D individuals who administer insulin injections or use medications that increase insulin and insulin-like growth factor I (IGF-I) and insulin-like growth factor II (IGF-II) have a higher frequency of cancer than those who have low insulin levels, and reduced IGF-I and IGF-II. The former group of subjects also has increased cancer mortality. IGF-I and IGF-II were discussed at length in Chapter 4. In summary IGF-I and II activate the PI3K/Akt pathway (discussed in Chapter 4) resulting in the transport of glucose into cells.

Cancer cells metabolize large quantities of glucose for their survival and especially for their multiplication. In the cancer cell the surplus glucose is converted to fatty acids and thence used to build plasma membranes, DNA, and proteins for new cancer cells. In fact T2D patients treated with metformin to lower insulin levels have a lowered incidence of the occurrence of cancer. The lower insulin levels reduce the uptake of glucose into the cancer cell. Metformin is an insulin-lowering drug that works through a mechanism to be discussed at length in Chapter 10. Cancer cells have two to three times more receptor sites for IGF than normal cells. Thus cancer cells are more responsive to IGF. In fact when IGF receptor sites are removed in mutant mice transplanted cancer cells grow and metastasize more slowly (42).

PI3K is associated with a tumor suppressor gene, phosphatase and tensin gene (PTEN). PTEN normally regulates PI3K, negatively, by dephosphorylating the 3 position of phosphoinositides (Fig. 8.7) Refer to Chapter 4 to review phosphatidyl

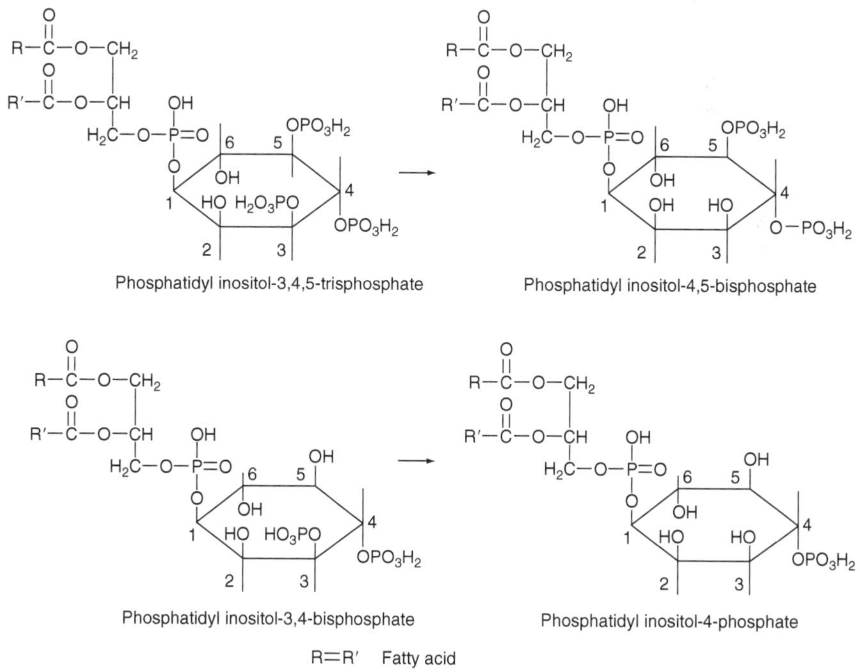

Phosphatidyl inositol-3,4,5-trisphosphate   Phosphatidyl inositol-4,5-bisphosphate

Phosphatidyl inositol-3,4-bisphosphate   Phosphatidyl inositol-4-phosphate

R=R′   Fatty acid

Figure 8.7 Dephosphorylation of phosphatidylinositol.

inositol-3,4-bisphosphate (PIP2), and phosphatidyl inositol-3,4,5-trisphosphate (PIP3). In many cancer types the PTEN gene is mutated and made inoperative usually by having stretches of it deleted. PTEN deletions occur in cancers such as prostate, endometrial, renal, small cell lung carcinoma, melanoma, meningioma, and glial cancers (43). The mutated PTEN gene is unable to inhibit the PI3K/Akt pathway. Thus insulin and IGF-I and II are increased and the above scenario occurs.

Summary Box 8.5

- Acetoacetate, β-hydroxybutyrate, and acetone are called the *ketone bodies* and are produced from a normally minor pathway.
- When produced in surplus the ketone bodies create an acidotic condition known as *DKA*.
- DKA occurs predominantly in T1D.
- Hospitalizations for HHNC syndrome occurs in less than 1% of diabetics.
- Hypoglycemia occurs frequently in diabetics who are treated intensively to keep their blood glucose levels near normal.
- Diabetics have an increased frequency of infection.
- Increased mortality for upper respiratory infections occurs more frequently in diabetics than for nondiabetic individuals.
- Sixty percent of nontraumatic lower limb amputations were in diabetics.
- As early as the 1980s Alzheimer Disease was linked with those suffering from diabetes.
- There have been at least four hypotheses to explain the diabetes link to AD.
- Diabetics and obese individuals have a higher risk of getting cancer.
- Mutations in the PTEN gene may explain the connection between cancer and diabetes.

PATHOPHYSIOLOGY OF DIABETIC COMPLICATIONS

Currently there are three different hypotheses that explain the underlying pathology of diabetic complications. They are glycation, sorbitol accumulation, and oxidative stress. We will discuss each in turn in the following discussions.

Glycation

Refer to Chapter 7, Part 2, for the glycation of hemoglobin. The globin of hemoglobin is not the only protein that is glycated by glucose. As with hemoglobin the degree of glycosylation (or glycation) of other proteins is in proportion to the glucose content of the blood during the life span of the protein. The carbonyl groups of glucose and other monosaccharides react with protein free amino groups such as the N-terminal amino group and internal lysine ε- amino groups. As shown in Figure 7.2, Part 2, this reaction produces an aldimine or Schiff base. The reaction is often called the *Maillard reaction.* The aldimine rearranges to a ketoamine (the Amadori rearrangement). Thus serum albumin has been studied as to its degree of glycosylation, and glycated albumin has been found to be proportional to blood glucose levels. However the half-life of albumin is only 15 days and thus glycated albumin cannot be used to determine long-term blood glucose concentrations.

Glycated proteins are involved in subsequent reactions such as dehydration, oxidation, and cross-linking with collagen. In addition to proteins they also derive from the formation of bonds between the aldehyde groups of various carbohydrates other than glucose and the free amino groups of lipids and nucleic acids. The many different compounds that are formed are called *advanced glycation end products* (AGEs). A receptor site for AGE has been characterized and is called *receptor for advanced glycation end products* (RAGEs). Several AGEs have been isolated and characterized.

Imidazolone has been found to be significantly increased in the red cells of diabetic patients. It is the result of the reaction between the guanidino group found in arginine and 3-deoxyglucosone (Refer to Fig. 8.8). It was also found in the mesangial matrix, the matrix between the capillaries and glomeruli in the kidneys, in diabetic nephropathy, and in the aortas of diabetics (44). Pentosidine is a compound that is derived from cross-linking of ribose with guanidino groups of arginine and the ε-amino groups of lysine. Skin pentosidine was elevated in T1D relative to nondiabetics (45). Urinary excretion of N^ε-carboxymethyl lysine (CML) is significantly increased in diabetic nephropathy (46). Derivatives of glyceraldehyde and dihydroxyacetone such as glyoxal and methylglyoxal also form AGEs (Fig. 8.9). Methylglyoxal is probably the principal AGE found in endothelial cells.

Activation of RAGE by AGE results in the production of cytokines, including TNFβ (refer to Chapter 5 for the THF group of cytokines), which results in abnormal production of the mesangial matrix. In subjects with high levels of AGE and hyperglycemia abnormal overproduction of the mesangial matrix leads to decreased kidney function (nephropathy). AGE deposition in arteries decreases arterial elasticity. High levels of RAGE have been associated with CHD in diabetics (47). AGE has also been implicated with several substances that promote the deterioration of the kidney and result in diabetic nephropathy. The substances elicited by AGE are cytokines,[14] chemokines,[15] growth factors, adhesion molecules, and oxidant stress substances (referred to later in this chapter). In diabetic animals investigators have found accumulations of AGE in the kidneys

---

[14]Cytokines are intercommunicative proteins secreted by cells.
[15]Chemokines are a class of cytokines that direct the movement of cells.

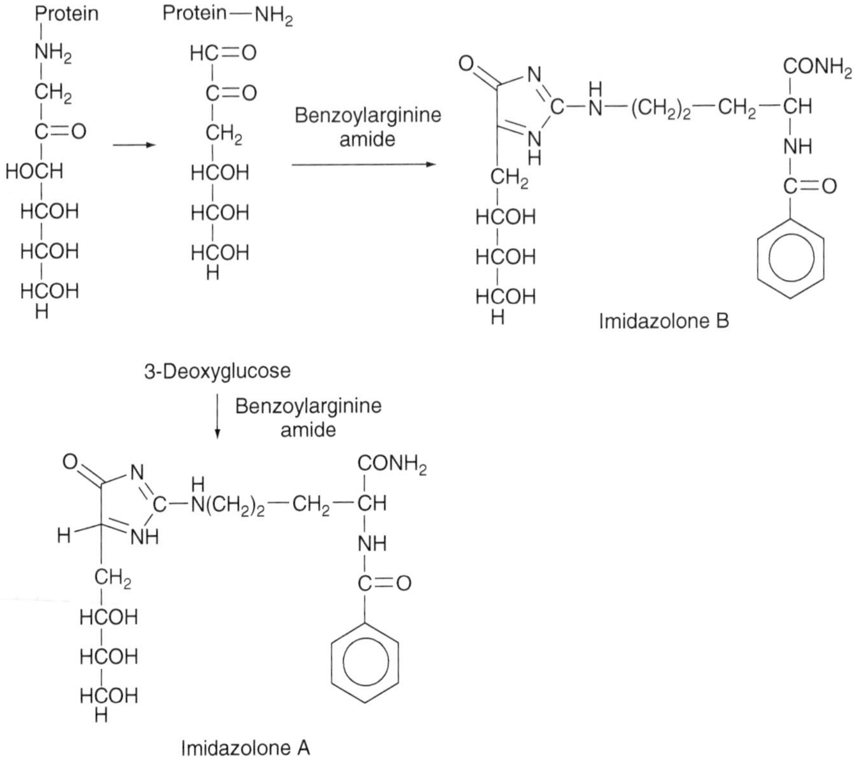

Figure 8.8 Synthesis of imidazolones.

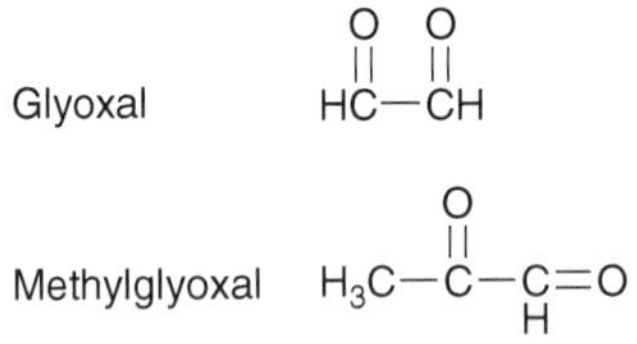

Figure 8.9 Structures of (a) glyoxal and (b) methylglyoxal.

associated with increased mesangial matrix, podocytes (cells that are wrapped around the capillaries of the glomerulus),[16] and renal tubular cells (48).

AGE induces the same substances described earlier for DR. AGE deposits on retinal blood vessels increases with the severity of the retinopathy. Glycation of

---

[16]Blood is filtered through spaces located between the processes (sometimes called *feet* and therefore the name podocytes). Large molecules do not filter through these spaces called *slit diaphragms*. Only small molecules, water, and electrolytes pass into the developing urine.

α-crystallin results in cataracts (49). Pentosidine, CML, and imidazolones were reported (50) in higher amounts in lenses with cataracts than in normal lenses.

In summary naturally occurring aldehydes such as glyceraldehyde, glucose, and glucose 6-phosphate can form Schiff bases with free amino groups of various proteins and then through a series of rearrangements, dehydrations, and oxidations produce AGEs. AGEs directly cause modifications of proteins, and indirectly bind to receptor sites for AGE called *RAGEs* to cause the release of chemical substances that modify the actions of nearby molecules. Inflammation may be promoted through the release of cytokines and growth factors, which may then promote vascular damage, nephropathy, and retinopathy.

Sorbitol Accumulation

The sorbitol pathway (51) is another pathway like the one that produces ketone bodies that is normally of minor significance. As you may see from Figure 8.10, glucose is reduced at the aldehyde position by aldolase reductase to sorbitol. NADPH is simultaneously oxidized to $NADP^+$. Next sorbitol is oxidized to fructose by sorbitol dehydrogenase. The cofactor in this reaction is NAD^+, which is reduced to NADH. You may recognize that the conversion of glucose to sorbitol is governed by the concentration of glucose, which is high in diabetics. Sorbitol does not diffuse through plasma membranes easily and when synthesized in excess it accumulates within the cell. Fructose does permeate the plasma membrane but when glucose levels are high sorbitol synthesis outstrips the conversion of sorbitol to fructose and sorbitol accumulates within the cell. Sorbitol and glucose intracellular accumulation results in osmotic swelling and injury to the cells affected. The cells affected are primarily nerve, lens, and glomerulus cells.

There are several other hypotheses with regard to the sorbitol pathway in diabetics. NADPH is a cofactor in this reaction. It is also a cofactor in the reduction of glutathione. When NADPH is used in excess in the sorbitol pathway, which is the case in diabetics, reduced glutathione becomes depleted. Reduced glutathione depletion leads to increased oxygen-free radicals in the cell. As we shall learn in the next section oxygen-free radicals are instrumental in cell damage. Furthermore NADH can pass from the cytosol into the mitochondria. In the mitochondria NADH is reoxidized to NAD^+-causing formation of superoxide radicals (52). Another result of the increase in the $NADH/NAD^+$ is the accumulation of triose phosphates, which leads to the formation of methylglyoxal and therefore AGEs.

Reactive Oxygen Species (ROS) in Diabetes

You may wish to review ROS discussed in Chapter 5. ROS refers to the generation of free oxygen radicals. They include superoxide anion ($\cdot O_2^-$), hydroxyl anion ($\cdot OH^-$), peroxyl ($\cdot RO_2$ where R is an alkyl or phenyl group), and hydrogen peroxide (H_2O_2). Free oxygen radical formation occurs to a greater extent in diabetics than in normal individuals. A free radical is an atom or group of atoms possessing an unpaired electron. Free radicals are very reactive as they seek to

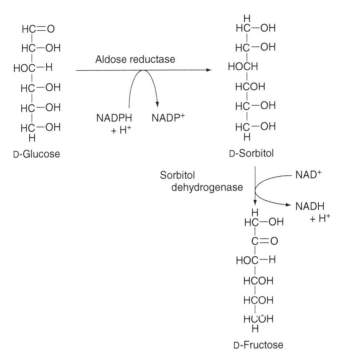

Figure 8.10 The sorbitol pathway.

pair their electron. $\cdot O_2^-$ is an ROS, which is formed when a single electron is added to O_2 and is the predominant free radical that will be addressed in the following discussion.

As you may remember from Chapter 3 when glucose is metabolized by the tricarboxylic acid cycle (TCA) via glycolysis, NADH and $FADH_2$ are generated. In the electron transport system, NADH donates electrons to Complex I and $FADH_2$ donates electrons to Complex II. In diabetics more glucose goes through the glycolytic and TCA pathways resulting in increased electron flow through Complex I and Complex II. A voltage threshold is reached at Complex III thus blocking the passage of electrons and instead resulting in a backup of electrons. The surplus electrons are diverted to the formation of $\cdot O_2^-$ (53). As shown later, superoxide dismutase catalyzes the conversion of the superoxide to hydrogen peroxide and then catalase, glutathione peroxidase, or peroxiredoxin converts hydrogen peroxide to oxygen and water. This pathway is a minor pathway in nondiabetics.

$$2\,\dot O_2^- + 4H^+ \xrightarrow{\text{Superoxide dismutase}} 2H_2O_2 \xrightarrow{\text{Catalase}} 2H_2O + O_2$$

The ROS pathway is an important signaling pathway for the secretion of insulin. Several investigators have provided evidence for this. In one study cultured mouse β-islets were stimulated to produce intracellular H_2O_2 from $\cdot O_2^-$

by the addition of H_2O_2 or diethyl maleate. The accumulation of H_2O_2 results in the stimulation of insulin secretion. Inhibitors to the synthesis of H_2O_2 result in reduced insulin secretion. Excessive ROS synthesis such as that occurs when H_2O_2 production is inhibited results in oxidation of macromolecules like proteins, DNA, and lipids. It is also hypothesized that $\cdot O_2^-$ accumulation disrupts the signal for insulin secretion. A defense against ROS damage is transcription factor NF-E2-related factor 2 (Nrf2), which protects cell components against oxidative damage (54).

The involvement of ROS in insulin secretion is further supported by experiments with cultured aortic endothelial cells, which were subjected to a hyperglycemic environment. The hyperglycemia increased the production of ROS. Increased ROS production was inhibited by the addition of an inhibitor of the electron transport chain at Complex II, an inhibitor of oxidative phosphorylation (formation of ATP from ADP), and by manganese superoxide dismutase (55). Furthermore, this hypothesis was supported by a study in which endothelial cells were cultured to lack the mitochondrial electron transport system. The mutant endothelial cells did not show an increased production of ROS when they were grown in a glucose milieu (53).

A hypothesis that unifies glycation, sorbitol, and ROS has been developed. It suggests that overproduction of mitochondrial $\cdot O_2^-$ is the activation factor that causes damage in nephrons arterials, nerve cells, and ocular tissues (53, 56).

In summary H_2O_2 has been shown to stimulate insulin secretion. Insulin secretion is turned off when $\cdot O_2^-$ is produced in excess and H_2O_2 is diminished. Essentially any agent that can increase the electron transport system can generate $\cdot O_2^-$. In T2D, patients' hyperinsulinemia may be caused by the overproduction of ROS.

Summary Box 8.6

- The principal hypotheses that explain the complications of diabetes are glycation, sorbitol accumulation, and oxidative stress.
- Glycosylation of proteins occurs similar to the glycosylation of hemoglobin.
- AGEs form due to subsequent dehydration, oxidation, and cross-linking to collagen.
- Receptor sites for AGE have been found, named RAGEs.
- AGE products include imidazolone, pentosidine, and methylglyoxal.
- Activation of RAGE by AGE produces cytokines.
- AGE produces decreases of kidney function and arterial elasticity.
- AGE is a possible cause of retinopathy and nephropathy.
- A minor pathway that increases in diabetes is the sorbitol pathway.
- Sorbitol and glucose intracellular accumulation results in retinopathy and nephropathy.

- NADPH formation in diabetics reduces glutathione and leads to the formation of free radicals.
- ROS refers to free oxygen radicals.
- ROS is a signaling pathway for insulin secretion.
- The formation of superoxide ($\cdot O_2^-$) inhibits the signal for insulin secretion.
- Hyperinsulinemia can be caused by surplus ROS.

GLOSSARY

Albuminuria   Is defined as urinary albumin concentrations of $\geq 208\,\mu g/min$.

Anion gap   Equals $[Na^+] + [K^+] - ([Cl^-] + [HCO_3^-])$

Apolipoprotein E4 (ApoE4)   Is found in approximately 15% of the population. Individuals who inherit this gene from just one parent have a susceptibility of AD. The gene dictates the synthesis of apolipoprotein E4 (ApoE4).

Chemokines   A class of cytokines that direct the movement of cells.

Cytokines   Intercommunicative proteins secreted by cells.

Dystrophic neurites   Are aggregated hyperphosphorylated tau proteins associated with amyloid plaques.

Hazard Ratio   Is defined as the ratio of an event occurring in an experimental population divided by the event occurring in a control population.

Microalbuminuria   Is defined as urinary albumin concentrations of ≥ 28 $\mu g/min$.

Neuritic amyloid plaques   Are similar to neurofibrillary tangles but do not have the tangled appearance of them.

Neurofibrillary tangles   Are also aggregated hyperphosphorylated tau proteins. They are found inside nerve cells. They consist of tau protein filaments wound about each other.

Podocytes   Blood is filtered through spaces located between the processes (sometimes called *feet* and therefore the name podocytes).

Relative risk   Is the probability of developing a disease after being exposed to it for longer and longer time periods.

Tau protein   Is a protein that normally stabilizes microtubules.

Vascular dementia   Occurs after a stroke.

REFERENCES

1. Madineh SMA. Avicenna's canon of medicine and modern urology part I: bladder and its diseases. Urol J 2008;5:284.
2. National Eye Institute. *Diabetes Retinopathy: What You Should Know*. Bethesda (MD): NIH; 2003. NIH Publication no 03-2171.

3. Teaching Topics. s.l.: N Engl J Med 2011. NEJM Resident E-Bulletin.

4. Sun JK, Keenan HA, Cavallerano JD, et al. Protection from retinopathy and other complications in patients with type 1 diabetes of extreme duration: the joslin 50-year medalist study. Diabetes Care 2011;34:968.

5. Javadi M-A, Zarei-Ghanavati S. Cataracts in diabetic patients: a review article. J Opthalmic Vis Res 2008;3:52.

6. Shiokawa A, Fukushima H, Kato S. Glycemic control and lens transparency in patients with type 1 diabetes mellitus. Am J Opthalmol 2001;131:301.

7. CDC. MMWR Morb Mortal Wkly Rep 2011;60(45):1549.

8. NDIC. Diabetic Neuropathies: The Nerve Damage of Diabetes, NIH Pub #08-3185. 2009. Available at http://www.diabetes.niddk.nih.gov. Accessed January 2011.

9. Vinik A, Nakavi A, del Pilar Silva Chuecos M. A break in the brake mechanism in diabetes: a cause of postprandial hyperglycemia. Diabetes Care 2008;31:2410.

10. Woerle HJ, Albrecht M, Linke R, et al. Impaired hyperglycemia-induced delay in gastric emptying in patients with type 1 diabetes deficient for islet amyloid polypeptide. Diabetes Care 2008;31:2325.

11. Feldman HA, Goldstein DG, Hatzichristou RJ, et al. Impotence and its medical and psychological correlates: results of the Massachusetts male aging study. J Urol 1994;151:54.

12. NDIC 2008. Sexual and Urologic Problems of Diabetes. National Diabetes Information Clearinghouse. s.l.: NIH Pub # 09-5135.

13. The Diabetes Control and Complications (DCCT) Research Group. Effect of intensive therapy on the development and progression of diabetic nephropathy in the diabetes control and complications trial. Kidney Int 1995;47:1703.

14. The Writing Team for the Diabetes Control and Complications Trial/Epidemiology of Diabetes Interventions and Complications Research Group. Sustained effect of intensive treatment of type 1 diabetes mellitus on development and progression of diabetic nephropathy: the Epidemiology of Diabetes Interventions and Complications (EDIC) study. JAMA 2003;290:2159.

15. de Boer IH, Rue TC, Hall YN, et al. Temporal trends in the prevalence of diabetic kidney disease in the United States. JAMA 2011;305:2532.

16. Levey AS, Stevens LA, Schmid CH, et al. A new equation to estimate glomerular filtration. Ann Int Med 2009;150:604.

17. Burrows NR, Hora I, Gerzoff RB, et al. Incidence of end-stage renal disease attributed to diabetes among persons with diagnosed diabetes – United States and Puerto Rico, 1996–2007. MMWR Morb Mortal Wkly Rep 2010;59:1361.

18. Fox CS, Coady S, Sorlie PD, et al. Increasing cardiovascular disease burden due to diabetes mellitus: the Framingham Heart study. Circulation 2007;115:1544.

19. Roger V, Go RA, Lloyd-Jones DM, et al. Heart disease and stroke statistics – 2012 update: a report from the American Heart Association. Circulation 2012;125:e160. DOI: 10.1161/CIR.0b013e31823ac046.

20. Chen G, McAlister RL, Walker RL, et al. Cardiovascular outcomes in Framingham participants with diabetes: the importance of blood pressure. Hypertension 2011; 57:891.

21. Kissela BM, Khoury J, Kleindorfer D, et al. Epidemiology of ischemic stroke in patients with diabetes. Diabetes Care 2005;28:355.

22. Cines DB, Pollak ES, Buck CA, et al. Endothelial cell in physiology and in the pathophysiology of vascular disorders. Blood 1998;91:3527.

23. PF D. Flow-mediated endothelial mechanotransduction. Phys Rev 1995;75:519.

24. Wei X, Schneider JG, Shenouda SM, et al. The FAS-NOS Connection: de novo lipogenesis maintains homeostasis through endothelial nitric oxide synthase (eNOS) palmitoylation. J Biol Chem 2011;286:2933.

25. Chatzizisis YS, Coskun AU, Jonas M, et al. Role of endothelial shear stress in the natural history of coronary atherosclerosdis and vascular remodeling:Molecular, cellular, and vascular behavior. J Am Coll Cardiol 2007;49:2379.

26. CD W. Case study: diabetic ketoacidosis complications in type 2 diabetes. Clin Diabetes 2000;18:2.

27. Prevention, Centers for Disease Control and Diabetic ketoacidosis as first-listed diagnosis, United States 1980–2005. [Online] 2008. Available at http://www.cdc.gov /diabetes/statistics/dkafirst/fig1.htm. Accessed 2012 Jan 25.

28. Kitabchi AE, Nyenwe EA. Hyperglycemic crises in diabetes mellitus: diabetic ketoacidosis and hyperglycemic hyperosmolar state. Endocrinol Metab Clin North Am 2006;35:725. DOI: 10.1016/jecl.2006.09.006.

29. Zoungas S, Patel A, Chalmers J, et al. Severe hypoglycemia and risks of vascular events and death. N Engl J Med 2010;363:1410.

30. Shah BR, Hux JE. Quantifying the risk of infectious disease for people with diabetes. Diabetes Care 2003;26:510.

31. Bertoni AG, Saydah S, Brancati FL. Diabetes and the risk of infection-related mortality in the U.S. Diabetes Care 2001;24:1044.

32. National Diabetes Information Clearinghouse. Complications of Diabetes in the United States. 2011. Available at http://diabetes.niddk.nih.gov/DM/PUBS/statistics/. Assessed February 2011.

33. Hoyer S, Oesterreich K, Wagner O. Glucose metabolism as the site of the primary abnormality in early onset dementia of Alzheimer type. J Neurol 1988;235:143. DOI: 10.1007/BF00314304.

34. Matsuzaki T, Sasaki K, Tanizaki Y, et al. Insulin resistance is associated with the pathology of Alzheimer disease. Neurology 2010;75:764. DOI: 10.1212/WNL .0b013e3181eee25f.

35. Ohara T, Doi Y, Ninomiya T, et al. Glucose tolerance status and risk of dementia in the community. Neurology 2011;77:1126. DOI: 10.1212/WNL.0b03e31822f0435.

36. de la Monte SM, Wands JR. Review of insulin and insulin-like growth factor expression, signaling, and malfunction in the central nervous system: relevance to Alzheimer's disease. J Alzheimers Dis 2005;7:45.

37. Steen E, Terry BM, Rivera EJ, et al. Impaired insulin and insulin-like growth factor expression and signaling mechanisms in Alzheimer's disease—is this type 3 diabetes? J Alzheimers Dis 2005;7:63.

38. Burns JM, Donnelly JE, Anderson HS, et al. Peripheral insulin and brain structure in early Alzheimer disease. Neurology 2007;69:1094. DOI: 10.1212/01.wnl.0000276952 .91704.af.

39. Burdo JR, Chen Q, Calcutt NA, et al. The pathological interaction between diabetes and presymptomatic Alzheimer's disease. Neurobiol Ageing 2009;30:1910. DOI: 10.1016/j.neurobiolaging.2008.02.010.

40. Zhao W-Q, De Flice FG, Fernandez S, et al. Amyloid beta oligomers induce impairment of neuronal in sulin receptors. FASEB J 2008;22:246. DOI: 10.1096/fj.06-7703com.

41. De Felice FG, Vieira MNN, Bomfim TR, et al. Protection of synapses against Alzheimer's-linked toxins: insulin signaling prevents the pathogenic binding of Abeta oligomers. Proc Nat Acad Sci U S A 2009;106:1971.

42. Gallagher EJ, LeRoith D. Minireview: IGF, insulin, and cancer. Endocrinology 2011;152:2546.

43. Cantley LC, Neel BG. New insights into tumor supression: PTEN suppresses tumor formation by restraining the phosphoinositide 3-kinase/AKT pathway. Proc Nat Acad Sci U S A 1999;96:4240.

44. Niwa T, Katsuzaki T, Miyazaki T, et al. Immunohistochemical detection of imidazolone, a novel advanced glycation end product, in kidneys and aortas of diabetic patients. J Clin Invest 1997;99:1272.

45. Sell DR, LaPolla A, Odetti P, et al. Pentosidine formation in skin correlates with the severity of complications in individuals with long-standing IDDM. Diabetes 1992;41:1286. DOI: 10.2337/diabetes.41.10.1286.

46. Coughian MT, Forbes JM. Temporal increases in urinary carboxymethyllysine correlate with albuminuria development in diabetes. Am J Nephrol 2011;34:9.

47. Colhoun HM, Betteridge J, Durrington P, et al. Total soluble and endogeneous secretory receptor for advanced glycation end products as predictive biomarkers of coronary heart disease risk in patients with type 2 diabetes. Diabetes 2011;60:2379. DOI: 10.2337/db11-0291.

48. Peppa M, Vlassara H. Advanced glycation end products and diabetic complications: a general overview. Hormones 2005;4:28.

49. Cerami A, Stevens VJ, Montier VM. Role of nonenzymatic glycosylation in the development of the sequelae of diabetes mellitus. Metabolism 1979;28:431.

50. Franke S, Dawczynski J, Strobel J, et al. Increased levels of advanced glycation end products in human cataractous lenses. J Cataract Refract Surg 2003;29:998.

51. Dods R. Diabetes. In: Pesce A, Kaplan L, editors. Volume 38,. *Clinical Chemistry: Theory, Analysis, Correlation*. 5th ed. St. Louis: Mosby; 2010. p 729.

52. Schalkwijk CG, Stehouwer CDA. Vascular complications in diabetes mellitus: the role of endothelial dysfunction. Clin Sci 2005;109:143.

53. Brownlee M. Banting Lecture 2004: the pathobiology of diabetic complications: a unifying mechanism. Diabetes 2005;54:1615.

54. Pi J, Bai Y, Zhang Q, et al. Reactive oxygen species as a signal in glucose-stimulated insulin secretion. Diabetes 2007;56:1783.

55. Nishikawa T, Edelstein D, Du XL, et al. Normalizing mitochondrial superoxide production blocks three pathways of hyperglycaemic damage. Nature 2000;404:787. DOI: 10.1038/35008121.

56. Forbes JM, Coughlan MT, et al. Oxidative stress as a major culprit in kidney disease in diabetes. Diabetes 2008;57:1446. DOI: 10.2337/db08-0057.

CHAPTER 9

HEREDITARY TRANSMISSION OF DIABETES MELLITUS

In many cases he (Friedrich Theodor von Frerichs 1819–1885) proved a hereditary transmission from the distant, and yet at the same time directly related, members of one family; for instance, supposing an uncle and a nephew to be afflicted by this disease, he inferred from this that the grandfather or grandmother, and eventually the great-grandfather or great-grandmother, were the original sources of diabetes. This constitutes the basis, on the strength of which, after lengthy examination extending over 20 years, I (Emil Schnee) was to make the following important discovery: Diabetes is a hereditary, constitutional disease; and the etiological element of this disease is lues[1] contracted by some ancestor! From Emil Schnee "Diabetes its cause and permanent cure from the standpoint of experience and scientific investigation" in Arthur Scott Donkin "On the relation between diabetes and food and its application to the treatment of the disease."

—Translated from German by L. Tafel, published by P. Blakiston, Son & Co.,
Philadelphia, 1889.

As can be seen from the above quotation, the familial characteristics of diabetes mellitus were inferred long before the origins of the disease were discovered. The early observations of the connection between heredity and diabetes came from physicians who recognized that diabetes was found more commonly in families. Modern day investigations focused on identical twins and the offspring and siblings of diabetics. They demonstrated clearly that diabetes developed from a complex interplay between the environment and genetic factors.

[1] Translated from the Latin—a plague.

Understanding Diabetes: A Biochemical Perspective, First Edition. Richard F. Dods.
© 2013 John Wiley & Sons, Inc. Published 2013 by John Wiley & Sons, Inc.

INHERITANCE OF T1D IN MONOZYGOTIC AND DIZYGOTIC TWINS

Refer to Figure 9.1 to understand how monozygotic (MZ) (identical) and dizygotic (DZ) (fraternal) twins are produced. As you can see from the figure, MZ twins originate from a single fertilization and therefore their gene complement is identical. DZ twins are derived from two separate fertilizations and therefore share half of their genes in common. In MZ twins, T1D in one twin presaged the disease in the second 27–50% of the time. If environment did not take part in the expression of the T1D gene(s) then both twins would be expected to be diabetic. For DZ twins both twins being afflicted with diabetes (concordance) is 4–10%. If environment did not take part in the expression of the T1D gene(s) then 50% would be expected to be diabetic.

Pairwise and Probandwise Concordance in T1D

First we need to define some terms used by geneticists to represent the effect of inheritance and environment in twins (1). Concordance is when both siblings (sibs) of a pair of twins have T1D. Discordance (or disconcordance) is when only one sib of a pair of twins has the disease. Pairwise concordance is defined as the number of twins who both have T1D divided by the sum of the number of twins in which only one has T1D and the number of twins who both have T1D. This can be represented as $C/(C + D)$ where C is the number of concordant twins and D is the number of discordant twins.

Pairwise concordance is the probability that both twins have T1D in a population in which at least one twin has T1D. For example, if 8 pairs of twins both have T1D and 12 pairs have only one twin with T1D, pairwise concordance

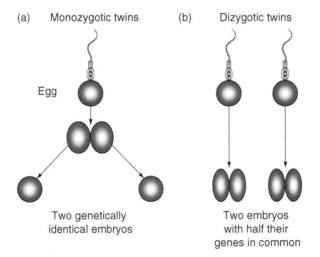

Figure 9.1 Depiction of the fertilization of (a) one egg producing MZ twins (identical) and (b) two eggs producing DZ twins.

would equal to 8/(8+12) or 8/20 or 40%. Probandwise concordance determines the proportion of twins in which only one has T1D. Probandwise concordance is calculated as two times the number of twins that both have T1D divided by the sum of two times the number of twins both having T1D and the number in which only one twin has T1D. This relationship can be determined from the equation $2C/(2C + D)$. For example, if 8 twins where both have T1D and 12 twins have only one twin who has T1D the probandwise concordance would equal 2*8 divided by (2*8) + 12, which equals 16/(16 + 12) or 16/28 or 57%.

Both terms are measures of the expression of the T1D gene. Or in other words they are a measure of the role environmental factors play in the expression of T1D.

Several studies concur with the conclusion that although heredity provides the genes for T1D it takes environment to express them. The problem that geneticists face is that few studies have the numbers of twins that would provide statistically significant results. Let us look at two of the more recent studies that do have a significant number of participants. Finland has the world's highest incidence of T1D. In a study (2) of 22,650 twins born between 1958 and 1986, the study identified 228 twin pairs with T1D who were MZ and 183 with T1D who were DZ. Table 9.1 presents the pairwise and probandwise concordances for the Finnish study. As you can see from the table Finnish investigators found that for MZ twins with T1D the pairwise concordance was 27.3% and the probandwise concordance was 42.9%. For DZ twins the former was 3.8% and the latter was 7.4%. Furthermore the pairwise concordance and the probandwise concordance are significantly greater for those who had the onset of T1D occurring at ≤10 years as contrasted with those with onset greater than 10 years. In addition the study shows that early onset in one sib increases the risk in the second. Nonetheless the majority of MZ sibs have only one sib with T1D. It should be noted that MZ twins can become concordant well after diagnosis in the first twin and after the limits of the study. This study suggested [using a best fitting liability

TABLE 9.1   Pairwise and Probandwise Concordance from the Finnish Study of Diabetes Mellitus in MZ and DZ Twins.

		Concordance, %	
	MZ or DZ	Pairwise	Probandwise
Males and females	MZ	27.3	42.9
	DZ	3.23	6.25
Age			
≤10 yr	MZ	50.0	66.7
	DZ	7.32	13.6
>10 yr	MZ	16.7	28.6
	DZ	3.83	7.37

Adapted from Hyttinen et al. (2).

model (3)] that 88% of the phenotypic variance[2] was due to genetic factors and the remaining (12%) was due to environment.

The second large-scale study (4) involved 187 MZ twins of subjects who had T1D. The subjects in this study were from the United Kingdom (134 pairs) and the United States (53 pairs). Many of the subjects were followed for up to 40 years. Initially all of the twins were discordant, that is, only one sib had T1D. After 14 years of observation 20 percent of the twins became concordant. The progression of the diabetes to the second twin was similar for both countries. Furthermore the study found that the rate of progression to concordance occurred more frequently when the initial sib contracted T1D at an earlier age. When one sib was diagnosed with T1D at ≤24 years of age 38% became concordant within 30 years. After 24 years of age this percentage decreased to 6%. This study emphasizes that there are environmental factors that initiate the onset of T1D in individuals who have the genes for T1D.

We will come back later in this chapter to discuss the genes that cause susceptibility (and resistance) for an individual to T1D.

Pairwise and Probandwise Concordance in T2D

An early study (5) used a relatively small number of T2D subjects to determine the concordance and discordance of MZ twins with T2D. The results indicated that pairwise concordance for MZ twins was as great as 90.6% and probandwise concordance was 95.0%. Incidentally the same study showed the pairwise concordance for MZ twins with T1D as 54.4% and probandwise concordance as 70.5%.[3] The subjects for this study came from King's College Hospital in London, "physicians at other hospitals," the British Diabetic Association, and "self-referral after programs on radio or television." The authors maintain that the population studied is reliable because medical history and oral glucose tolerance testing determined whether or not they were diabetic and that monozygosity was established by typing for blood groups. The oral glucose tolerance test identified some sibs who otherwise would have been determined to be normal thus increasing the concordance frequency.

Later articles presented somewhat more modest results for the concordance of T2D. For example, a Danish study (6) used the 75 g oral glucose tolerance test to identify 62 pairs in which T2D was concordant or discordant, that is, one or both twins had T2D. This study made a distinction between IGT and T2D using the criteria put forth in Chapter 6. As can be seen in Table 9.2, the probandwise concordance for both MZ and DZ twins was increased when IGT twins were included. The pairwise concordance also increased when IGT twins

---

[2]Phenotype refers to the expression of a specific trait due to genetic and environmental influences. Phenotypic variation refers to the variance of the phenotype due to the sum of genotype and environmental factors.

[3]The concordances were calculated from the raw numbers for concordant and discordant pairs from the original paper.

TABLE 9.2 Concordances for T2D from a Danish Study of MZ and DZ Twins Using the New Criteria by WHO*

MZ or DZ	Concordance (%)	
	Pairwise	Probandwise
T2D without an OGTT		
MZ	22.7	37.0
DZ	1.00	18.2
T2D + IGT as determined by an OGTT		
MZ	38.9	56.0
DZ	21.6	35.6

*Adapted and recalculated to express the percent concordances to three significant figures by this author.

were included. While this study was in progress the WHO definition for IGT and T2D was modified (refer to Chapter 6). Thus only the results for the new criteria are shown in the table although both are presented in the paper.

Another term that geneticists often use is heritability. Heritability (h^2) refers to the proportion of the total variation of T2D attributable to genetic variation. It is a comparison of the similarity of T2D within MZ twins with the similarity within DZ twins. Heritability is calculated using the equation $h^2 = 2(r_{MZ} - r_{DZ})$ where r represents a specific inherited body measurement or metabolic variable (phenotype). For example, the *r* value for weight for MZ twins was measured in this study as 78% and for DZ twins was 39%. $H^2 = 2(78\% - 39\%) = 78\%$. This indicates a major genetic contribution (78%) to variation of weight. BMI also showed a strong genetic contribution whereas waist to hip ratio, fasting plasma glucose, and fasting plasma insulin indicated a weak contribution. The 2-h OGTT plasma glucose and 30 min OGTT plasma glucose indicated a moderate contribution. You may recall that MZ twins have identical genes, and differences in a given trait are caused by environmental factors. DZ twins carry a 50% complement of genes. Thus the extent to which MZ twins are more similar than DZ twins reflects the genetic influence. Note that the greater the concordance for MZ and/or the lower the concordance for DZ, the greater the heritability.

Other studies (7–9) present concordance data from which I calculated the pairwise concordance as 39–68%.

Studies of T2D twins suggest that both genetic and environmental factors contribute to the disease with weight and BMI being the principal nongenetic contributors.

Summary Box 9.1

- The length of the study and the age of the subjects at the time they were diagnosed affect pairwise and probandwise concordances.

- For T1D pairwise and probandwise concordances, MZ twins averaged about 27% and 43% and DZ twins averaged about 3% and 6% respectively. If the disease is diagnosed earlier than 10 years the values are about three times higher than if diagnosed after 10 years (Finnish study).
- For T2D pairwise and probandwise concordances, MZ twins who did not take an OGTT (and thus IGT were not included) were about 23% and 37% and for DZ twins were about 1.0% and 18%. If IGT patients were added, MZ twins averaged about 38.9% and 56%; 22% and 36% for DZ twins (Danish study).
- Heritability calculations for T2D patients show that there is a major genetic contribution for weight: 78%. Strong genetic contributions also were attributable to BMI and moderate contribution from 2-h OGTT and 30-min OGTT plasma level.

Diabetes in Offspring of One or Two Diabetic Conjugal (Biological) Parents

Since the 1950s and possibly earlier there have been numerous studies that show that T1D and T2D follow a familial pattern. Some of these studies are discussed below.

A thorough but early review (1977) (10) of the literature on genetic factors in the etiology of diabetes mellitus concluded that the frequency of diabetes was 6–10%. Conjugal parents (biological parents) who have T2D (the term used in the review is "mild maturity-onset type diabetes") produce offspring that, if diabetic, have T2D. This is often termed *producing true to type*. T1D (the term used in the review for these diabetics was *classical juvenile-type diabetic*) parents were "uncommon" and those with both parents possessing T1D were "rare." The frequency of T1D offspring when one parent had the disease was not different from the frequency of T2D offspring of conjugal parents who both had T2D. When OGTT was administered the review found 25–40% of the offspring of diabetic parents had IGT (called *chemical (latent) diabetes* in the review). Repeated tolerance testing increased this value to 55–60%.

In an (2) ongoing study 700 offspring of 205 diabetic parents led to the conclusion that when the offspring would reach 85 years of age approximately 33% would have overt T2D and 50% of them would have abnormal OGTTs.

In a study by the same principal authors (11) involving 274 offspring of 80 conjugal diabetic parents repeated administration of the OGTT determined a frequency for IGT (called *chemical diabetes* in the paper) of 41–62% depending on the age and weight of the offspring as compared to 8.8% for overt diabetes (T2D). The reason for repeating the OGTT is that an abnormal test was frequently followed by one or more normal tests and vice versa.

A study (12) from the Karlsburg Clinic for Diabetes and Metabolic Diseases in Germany, 58 offspring from 46 parents both with T1D were followed from

1955 to 2001. Overall 25 (43%) of them developed diabetes at between 1 and 42 years. Prevalence of diabetes in this study suggests a genetic aspect to T1D development.

Numerous studies in respect to the offspring of two T2D conjugal parents have been reported. A study (13) with 199 offspring of 37 diabetic patients had a prevalence of 11.5%. However when 123 offspring were tested with an OGTT, 28 were determined to have what was then called *latent diabetes* (now called *IGT*). If extrapolated to 60 years of age, 36.5% would have an abnormal OGTT.

A Framingham offspring study (14) of 2527 offspring from 1303 nuclear families concluded that risk ratios for T2D offspring were consistent with an additive risk model. That is, when both father and mother have T2D the risk of their child developing T2D is equal to the sum of the risks when either parent has T2D. Also a conclusion of this study was that when the mother is diabetic, perinatal exposure to diabetes increases the T2D risk for the offspring.

Diabetes in Siblings of Diabetics

In the 1977 study cited earlier (10) the prevalence of diabetes in siblings of diabetics was 12% for T1D and 10% for T2D.

A recent study (15) of 701 siblings of T1D patients found that 47 (6.70%) were diagnosed with T1D during the 15-year period of the study. T1D was identified in some cases by the presence of autoantibodies to β-islets. The autoantibodies measured were ICA, GADA, IAA, and IA-2A. Refer to Chapter 7 for a discussion about β-islets autoantibodies in T1D. Also an intravenous glucose tolerance test and gene testing for HLA DR_3/DR_4 were used to detect T1D.

Summary

T1D

- Diabetes mellitus is transmitted true to type. Diabetic offspring of T1D are usually T1D. Offspring of T2D are usually T2D.
- The risk for T1D for an individual in the general population is approximately 0.4%.
- T1D in MZ twins has a concordance (pairwise or probandwise) of less than 100% (about 20–30%) thus suggesting the importance of environmental factors in its causation.
- Children whose fathers have T1D have a frequency of about 6% of developing T1D. Children whose mothers have T1D have a frequency of 4% if born before the mother is 25 years of age and 1% after 25 years of age.
- The risk to a nontwin sibling if one of his/her siblings has T1D is about 6%.
- T2D shows familial clusters. There is an increased frequency for children of diabetic parents and an increased frequency for children that have sibs that have T1D or T2D.

- When only one parent has T2D the frequency of the child developing T2D is about 14% if the parent was diagnosed before age 50 and about 8% if diagnosed after 50 years. The frequency is higher when the parent is the mother.
- When both parents are T2D the frequency of their child having T2D increases to about 50%.
- The OGTT adds subjects that would otherwise be considered normal to the total number of T2D. In other words IGT subjects identified by the OGTT add to the overt T2D totals.

Summary adapted and modified in part from the American Diabetes Association, www.diabetes.org/diabetes-basics/genetics-of-diabetes.html, accessed on March 14, 2012.

Summary Box 9.2

- Studies show that T1D and T2D are passed on from parents to their offspring, true to type, but at low frequencies.
- The risk of inheriting T1D when a sib has the disease is higher than if the sib does not have the disease. The same correlation is found when one sib has T2D.

THE GENETIC COMPONENT OF DIABETES MELLITUS

The Major Histocompatibility Complex Proteins or Human Lymphocyte Antigens and Disease

Before we discuss the genes that provide susceptibility (and in some cases resistance) to T1D and T2D, we have to understand the nature of the major histocompatibility complex proteins (MHC). In humans MHC proteins are called *human lymphocyte antigens*[4] (HLA). HLAs are proteins found in nucleated cells, which are responsible for the rejection of tissue transplanted to an individual from another unrelated individual. As you can see from Figure 9.2, the genes that code for the HLA proteins are on chromosome 6 and have been localized at loci A, B, C, D, and DR. Each locus gives rise to two proteins. This is derived from the pairing of chromosomes, one from each parent. The paired genes are called *alleles*. As of 2012, there have been 7527 alleles determined on the HLA complex. Unlike most genes both HLA alleles are expressed (i.e., both alleles produce proteins). The HLA region contains genes in close proximity to each other that tend to be inherited together, a phenomenon called *linkage disequilibrium*.

Shortly after discovery of the HLAs, a startling correlation between disease and HLAs was found. Specific HLA antigens were found to occur at a high

---

[4]Human lymphocyte antigens are also called human leukocyte antigens.

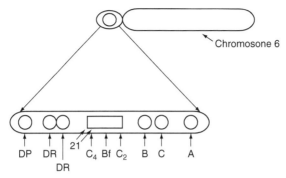

21 = 21 hydroxylase

Figure 9.2 A simplified map of chromosome 6 showing the HLA complex. A complete mapping of the HLA region on chromosome 6 can be found at http://hla.alleles/index.html.

frequency in certain disease states. Individuals who carry the B27 HLA antigen develop a spine deformity (ankylosing spondylitis) 100 times more frequently than individuals who lack this HLA antigen. The 10th International Histocompatibility Workshop held in 1987 divided the HLA antigens into three classes: Class I occurs on all nucleated cells and is composed of HLA A, B, C, E, F, and G antigens; Class II is found on B-lymphocytes, macrophages, activated T-lymphocytes, and epithelial cells of β-islets and is composed of HLA-DR (D-Related), -DM, -DO, -DP and -DQ antigens; and Class III, which is a nonHLA region that produces factors C2, C4, and Properdin factor B (Bf) that are members of the complement system, the enzyme 21-hydroxylase, tumor necrosis factor α (TNF-α), and tumor necrosis factor β (TNF-β). Class I molecules are recognized by CD8 T cells. Class II proteins are recognized by CD4 T cells. Class I HLA molecules are composed of two polypeptide chains, the α chain, which is the longer of the two, and the β chain. The Class II HLA molecules are composed like the Class I molecules of two chains, α and β, but of nearly equal length.

What is the role of HLAs? The answer is that they are involved in the immune response of the body to the presentation of a foreign antigen (a protein, carbohydrates, nucleic acids, or lipid from a virus, bacteria, etc.). As you can see from Figure 9.3, the first step in the immune response is hydrolysis of the antigen by proteosomes[5] to smaller fragments. The antigen (for purposes of illustration a protein is used as the antigen and the antigen presenting cell (APC) is a macrophage) is hydrolyzed to produce products (peptides) that bind to specific HLAs. Class I bound peptides are 8–10 amino acids long and Class II are 13–25 amino acids in length.

HLAs have regions that are highly variable. The peptide that an HLA binds has specific amino acids at particular positions in its sequence. A specific HLA will not bind to a peptide with a different sequence although there may be other HLAs

---

[5]A proteasome is a complex of proteases that degrade proteins to peptides.

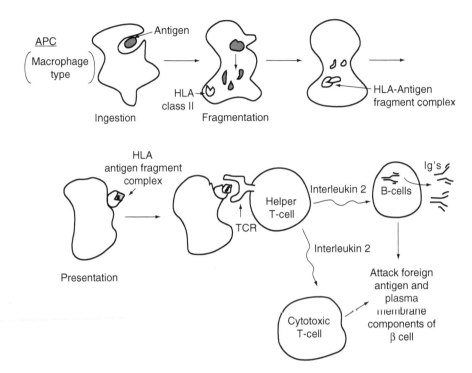

HLA denotes Human Lymphocyte Antigen (called major histocompatibility complex (MHC) in animals other than humans)
APC denotes antigen presenting cells (include macrophages, B cells, dendritic cells)
TCR denotes T cell antigen receptor

Figure 9.3   Depiction of a macrophage (an APC) ingesting a foreign antigen, fragmenting it, binding the fragment to a HLA class II molecule, presenting it on the exterior of the APC cell to a helper T-cell at its TCR, releasing interleukin 2 and thus preparing B-cells to release immunoglobulins and/or preparing Tc to attack foreign antigens such as plasma membrane components of β-islet cells. HLA denotes human lymphocyte antigen [called *major histocompatibility complex* (MHC) in animals other than humans]. APC denotes antigen presenting cells (include macrophages, B cells, dendritic cells). TCR denotes T cell antigen receptor.

that would bind the peptide. This is similar to the manner in which immunoglobulins bind antigens. The mechanism by which the variable regions are produced is also similar to the production of the variable regions of immunoglobulins. Therefore a given antigen always finds an HLA with a homologous epitope.[6] Next the HLA-antigen fragment moves to the periphery of the cell and if it is a Class I HLA-antigen, it binds noncovalently with the T cell receptor (TCR) of a cytotoxic T cell (Tc) and if the HLA-antigen is Class II, the complex binds (also noncovalently) to the TCR of a helper T-cell (Th) (as shown in the figure). The

[6]Epitope refers to a sequence of amino acids.

Th is induced to produce cytokines such as interleukin 2 (IL-2), which activate B-cells to proliferate and produce immunoglobulins directed toward the antigen. Tc attack foreign antigens, in some cases plasma membrane components of β-islets. As mentioned earlier, Class I HLA-antigens bind only to Tc, in which the TCR recognizes its epitope. When this happens, the Tc induces cell self-destruction (apoptosis). The binding of HLA to TCR is relatively weak. The forces that bind are ionic, H-bonding, and Van der Waals forces. Thus accessory molecules are necessary to stabilize the interactions. The principal accessory molecules are CD8 for Class I HLA and CD4 for Class II.

How susceptibility and resistance (protective) HLAs affect T1D is not perfectly clear. One model is based on the finding that HLA molecules bind to antigens in a wide spectrum of affinities (16). T1D antigens (those antigens that cause T1D) bind to Class II HLAs. T1D is more probable if the HLA carrying the susceptibility gene binds more tightly to the T1D antigen than other HLAs not carrying a susceptibility gene. When the HLA that binds to the antigen does not carry susceptibility genes, then T1D is less likely.

The second model is almost the exact opposite of the aforementioned model (17). It focuses on the inability of the immune system to maintain tolerance to β-islet cells. When Class II HLAs bind a tolerance-producing antigen (the tolerogen) tightly and in the correct orientation (during presentation to Th cell) resistance to T1D is provided. Those that bind the tolerogen weakly or in the incorrect orientation are less resistant to T1D. Class II HLAs that bind the tolerogen tightly but in the wrong orientation are nonresistant. The tolerogenic agent is a fragment of a β-cell protein. It is only tolerogenic if an HLA can bind it in the correct orientation. Remember these are only models and may or may not reflect what is really going on at the molecular and cellular level.

I suspect that there is a competition between HLAs that confer susceptibility and resistance. The more HLAs that carry the susceptibility genes in a specific individual the more likely one or more of them will be predominant and cause T1D. But as we shall see later in this chapter, there are also environmental factors that play an important role in T1D.

This was a brief review of HLA biochemistry. There are some interesting aspects of the HLA complex and its function, which we are unable to discuss because that would be best left to a book on immunology. I suggest to the reader that HLA, TCRs, the role of the thymus and cell-mediated immunity provide more beautiful concepts that would make for interesting reading.

Online Mendelian Inheritance in Man

There is one other prerequisite before we discuss the relationship between genes and diabetes mellitus. That is the manner in which we catalog human genes and the diseases of the genes. The catalog used is the Online Mendelian Inheritance in Man (OMIM), which is compiled at the McKusick-Nathans Institute of Genetic Medicine at the Johns Hopkins University School of Medicine, Baltimore, MD. As shown in Table 9.3 OMIM entries use a six-digit

TABLE 9.3   The Online Mendelian Inheritance in Man Catalog System

- OMIM uses a 6-digit number with a symbol preceding it. The first number in the sequence designates the following:
  - 1 and 2—Autosomal loci or phenotype created before May 15, 1994
  - 3—X-linked loci or phenotype
  - 4—Y-linked loci or phenotype
  - 5—Mitochondrial loci or phenotype
  - 6—Autosomal loci or phenotype created after May 15, 1994
- Allelic variants are designated by a decimal point followed by a 4-digit entry. For example 3XXXXX.XXX1
- Symbols preceding the number designate the following:
  - * a gene
  - # a descriptive entry, usually a phenotype
  - + entry contains the description of a gene of known sequence and phenotype
  - % entry describes a confirmed Mendelian phenotype or phenotypic locus for which the underlying molecular basis is not known
  - No symbol indicates a description of a phenotype for which the Mendelian basis has not been clearly established or that the separateness of this phenotype from that in another entry is unclear
  - ^ entry no longer exists because it was removed or moved to another entry location
- A mutation is cataloged in the allelic variants section

numbering system with a symbol preceding it. This nomenclature will be used for genes outside of the HLA region.

HLA Nomenclature

For HLA genes the nomenclature generally in use is quite different from that for genes outside of the HLA region, although the OMIM catalog numbering system can be used.

The rules for HLA nomenclature were modified in 2010. The old system assigned significance to pairs of digits. For example, HLA-A*0201 represents

- HLA identifies the allele as an HLA allele
- - is a separator
- A identifies the gene
- The first two digits 02 represent the allele group
- The third and fourth digits 01 represent the specific allele

There can be more digits designating silent mutations, mutations in introns, or regions that are not translated in exons. But the latter two instances will not concern us in this book. In the new nomenclature system a colon is used as a separator between pairs of numbers. Thus in the new system HLA-A*0201 becomes HLA-A*02:01. The change is that as new alleles are discovered the number of alleles for a given group becomes greater than 99. If HLA-A*02:99 is discovered, then the next one is HLA-A*02:100 in the new nomenclature. The / between alleles designates the paternal and maternal chromosomes, for example, DRB1*02:01/03:02. Whether the new nomenclature will become popular among epidemiologists and geneticists is yet to be determined. Since the new nomenclature has not as yet replaced the older nomenclature this book will use the older nomenclature.

The HLA complex, HLA nomenclature, and the OMIM classification system are featured prominently in the following discussion of the genetics of diabetes mellitus.

Summary Box 9.3

- HLAs are located on chromosome 6.
- HLAs are associated with many diseases including T1D.
- Proteosomes found in APCs hydrolyze foreign proteins to fragments. A specific HLA molecule binds to an antigen fragment. The antigen fragment-HLA molecule travels to the surface of the APC and presents itself to a TCR site on a helper T-cell. Accessory molecules, CD4 and CD8, help in the binding to the TCR site. Susceptibility and resistant (protective) HLAs are involved. How susceptibility and resistant HLA molecules determine whether an individual develops T1D is not well understood.

HLAS AND DIABETES MELLITUS

T1D and Class II Genes

HLA Class II DR and DQ genes are the most important genetic factors in T1D. Recently (as we shall learn later) HLA Class I have been shown to have associations with T1D. Class II HLAs DR3 and DR4 (both or one alone) occur in 90–95% of T1D. The locations of these loci have been refined to allele DQB1*0302[7] at the DR4 haplotype[8] and allele DQB1*0201 at the DR3 haplotype (18). Susceptibility to T1D is greater when DQB1*0302 is produced in conjunction with DQB1*0201. The DQB1*0201 and DQB1*0302 both lack an aspartic acid residue at position 57 of the DQ β-chain.

Susceptibility to T1D is further increased when the DQ β-chain lacks aspartic acid at position 57 and has arginine present at position 52 of the DQ

---

[7]The B in the name DQB1 stands for its being a product of the β-chain. An A would stand for being a product of the α-chain.

[8]Haplotype refers to a group of alleles at adjacent locations (loci) on the same chromosome that are transferred together.

α-chain. The hypothesis has been advanced that these positions on the DQ α and β chains are crucial to the binding of antigenic peptides for presentation to T-cells (19). The highest risk is derived from DRB1*03-DQA1*0501-DQB1*0201/DRB1*0401-DQA1*0301-DQB1*0302. Those in the general population who have this genotype have a 5% absolute risk[9] of having T1D by 15 years of age (20).

Another interesting study (21) found an association between the appearance of islet cell antibodies, HLA-DR3/4 DQ8, and the eventual occurrence of T1D. The Diabetics Autoimmunity Study in the Young (DAISY) followed 1972 children for islet autoantibodies (see Chapter 7 for ICA, GADA, IAA, IA-2A) and T1D. Thirty-six of the children tested persistently positive for at least one autoantibody. In this study only the HLA-DR3/4 DQ8 genotype was predictive of progression to T1D. The high risk genotype (T1D by 20 years of age) was determined to be DRB1*04, DQB1*0302/DRB1*0301, DQB1*0201, (DR3/4DQ8). The follow-up to this study was an average of 4.06 years. The progression to T1D from the first determined autoantibody is years if not decades. Therefore individuals found positive for one or more autoantibodies although nondiabetic currently may still progress to T1D. The predictive value of the HLA-DR 3/4 DQ8 genotype and the presence of autoantibodies may be used to identify those who may develop T1D later in life.

Resistance to T1D is found on DR2 at haplotype DQA1*0102/DQB1*0602. Approximately 20% of Americans and Europeans have this protective allele whereas greater than 99% of individuals with T1D do not have this allele. The resistance to T1D conveyed by DQA1*0102/DQB1*0602 is dominant over risk conveyed by several susceptibility haplotypes (22).

The association of some HLA haplotypes with T1D as conveying high, weak, or medium susceptibility and high, weak, or medium resistance (protection) are listed in Table 9.4.

Summary Box 9.4

- HLA Class II DR and DQ genes are strongly associated with T1D.
- HLA Class I have been recently associated with T1D.
- T1D susceptibility is enhanced by lacking aspartic acid at position 57 of the HLA DQ β-chain and has arginine at position 52 of HLA DQ α-chain.
- Islet autoantibodies appear in children before T1D.
- The protective HLA DQA1*0102/DQB1*0602 is rare in individuals who have T1D.

---

[9]Absolute risk is the probability that a specified event (occurrence of T1D) will occur in a defined population (those carrying the genotype DRB1*03-DQA1*0501-DQB1*0201/DRB1*0401-DQA1*0301-DQB1*0302). Thus where Cl is the total number of cases of T1D within an age range, absolute risk is Cl times the ratio of the frequency of the above genotype in those who have T1D to the frequency of this genotype in the general population (controls).

TABLE 9.4   HLA Haplotypes and High, Weak, or Medium Susceptibility and Resistance for Type 1 Diabetes

High susceptibility
DQA1*0301-DQB1*0302-DRB1*0401 (or 0402, 0405)
DQA1*0501-DQB1*0201-DRB1*0301

Medium susceptibility
DQA1*0101-DQB1*0501-DRB1*0101
DQA1*0301-DQB1*0303-DRB1*0901
DQA1*0401-DQB1*0402-DRB1*0801

High resistance
DQA1*0101-DQB1*0503-DRB1*1401
DQA1*0102-DQB1*0602-DRB1*1501
DQA1*0201-DQB1*0303-DRB1*0701

Weak or medium resistance
DQA1*0201-DQB1*0201-DRB1*0701
DQA1*0301-DQB1*0302-DRB1*0403
DQA1*0501-DQB1*0301-DRB1*1101

Modified from Reference 19.

T1D and Class I Genes

Although Class II alleles are the principal effectors for T1D recently T1D association with Class I loci has been discovered. A large international collaborative study (23) from the Type 1 Diabetes Genetics Consortium has strongly associated B*5701 and B*3906 with T1D. Other alleles also associated with T1D are A*2402, A*0201, B*1801, and C*0501. These alleles are susceptibility alleles. Resistant (protective) alleles are A*1101, A*3201, A*6601, B*0702, B*44023, B*3502, C*1601, and C*0401. The subjects were from the Asia Pacific, Denmark, Europe, United States, Sardinia, and the United Kingdom. These HLA Class I alleles were associated with T1D independent of HLA Class II alleles.

Another study (24) showed a protective effect of A*03 whereas B*39 had a promoting effect. In this study from Finland, B*39 had a strong effect on the progression from β-cell autoimmunity to clinical disease only when the subjects also carried the Class II haplotypes DR3/DR4 genotype. The subjects were chosen as follows:

- The subjects were Finnish newborns.
- They were screened for HLA-DQ-associated-genetic risk, that is, HLA Class II alleles that carry susceptibility genes for T1D.
- Those that were genetically susceptible to T1D tested for ICA.
- Those who tested positive for ICA were tested for IAA, GADA, and IA-2A.

- Those who tested positive for two or three of the autoantibodies were then subjected to testing for HLA-A and -B alleles.

HLA-A*03 was associated with resistance and HLA-B*39 with progression to T1D in children with the HLA-DR3/DR4, but was not associated with T1D in the absence of this genotype. This study also demonstrated that individuals carrying Class I and II HLA susceptibility alleles progress in stages from the presence of one autoantibody to 2–3 autoantibodies before manifesting T1D.

NON-HLA T1D PROMOTING ALLELES

During the Human Genome Project, regions of chromosomes, which were associated with T1D, were identified. They were called *IDDM* (the outdated name for T1D, insulin-dependent diabetes mellitus). The HLA region was called *IDMM1*. At least 15 IDDMs were named. Although not much used today, they do appear currently in some publications. In Table 9.5 genes that have been associated with T1D are listed with the name of their locus, the chromosome, their location, and where possible their IDDM number.

As we have already discussed, genes play an important role in development of T1D. Included in Table 9.5 are genes for which there is strong evidence that they are associated with T1D. Those genes that are "candidates" or in the status of "possibly" or "maybe" for T1D associations are left out until stronger evidence is available to link them to T1D. The references used to compile the table were (4, 18, 19).

GENETICS OF T2D

Numerous studies have been reported that link gene clusters to increased risk of T2D. Table 9.6 lists some of the genes that have been associated with T2D. The listing does not include the newly found genes in the three studies we will discuss in this chapter.

Recently geneticists have used a new tool in their search for genes that are associated with diabetes. This tool enables researchers to discover new genes that promote T2D. The tool is a genome-wide association study (GWAS) (sometimes called a *whole genome association study*, WGAS). It is an examination of common genetic differences among individuals (called *genetic variants*) to learn if any (from our perspective) is associated with T2D. When the variant (usually a gene) is frequently associated with T2D then it is considered to be related to the disease. Previously studies concentrated on genetic regions and related them to the disease. GWAS focuses on the entire genome.

TABLE 9.5 The Gene Name, Chromosome, Location, IDDM Number, and Function for Non-HLA Genes Associated with T1D

Name	Chromosome	Location	IDDM	Function
Cytotoxic T-lymphocyte antigen (CTLA-4)	2	q33	12	Codes for a receptor on the surface of activated T-cells that controls (CTLA-4) proliferation of activated T-cells. Alanine substituted for threonine at codon 17 has been associated with susceptibility for T1D. Disruption of this gene may result in activated T-cells attacking self-antigens (such as β-islet cells)
INS	11	p15.5	2	This locus has been linked to susceptibility for T1D. It contains a VNTR in the promoter region that controls the quantity of insulin produced. VNTR class 1 (26–63 repeats) determines high risk for T1D. Class 3 (140–200 repeats) determines resistance to T1D. Most important susceptibility factor after HLA
Protein tyrosine phosphatase non-receptor type 22 (PTNP 22)	1	p13	—	Codes for lymphoid protein tyrosine phosphatase and is associated with susceptibility to T1D. A single nucleotide polymorphism (SNP)* at base pair 1858 is associated with susceptibility to T1D
Interleukin 2 receptor (IL2RA)	10	p15	—	IL2RA (see IL2 in Fig. 9.3) is found on Tc and Th cells and has been associated with T1D. The SNP ss† 52580101 is closely associated with T1D susceptibility to T1D
Interferon-induced helicase C containing protein 1	2	q24.3	—	IFIH1 is strongly associated with T1D susceptibility. A strong association between T1D and SNP rs‡ 1990760 has been found. IFIH1 releases interferon-γ which promotes the apoptosis of cells which are infected by viruses. The importance of this effect will be discussed later in this chapter

VNTR, variable number tandem repeat region.; INS, insulin

*SNP refers to a substitution of a base for another base in DNA. For example, thymidine (T) substitutes for adenine (A).

†SNP ss followed by an id number refers to a submitter id.

‡SNP rs followed by an id number refers to a reference id. These numbers are catalogued by the National Center for Biotechnology Information (NCBI). SNPs cataloged in the NCBI database have an rs id assigned to them.

TABLE 9.6 T2D Susceptibility and Insulin Resistance Genes

Phenotype	Gene/Locus	Gene/Locus MIM Number	Location	Comment
{Diabetes, type 2, susceptibility to}	GPD2	138430	2q24.1	Codes for mitochondrial glycerophosphate dehydrogenase
{Diabetes mellitus, noninsulin-dependent}	NEUROD1	601724	2q31.3	Caused T2D in two families
{Diabetes mellitus, noninsulin-dependent}	IRS1	147545	2q36.3	Mutation in IRS1 gene
{Diabetes, type 2}	PPARG	601487	3p25.2	Susceptibility gene Promotes obesity and insulin resistance
{Diabetes mellitus, noninsulin-dependent, susceptibility to}	IGF2BP2	608289	3q27.2	Susceptibility gene Insulin like growth factor 2
{Diabetes mellitus, noninsulin-dependent, association with}	WFS1	606201	4p16.1	Wolfram syndrome
{Diabetes mellitus, noninsulin-dependent}	NIDDM4	608036	5q34-q35.2	Susceptibility gene
{Diabetes mellitus, noninsulin-dependent, susceptibility to}	CDKAL1	611259	6p22.3	CDK regulatory unit
{Diabetes mellitus, noninsulin-dependent, susceptibility to}	HMGA1	600701	6p21.31	
{Diabetes mellitus, noninsulin-dependent, susceptibility to}	ENPP1	173335	6q23.2	Susceptibility gene Phosphodiesterase
Diabetes mellitus, noninsulin-dependent, late onset	GCK	138079	7p13	Hexokinase 4, MODY2
Diabetes mellitus, type 2	PAX4	167413	7q32.1	MODY4
{Diabetes mellitus, noninsulin-dependent, susceptibility to}	SLC30A8	611145	8q24.11	Zinc carrier
{Diabetes mellitus, type 2, susceptibility to}	TCF7L2	602228	10q25.2-q25.3	Susceptibility gene Transcription factor 7
{Diabetes mellitus, type 2, susceptibility to}	KCNJ11	600937	11p15.1	Potassium channel

Disorder	Gene	MIM	Location	Comment
Diabetes mellitus, noninsulin-dependent	ABCC8	600509	11p15.1	Sulfonylurea receptor
{Diabetes mellitus, noninsulin-dependent}	MAPK8IP1	604641	11p11.2	T2D in 4 successive generations Mitogen activated protein kinase 8
{Diabetes mellitus, type 2, susceptibility to}	MTNR1B	600804	11q14.3	Susceptibility gene Melatonin receptor
{Diabetes mellitus, noninsulin-dependent, 2}	HNF1A	142410	12q24.31	Glv→Ser at position 319 Susceptibility gene MODY3
{Diabetes mellitus, type II, susceptibility to}	IPF1	600733	13q12.2	RNA polymerase, MODY4
{Diabetes mellitus, noninsulin-dependent}	IRS2	600797	13q34	IRS2
{Diabetes mellitus, noninsulin-dependent}	LIPC	151670	15q21.3	Hepatic lipase gene IGT progresses to T2D
{Diabetes mellitus, noninsulin-dependent}	SLC2A4	138190	17p13.1	Mutation at GLUT4 gene
Diabetes mellitus, noninsulin-dependent	HNF1B	189907	17q12	Found in Japanese patient Transcription factor 2
{Diabetes mellitus, noninsulin-dependent}	GCGR	138033	17q25.3	Glucagon receptor
{Hypertension, insulin resistance-related, susceptibility to}	RETN	605565	19p13.2	Susceptibility (in Chinese) Resistin
{Diabetes mellitus, noninsulin-dependent, susceptibility to}	RETN	605565	19p13.2	
Diabetes mellitus, type II	AKT2	164731	19q13.2	Murine thymoma viral
{Diabetes mellitus, noninsulin-dependent}	NIDDM3	603694	20q12-q13.1	NIDDM3
{Diabetes mellitus, noninsulin-dependent}	HNF4A	600281	20q13.12	Mutation in Hepatocyte nuclear factor 4-alpha MODY8
{Insulin resistance, susceptibility to}	PTPN1	176885	20q13.13	Protein tyrosine phosphatase non-receptor-type 1

Adapted from www.omim.org/entry/125853?search=niddm.highlight=niddm.

Another tool often accompanies GWAS studies. The tool is meta-analysis. Specifically from our perspective, meta-analysis involves the pooling of several studies relating a variant (gene) to a disease such as T2D. The principal weakness in meta-analysis is the manner in which the studies included have been selected. Bias can enter the overall study from the selection of studies that are included. Although this text does not present the statistical approaches to meta-analysis or the criteria that determine a good meta-analysis from a poor one, there are many excellent statistics textbooks that do.

In studies that relate genes to T1D or T2D, the term SNP appears frequently. SNP is an acronym for single nucleotide polymorphism (sometimes but not always correctly the expression point mutation is used). An SNP occurs in a gene or regulatory region and substitutes in DNA a single base for another base. For example, CCTAATAG to CCTAACAG the T → C is an SNP. Depending on the change, the protein coded for this stretch of DNA could be aberrant or could be unaltered in properties. For example, a nonpolar amino acid, say glycine, could be replaced by a nonpolar amino acid, alanine, resulting in little change to the conformation of the protein. The term missense mutation is used if the base changes results in a nonfunctional protein.

How are the SNPs determined? The answer is a "microarray." The same technology that was developed to produce computer semiconductors is used to build a square of glass (a chip) approximately the size of a thumbnail, which has greater than 400,000 squares etched onto it. Each square is 8 μm in length and contains millions of identical copies (a probe) of one DNA sequence that is about 25 bases in length. There are several commercial companies that produce the SNP microarrays. For most microarrays there are about 10,000 SNPs that can be assayed per chip. As there are some 400,000 squares, these SNPs are duplicated about 40 times thus permitting an exceptional degree of certainty as to whether the SNP matched a portion of the DNA sample. In general the procedure for determining the presence of specific SNPs in human DNA is as follows.

- SNPs are produced as probes to study their presence in human DNA. The International HapMap Project of the National Human Genome Research Institute has developed a haplotype map to find genes and genetic variations that affect health and disease. SNPs that are found on a single chromosome are inherited in clusters. The HapMap is a map of these clusters. The clusters of SNPs are associated with disease in different ways (risk, susceptibility, protection, etc.).
- The piece of DNA to be studied is extracted from a biological sample such as blood, cheek cells, and so on.
- The strands of DNA are isolated, purified, and duplicated using the polymerase chain reaction (PCR).
- The DNA chains are fragmented enzymatically.
- Biotin is added to each strand.

- The DNA sample is washed over the chip(s). Complementary DNA[10] sticks to the biotin tagged probes.
- The DNA strands that were not complementary to the probes are washed away.
- A fluorescent dye is washed over the chip(s). The fluorescent dye sticks to the biotin of the DNA strands that had adhered themselves to the probes by virtue of their complementary structure.
- A scanner is used to identify the fluorescence of the DNA that is bound. Many manufacturers produce microarray scanners with various features.

Now that we are further armed with the necessary understanding of the techniques and language used in genetics we can discuss three recent studies that explored an aspect of T2D and heredity. The first study (25) presented the meta-analysis of GWAS tests of approximately 2.5 million SNPs with respect to fasting proinsulin production in 10,701 nondiabetic adults of European ancestry. The study followed up on 23 loci in up to 16,378 adults. To give you an idea of how meta-analysis is conducted, I will list individual studies used.

- Framingham Study, Precocious Coronary Artery Disease, the Finland study, Diabetes Genetics Initiative. Total 10,701 participants.
- Metabolic Syndrome on Men, Botna Prevalence, Prediction and Prevention of Diabetes, Helsinki Birth Cohort Study, the Ely study, the Hertfordshire study, Uppsala Longitudinal Study of Adult Men, Relationship between Insulin Sensitivity and Cardiovascular Disease, Prospective Investigation of the Vasculature in Uppsala Seniors, Segovia, the Greek Health Randomized Aging Study, and Stockholm Diabetes Prevention Program. Total: 16,378 participants.

The study used four genome-wide commercial microarrays for genotyping. Nine SNPs at eight loci were associated with fasting proinsulin concentrations. Two were new, LARP6 (OMIM 611300) and SGSM2 (OMIM 611418), one MADD (OMIM 231680) was also associated with fasting glucose levels, one PCSK1 (162151) with obesity, four increase the risk for T2D (TCF7L2 (OMIM 602228), SLC30A8 (OMIM 611145), VPS13C/C2CD4A/B, ARAP1 (OMIM 606646).[11] ARAP1 was associated with low glucose levels, improved

---

[10]Complementary DNA is simply the reverse of the probe DNA. For example if a probe has the base sequence ATCTTAG and the strand of DNA TAGAATC a duplex structure will be produced held together by H-bonding, ionic forces, and van der Waals forces. If the DNA strand has a structure with even one mismatched base pair, such as TTGAATC, it will not form a duplex structure and will be washed away in the ensuing step.

[11]The names for these acronyms are: LARP6-La ribonucleoprotein domain family member 6, SGSM2-small G protein signaling modulator 2, MADD-multiple acyl-CoA dehydrogenase deficiency, PCSK1-proprotein convertase subtilism/kexin type 1, TCF7L2-transcription factor 7-like 2, SLC30A8-solute carrier family 30 member 8, VPS13C-vacuolar protein sorting 13 homolog C,

β-cell function, and a lower risk of T2D. This seems to adhere to the T2D resistant or protective function. PCSK1 codes for prohormone convertase 1/3, an enzyme found in both α and β cells and involved in the conversion of intestinal proglucagon to GLP-1 (26) and proinsulin to insulin and C-peptide (27). Hyperinsulinemia coupled with elevated levels of proinsulin and insulin denotes β-cell stress. In fact as you might expect destruction of β-islets occurs with the release of insulin and proinsulin.

The second study (28) of interest determined 12 new variants that confer susceptibility to T2D. Most of the variants involved defects in β-cell function and insulin action. The study utilized meta-analysis of eight GWAS studies of T2D. In the first phase of the study the data collected from several studies were utilized to determine the association of genetic variants and T2D. A total of 8130 T2D subjects were used with 38,987 controls. Subjects were of European origin. In the second phase, the investigators used 34,412 T2D and 59,925 controls (who did not have T2D). The associations between genetic variants and T2D that were found in phase 1 were confirmed in phase 2. Finally the study combined the participants in phases 1 and 2. Chromosomal regions that were not previously studied were the focus of the investigation. For example, a variant was found on the X-chromosome near the DUSP9 gene. Variants were found near the following genes, BCL11A (OMIM 606557) found in African-Americans, ZBED3,[12] KLF4 (OMIM 164177), TP53INP1 (OMIM 606185), CHCHD9,[13] KCHQ1 (OMIM 607542), CCNTD2,[14] HMGA2 (OMIM 600698), HNFIA (OMIM 142410), ZFAND6 (OMIM 610183), PRC1 (603484), and DUSP9 (OMIM 300134).[15]

In addition, the investigators found and confirmed T2D susceptibility with 13 previously reported loci. They found that of the newly found loci and the previously reported loci associated with T2D, four were related to higher fasting insulin probably emanating from "a primary effect on insulin action" and three loci were associated with decreased fasting insulin, thus indicating "beta cell dysfunction."

The last study that will be discussed involves the largest and most comprehensive study of genetic variants associated with T2D todate. The investigators performed a GWAS of 39 multiethnic studies that analyzed 50,000 genetic

---

ARAP1-ArfGAP with RhoGAP domain, ankyrin repeat and PH domain 1. OMIMs were added by the author.

[12]No known OMIM.

[13]" "

[14]" "

[15]The names for these acronyms are: BCL11A-β-cell lymphoma/leukemia 11A, ZBED3-zinc finger, BED type containing 3, KLF4-Kreuppal-like factor 4, TP53INP1-tumor protein 53-inducible nuclear protein 1, CHCHD9-coiled-coil-helix-coiled-coil helix domain containing 2 pseudogene 9, KCNQ1-potassium voltage-gated channel, KQT-like subfamily, member 1, CCNTD2- cyclin D2, HMGA2-high-mobility group AT-hook 2, HNFIA-Codes for MODY3, found in Oji-Cree people, hepatic nuclear factor-ia, ZFAND6- zinc finger, AN1-type domain 6, PRC1-protein regulating cytokinesis 1, and DUSP9- dual specificity phosphatase 9. OMIMs were added by the author.

variants located in or near 2000 genes that had been previously linked to cardiovascular, inflammatory, and metabolic functions. The study included 17,418 subjects and 70,298 controls. One of the unique aspects to this study was that it was across several ethnicities. The study's participants included 14,073 subjects and 57,489 controls of European origin, 1986 subjects, and 7695 controls of African-American origin, 592 subjects, 1410 controls of Hispanic origin, and 767 subjects, 3704 controls of Asian origin. Previous studies had consisted primarily of European ethnicities. SNPs were determined using the 50K SNP KITMAT-BroadCARe (IBC) gene analysis array (a type of chip). T2D was determined on the basis of one of six criteria: ADA criteria, fasting blood glucose $\geq$126 mg/dl ($\geq$7 mmol/L), 2h-PG (from OGTT), nonfasting blood glucose $\geq$200 mg/dl (11.1 mmol/l), or physician report or self-report of physician diagnosed diabetes. Age of diagnosis was $\geq$25 years to decrease the possibility that T1D would be accidently included.

The study (29) identified four loci associated with T2D in individuals of European origins ($p^{16} < 2.4 \times 10^{-6}$); GATAD2A[17], SREBF1 (OMIM 184756), TH/INS (OMIM 191290/176730), and BCL2 (OMIM 151430).[18] Sixteen previously described loci were confirmed as associated with T2D (p value significance also $<2.4 \times 10^{-6}$). Eight loci also were associated with T2D at less than 5.0×10^{-8}. HLA-DQ8 was previously associated with T1D. African-Americans showed associations to T2D within genes TCF7L2 and HMGA2. Previous studies have shown that genetic variations near or at TCF7L2 produce the greatest risk for T2D. Decreased β-cell function occurs in carriers of this gene (30). KCNJ11 (OMIM 600937) has been shown to reduce insulin release during OGTT suggesting impaired β-cell function (31). Additional T2D genes and their function in causing T2D are reviewed in (32) and (33). More information especially with regard to gene loci that were found spanning multiple ethnicities can be determined by reading the actual publication.

Modern techniques such as GWAS, meta-analysis, and SNP microarrays will probably identify many more loci associations with T2D in the near future. After all only about 10% of T2D genetic variation has been discovered at the time this book was written. Further discovery of genes contributing to T2D and investigations of how they fit into the diabetes puzzle will guide the development of new pharmaceuticals for the prevention and treatment of T2D.

Summary Box 9.5

- GWAS concentrates on the whole genome in determining genes that are associated with T2D.

---

[16]Consult a statistics textbook to understand what p (probability) values denote.

[17]No known OMIM.

[18]Names for these acronyms are: GATAD2A- GATA zinc factor domain containing 2A, SREBF1-sterol regulatory element-binding transcription factor 1, TH/INS- tyrosine hydroxylase/insulin, BCL2-B-cell lymphoma 2.

- Meta-analysis pools several studies in relating a gene to T2D.
- A SNP substitutes one base for another.
- SNP microarrays placed on chips serve as probes to study human DNA. Each chip contains thousands of identical DNA copies featuring an SNP that is under study.
- Using the techniques listed above new DNA loci associated with T2D were found.

T1D AND ENVIRONMENT

We started this chapter by alluding to the fact that even MZ twins have only a 27–50% risk of both being T1D not 100% as expected. Some aspect(s) of environment must play an important part in this story. The earliest recorded association between T1D and viruses dates to 1864 when a Norwegian physician, J, Stang reported (34, 35) that a patient of his had mumps shortly before developing T1D. He hypothesized that there may be a link between the two events. Since then there has been growing evidence (36) that there indeed is a connection between viral infections and the onset of T1D. However it was very difficult to obtain absolute proof of the association between viruses and T1D.

Epidemiological studies (37, 38) have shown a seasonal incidence for T1D, which correlates with the occurrence of certain viral infections such as mumps, measles (rubella, sometimes called *German measles*), influenza, and hepatitis. Animal studies have demonstrated that certain viruses such as mumps, rubella, Coxsackie 4B, influenza, adenoviruses, enteroviruses, and cytomegalovirus (CMV) can induce diabetes in some strains of mice. Studies conducted in humans during the 1970s and 1980s of viruses and T1D have been inconsistent and contradictory. In some studies the proverbial question arises; is it the chicken or the egg? Did the virus enter the body before the onset of T1D and trigger the diabetes or did it enter after the destruction of the β-islets by the autoimmune process?

How could a virus cause T1D? There are two plausible ways;

- The virus may directly infect and destroy the β-islet cells.
- The virus may promote an autoimmune response, which destroys the β-islet cells.

As to the former mode of action there is clear evidence that viruses infect and destroy β-islet cells in animals but in humans the evidence is murkier. As T1D has been well characterized as an autoimmune disease, we will concentrate on the latter etiology.

How could a virus cause an autoimmune response from a human? There are several possible etiologies summarized below (adapted and modified from

(39)) but the mechanisms by which they activate autoimmunity currently are unknown:

- Virus epitopes may be similar to those on β-islet cells (termed *molecular mimicry*).
  - This results in Tc and antibodies that are directed at the virus and at β-islet cells.
- Primary antibodies directed toward the virus produce secondary antibodies. The secondary antibodies (called *antiviral antibodies*) recognize amino acid sequences (idiotypes) on the primary antibody. If the primary antibody was produced to react for a part of the virus coat that infects the β-islet cell, it would produce antiviral antibodies that will react with both the primary antibody and the β-islet cell protein.
- Epitope spreading refers to activation of virus specific T-cells that recognize viral epitopes. Cellular destruction results in release of antigens that activate T-cells and result in an autoimmune response.
- The virus could alter the β-islet cells in such a way that they are looked at as foreign by the individual's immune system.
  - Alteration of surface proteins on the β-islet cells.
  - Cause development of new proteins.
  - Release sequestered proteins during cell lysis.
  - Incorporate cell proteins in the viral package.
  - Activate HLA class I or II molecules on β-islet cells.
- The virus could alter the immune system.
  - Polyclonal B cell activation resulting in autoantibody production.
  - Release of lymphokines,[19] which recruit immunocytes[20] to β-islet cells.
  - Activation of immune cells resulting in decreased immune tolerance.
  - Disruption of Th1/Th2 immune balance. Th1 cells initiate an immune response to foreign invaders that attack cells by getting into their interiors and Th2 initiate an immune response to foreign invaders that are found outside the cell in the blood and other fluids. A balance between the two is a prerequisite for a good immune system. In animal studies T1D is promoted by a Th1/Th2 ratio that greater than 1 and protection by less than 1.

Below we shall discuss the evidence that implicates some viruses in the development of T1D.

---

[19]Cytokines released by lymphocytes.
[20]Such as B-cells, T-cells.

Enteroviruses (Coxsackie B Virus)

Grabbing the spotlight recently have been the enteroviruses. These very small viruses are found in the gastrointestinal tract and have been related to the onset of T1D. Enteroviruses are common viruses in newborns and young children. They are the agents that produce cold-like symptoms such as fever and muscle aches. A recent meta-analysis of 23 papers and 2 letters included 4448 subjects (40). The studies were classified as pre-diabetes and diabetes. Although study design varied greatly and there was high degree of statistical heterogeneity, meta-analysis showed a significant connection between enterovirus infection and T1D. Presence or recent presence of enterovirus infection was determined by the existence of enterovirus RNA or viral capsid protein in blood, stool, or tissue. There were nine studies that included only pre-diabetic children. The summary (or average of the nine studies) odd ratio was 3.7. For T1D (25 studies) the ratio was 10.0. All told the risk of T1D accompanying enterovirus infections was more than nine times and in children who did not have T1D but did have an autoimmune response, the risk of infection was three times.

Another recent study (41) supports the above research. Enterovirus RNA was found in T1D subjects more frequently than controls. In these cases there was an ongoing inflammation process in the gastrointestinal tract. The conclusion to the study was that T1D subjects have prolonged and persistent enterovirus infections associated with gastrointestinal tract inflammation.

Coxsackie B4 virus is a member of the enterovirus family. There have been several studies connecting T1D with Coxsackie B viral infections. But there also have been some studies that show no correlation between T1D and Coxsackie B. Among the studies that support a correlation between the two is the report of the death of a boy who was afflicted with T1D. The autopsy showed β-islet cell lymphocytic infiltration and β-islet cell death (necrosis) and the presence of Coxsackie B in his pancreas. The isolated virus induced T1D in a strain of mice but not others. A girl who died from T1D and myocarditis also had lymphocytic infiltration and β-islet cell necrosis. Her blood showed high titers of an antibody against the Coxsackie B virus. Other accounts similar in nature have been reported.

Coxsackie virus displays an epitope of its P2-C protein that is found on glutamic acid decarboxylase (GAD) (molecular mimicry) (see Fig. 9.4), which might explain the immune response that results in β-islet cell destruction. Some studies suggest that enteroviruses are less prevalent in countries with high incidences of T1D. In other countries the reverse is seen. This leads to the concept of "hygiene hypothesis." The hygiene hypothesis proposes that environmental exposure early in life to viruses promotes an immune response to the virus and thus protects the individual against a full-blown infection. If enterovirus does cause T1D then immunity to this virus results in fewer cases of T1D. In Western cultures the infant is protected against infection and this may produce children who do not build up immunity to enteroviruses, become infected with the virus, and thus are susceptible to T1D. The hygiene hypothesis explains how T1D diabetes can

Coxsackie virus protein P2-C	P 38	E	V	K	E	K 43
GAD$_{65}$	P 260	E	V	K	E	K 265
GAD$_{67}$	P 268	E	V	K	T	K 273

This sequence of amino acids represents the longest stretch that is similar in GAD$_{65}$ and GAD$_{67}$ and the protein P2-C found in Coxsackie virus. Nearby single amino acids in the three proteins also can be aligned. They are:

Protein	Amino acids	Position
GAD$_{67}$	Lysine	265
GAD$_{65}$	Lysine	257
Coxsackie	Lysine	38
GAD$_{67}$	Leucine	281
GAD$_{65}$	Leucine	273
Coxsackie	Leucine	50

P = Proline
E = Glutamic acid
V = Valine
K = Lysine
T = Threonine

Figure 9.4 Shown in this figure is the longest amino acid sequence common to Coxsackie virus protein P2-C and GAD$_{65}$ and GAD$_{67}$. Single amino acids of P2-C that align with amino acids present in GAD$_{65}$ and GAD$_{67}$ are also shown. Adapted from Kaufman DL, Erlander MG, Clare-Selzier M, et al. Autoimmunity to two forms of glutamate decarboxylase in insulin-dependent diabetes mellitus. J Clin Invest 1992;89:283.

be prevented by viral infection. This hypothesis may be applied to any of the viruses we shall discuss (42).

Rubella Virus (German Measles)

Congenital rubella syndrome (CRS) has been associated with T1D. ICAs and anti-insulin antibodies have been found in 50–80% of T1D infected with CRS. In contrast 20% of non-T1D patients were found to have these antibodies (43). Rubella virus has been shown in animal models to infect β-islet cells. The virus may induce T1D by molecular mimicry. In an investigation of this possibility, diabetic antibodies were absorbed on rubella virus capsid protein, then eluted and reacted with 52 kDa insulinoma[21] protein. The antibodies were absorbed by the 52 kDa insulinoma protein suggesting that the region to which the antibodies adhered to on the virus capsid protein were similar in structure to that of the β-islet cells (44).We speculate that with the advent of live attenuated rubella vaccinations, CRS has been eliminated as a cause of T1D.

Mumps Virus

The mumps virus is one of first viruses that were linked to the onset of T1D. It has been reported that T1D occurs 2–4 years after a mumps infection (45).There has been contradictory evidence as to the effect on T1D incidence due to the mumps vaccine, which limited the number of cases of mumps in young children.

Cytomegalovirus

CMV infection is common in humans. The CMV remains latent in tissues after an acute CMV infection (46). The strongest evidence for a connection between T1D and CMV infection is derived from a study (47) that showed that 22% of T1D patients had the presence of a CMV specific genome in their blood; but otherwise a connection between CMV and T1D needs further investigation.

Retrovirus

Retrovirus-like particles were observed by electron microscopy in the cyto-plasm of β-islet cells of deceased T1D patients. None were observed in non-diabetic patients. Insulitis was present in all the T1D patients accompanied by macrophages, CD8 Tc, and natural killer (NK) cells. None of the nondiabetic patients showed insulitis (39). This is the best evidence so far that human T1D is associated with a retrovirus. In addition molecular mimicry was suggested in a study in which anti-insulin autoantibody-positive sera was found to contain immunoglobulins that bind both insulin and a retroviral protein called *p73* (48). There have been many animal studies that suggest the association of T1D with retroviruses.

Reovirus and Rotavirus

Although reovirus has been related to T1D in animals, there is little evidence for its association with T1D in humans. Rotavirus, a genus within the reovirus family, may be linked to T1D because of the report that its protein coat contains

---

[21]52 kDa protein refers to 52 kilodalton protein. Insulinoma is a tumor that replicates β-islet cells and thus produces an excess of insulin.

amino-acid sequences in common with T-cell epitopes[22] of the autoantigens GAD and tyrosine phosphatase (IA-2). Again the mechanism of molecular mimicry is suggested. Repeated rotavirus infection resulted in increased insulin autoantibodies (most strongly), GAD65 antibodies, and tyrosine phosphatase autoantibodies (49). Rotaviruses are a common cause of gastroenteritis in children.

Epstein–Barr Virus

The Epstein–Barr virus (EBV) is the agent that is responsible for infectious mononucleosis. The GPPAA in the region of aspartic acid-57 of HLA-DQ8 on the β chain also occurs in six successive repeats in the BERF4-encoded EBNA3C protein of the EBV. Those persons who carry this epitope in the HLA-DQ molecule are susceptible to T1D through the mechanism of molecular mimicry. Two of seven individuals who were infected with EBV produced antibodies against an EBV-derived peptide, GPPAAGPPAAGPPAA. These two individuals developed T1D immediately after infection. The other five individuals who were infected but did not produce antibodies directed against these five amino-acid sequences did not develop T1D (50).

Viruses that Need More Evidence for the Assumption that They Promote T1D in Humans

Hepatitis A, Measles virus (not German measles), Influenza virus, Varicella Zoster, and Polio virus

Viruses That Produce T1D in Animals but so Far no Evidence in Humans

The following viruses have been implicated in producing T1D in animals but so far there is no evidence that they promote T1D in humans.

Encephalomycarditis virus, Kilham rat virus, Mengovirus, Bovine viral diarrhea-mucosal disease virus, and Ljungan virus.

Summary Box 9.6

- Even in monozygotic twins the risk of having T1D from heredity is 27–50%; the rest must be from environmental factors.
- The occurrence of T1D correlates seasonally with the occurrence of certain viral diseases such as measles and mumps.
- Lack of viral causation of T1D is proof of whether the virus entered the body before or after the onset of T1D.

---

[22]T-cell epitopes are present on the surface of antigen-presenting cells such as macrophages and dendritic cells. They are then bound to HLA molecules. The epitopes are amino-acid sequences of 8–11 for HLA class I and 13–17 for class II.

- Viruses could cause T1D by directly infecting and destroying β-islet cells or promoting an autoimmune response that destroys the β-islets. The former hypothesis is not as well proven in humans as it is in animals.
- There are several hypotheses as to how viruses can cause an autoimmune response in humans.
- The best-known hypothesis is the molecular mimicry where virus epitopes are similar to those on β-islet cells.
- Currently enteroviruses and coxsackie B virus have been focused on as a cause of T1D in genetically susceptible persons.
- Other viruses that have been implicated are rubella virus, mumps virus, CMV, retrovirus, reovirus, rotavirus, and EBV.

OTHER ENVIRONMENTAL FACTORS

In this section we will discuss several additional environmental factors that have been studied as precipitating factors for T1D. They are cow's milk and vitamin D.

Early Exposure to Cow's Milk as Opposed to Breast Milk

Only a few studies will be mentioned here among many. But the flavor of the arguments in favor and against a connection between cow's milk (or formula milk) and T1D will become evident. The association between cow's milk administered at an early age and T1D remains quite controversial. Cow's milk became associated with T1D in the middle 1980s, when researchers found that diabetes-prone rats on a diet free of cow's milk for the first 2–3 months of age did not contract the disease (51). In 1993, a Finnish study showed that the risk of T1D was increased to 1.5 in children for whom breast-feeding was terminated at less than 2 months of age and increased to 2 for those who were fed dairy products less than 2 months of age (52). A mechanism for the association of cow's milk with T1D was advanced with the discovery of a 17 amino-acid stretch (from 152 to 168) in bovine serum albumin that reacts with p69 protein found in a β-islet cell surface protein. The one letter abbreviation for this segment in bovine serum albumin is ADEKKFWGKYLYEIARR. The human amino-acid sequence is HDNEETFLKKYLYEIAR. The epitope was called *17-amino-acid bovine serum albumin peptide* (ABBOS). Thus the mechanism that was initiated by cow's milk to cause T1D was considered to be molecular mimicry (53). The corresponding amino acid sequence in breast milk does cause an antibody response. A Swedish study (54) used a questionnaire that was administered at birth, 1 year and 2½ years of age, and testing for β-islet cell autoantibodies (GADA and IA-2A) in 7208 2½ year old children. Six hundred fifty seven children had either IA-2A or GADA and thirty-eight had both. Cessation of breast feeding when the child was less than 2 months of age was associated with the occurrence of both IA-2A and GADA. The early introduction of cow's milk and the late introduction of gluten-containing food gave an odds ratio of 6.0 for the occurrence of at least one autoantibody at 1 and 21 years of age.

Although there are several other studies that affirm the association between early feeding of cow's milk (instead of breastfeeding) and T1D, there are many convincing studies that contradict this hypothesis. The DAISY (55) screened 253 children from families that had a first degree relative with T1D for β-islet cell autoimmunity. Eighteen cases showed at least one occurrence of insulin autoantibody, GADA, or tyrosine phosphatase autoantibody. There was no difference in the proportion of children that exhibited autoantibodies and those that did not with respect to exposure to cow's milk or foods containing cow's milk. An Australian study (56) using 317 children with a first-degree relative with T1D concluded that there was no association between duration of breast feeding or introduction of cow's milk and the development of islet autoimmunity. The researchers suggested that breast milk protects children from enteroviral infections and thus T1D.

Vitamin D

Vitamin D_3 (cholecalciferol) is a steroid similar to cholesterol that is produced from 7-dehydrocholesterol by sunlight (UV). Vitamin D_3 is hydroxylated in the liver to 25-hydroxy-D_3. Next in the kidneys 25-hydroxy-D_3 is hydroxylated further to 1,25-dihydroxyl-D_3. 1,25-Dihydroxyl-D_3 activates enzymes in the intestinal mucosa to produce calcium-binding protein, required for the intestinal absorption of calcium, stimulates reabsorption of bone calcium, and promotes reabsorption of calcium in the kidney. In animal studies the administration of vitamin D_3 protects against T1D. In humans a large-scale study[23] (57) (820 patients and 2335 controls) suggests that the administration of vitamin D_3 in early childhood protects against T1D. Breast milk does not contain enough vitamin D_3 to satisfy the bodies' requirements. Cow's milk is fortified with vitamin D_3. Unless supplements containing vitamin D_3 are given to infants fed breast milk, the infant will be deficient in the vitamin. In climates such as in the Scandinavian countries where sunlight is limited, vitamin D_3 is found at lower levels in the blood than in countries where the sunlight is more prevalent. It is therefore plausible that without the supplements of vitamin D_3 T1D risk would be higher especially in countries with less sunlight. It has been speculated that the constant increase of the incidence of T1D in Finland might be related to lower dose requirements (4000–5000 IU/day in the early 1960s to 400 IU/day in 1992) and a lack of compliance on the part of the citizenry (58).

---

[23] The countries that participated in the study were Austria, Bulgaria, Latvia, Lithuania, Luxembourg, North Ireland, and Romania.

Summary Box 9.7

- Several studies suggest that the early administration of cow's milk is associated with T1D.
- The mechanism of this association is speculated to be molecular mimicry due to ABBOS, a 17 amino-acid sequence found both in bovine serum albumin and β-islet cells.
- There have been several studies that contradict the above hypothesis.
- Vitamin D_3 has been associated as a preventive to the occurrence of T1D.
- Complicating studies of cow's milk and its association with T1D is the fact that vitamin D_3 is now added to cow's milk and formulas.

SUMMARY

The viruses listed earlier have been implicated in causing T1D in genetically susceptible individuals. The principal mode of action by these viruses appears to be molecular mimicry although there are other mechanisms that have been alluded to at the beginning of this section. Animal models show a whole range of the mechanisms. The direct causative effects in humans are hard to determine. On the other side of the coin, viral infections can build immunity to viruses and thus decrease the incidence of T1D.

Other environmental factors affecting the incidence of T1D include cow's milk and vitamin D_3. Complicating the situation in this instance is the fact that vitamin D_3 is believed to lend protection against T1D and is not present in breast milk. To adjust for this, newborns are given formulas to which vitamin D_3 is added. Also cow's milk is fortified with vitamin D_3. Nowadays ricketts is a health problem only for underdeveloped nations. For the most part, studies of T1D and its association with vitamin D_3 have not until recently taken the dietary supplementation of vitamin D_3 into account. Possibly new studies that have adequate vitamin D_3 in their diets will show definitively whether cow's milk (or formula) promotes the onset of T1D in genetically susceptible infants.

GENES AND OBESITY

The principal environmental factor promoting T2D is obesity. Like T1D and T2D, overweight and obesity are caused by many characteristics including poor nutrition, a sedentary lifestyle and heredity. Sushruta (see Chapter 2) recognized "obesity, voracity . . . increased soporific tendency and inclination for lounging in bed or on cushions". as characteristics of a second class of diabetes. But of course he didn't recognize there were genes that made an individual prone to obesity. In fact he didn't know that there were genes.

Recently two genes have been implicated in the onset of childhood obesity, the fat mass and obesity associated gene (FTO) and the Kruppel-like factor gene (KLF14).

The FTO Gene

Four studies implicating FTO were published within months of each other in 2007. In one study (59) SNPs at the first intron on the FTO gene located on chromosome 16q12.2 were found to be strongly associated with early childhood obesity in both children and adults of Europeans. They used 2900 obese individuals and 5100 controls.

A second study (60) included 38,759 Caucasian European subjects. These investigators studied the association of an FTO gene variation called *reference SNP* (rs) 9939609 and the BMI of the participants. Those individuals who were homozygous for this allele weighed approximately 3 kg (6 lb and 9.82 oz.) more than those not having the variant. This made their odds of obesity 1.67-fold greater than those without the variant. Sixteen percent of adults in the study were homozygous for the rs9939609 allele.

A GWAS of T2D associated another FTO variant, rs8050136, with obesity. The variant was found in a Finnish population (61). In a letter sent to the journal *Diabetologia*, researchers described (62) a study that associated the rs8050136 with a reduced insulin effect on β-islet activity (insulin resistance). You may recall that insulin resistance is a hallmark of T2D.

Thus far, the FTO variants studied were in Caucasian populations. The next study that we shall consider was a GWAS conducted in the Greater Philadelphia, Pennsylvania region of the United States and involved 418 Caucasian obese children and 2270 Caucasian controls. Also included in the study were 578 African-American obese children and 1424 African-American controls. Two SNPs were researched, rs3751812, which presented significant risk for obesity in African-Americans and Caucasians and rs8050136, which presented significant risk in Caucasians only. RS9939609 was not included in the study. When present, rs3751812 and rs 8050136 conferred a similar order of risk for obesity {odds ratios 1.27–1.3) as did rs9939609 (reference 60) (odds ratio 1.7). We can conclude from the above studies that polymorphisms occurring at the FTO gene locus promote obesity in children and adults across two ethnicities, Caucasian American and Europeans and African-Americans. The metabolic pathways that are affected by the FTO gene are not currently known but a clue to what they involve was provided by the next research study (63).

The FTO gene product (an enzyme) removes methyl groups from DNA and RNA (making it a demethylase). It reacts preferentially with 3-methylthymidine in single stranded DNA and 3-methyluracil in single-stranded RNA. The natural substrates for the FTO enzyme are unknown. What is known is that the FTO enzyme has a loop that competes with the unmethylated strand of double-stranded DNA for binding to FTO. This information alone can make FTO a target for pharmaceuticals that can affect the enzyme.

The KLF14 Gene

Previously an SNP rs4731702 located near the KLF14 gene has been implicated with T2D and high-density lipoprotein (HDL) cholesterol levels (28). You may recall that HDL cholesterol is the "good" cholesterol and the higher the levels in the blood, the less likely cardiovascular disease will develop in an individual. In a more recent paper (64) the investigators believe that they have found the "master switch" that links T2D, cholesterol, and obesity. It appears that they have discovered a specific gene that controls the complex of conditions that comprise the *metabolic syndrome* that was discussed in Chapter 6. Although this research still needs more collaborative studies, it does seem promising.

A little more background in genetics is needed to fully understand the study. It simply comes down to this: a cis-regulatory unit is a binding site on DNA for a protein (or other soluble molecule) that is produced by another site on the DNA called the *trans-regulatory site*.

The study showed that a cis-regulatory locus called the *cis-acting expression quantitative trait locus* (eQTL) found near the maternal KLF14 gene acts as a master trans-regulator of adipose expression. The study associated SNP rs4731702 and its expression in an array of 16,663 genes from the adipose tissue biopsies of 776 healthy female twins. RS4731702 is located close (about 14 kb) to the KLF14 locus. Ten genes showed genome-wide significant trans (GWST) associations driven by rs4731702. The expression levels of six genes were associated with BMI, the same six with HDL cholesterol, five with triacylglyerols (triglycerides), the same five with fasting insulin levels, the same five with HOMA-IR (a measure of insulin sensitivity), three with adiponectin,[24] and two with fasting blood glucose.

Researchers are presently working on defining how the gene affects these processes and how the information can contribute to the treatment of obesity and T2D.

Summary Box 9.8

- Four studies that came out in 2007 associated the FTO gene with early childhood obesity.
- Three of the studies associated variants near or at the FTO gene with obesity of the young.
- A fourth study discovered a variant of FTO that affected African-Americans.
- The KLF14 gene showed that a cis-acting eQTL acts as a master trans-regulator of adipose expression.

---

[24]Adiponectin is secreted as a 244 amino-acid protein from adipose tissue. It is involved in glucose regulation and fatty acid catabolism.

- Essentially a trans-regulator produces a protein that binds to the promoter region of another gene. For KLF14 it affects 10 genes that regulate metabolic pathways, which control BMI, HDL-cholesterol, triacylglycerols, insulin, glucose metabolism, and insulin resistance.

PROJECTS

9.1 Further investigate newly found metabolic pathways regulated by the FTO gene.

9.2 Search for new studies concerning the metabolic pathways regulated by the FTO gene region, especially with respect to HDL cholesterol, weight, triacylglycerol formation, insulin biosynthesis, glucose metabolism, and mechanisms of insulin resistance.

GLOSSARY

Absolute risk   Probability that a specified event will occur in a define population.

Antigen presenting cell (APC)   Proteosomes in APC hydrolyzes proteins to fragments that bind to HLA proteins. The fragment-HLA protein is presented on the surface of the cell. APCs include macrophages and dendritic cells.

Autoantibodies   Proteins produced by the immune system, which are directed against the individual's tissues and organs.

Autoantigens   An antigen normally found in an individual against which antibodies are produced.

Cis-regulatory unit   Binds proteins (from the trans-regulatory unit) that regulate its expression.

Concordance   Both siblings of a pair of twins have T1D.

Cytokines   Proteins secreted by cells to communicate with other cells.

Cytotoxic T-cell (Tc)   Cytotoxic T-cells are activated by cytokines released by Helper T-cells. They attack foreign antigens.

Disconcordance   One of a pair twins has T1D.

Discordance   See definition of disconcordance.

Dizygotic   Derived almost simultaneously from the fertilization of two eggs.

Epitope   Refers to a sequence of amino acids.

Genome   The genome constitutes all of the genes on all of the chromosomes in an organism.

Genome-wide association study (GWAS)   Study of variant genes across the entire genome.

Haplotype   Refers to a group of alleles at adjacent loci on the same chromosome that are transferred together.

Helper T-cells (Th)   Recognizes antigen on the surface of APC and binds them at the TCR site.

Heritability   Refers to the proportion of the total variation of T2D attributable to genetic variation. Calculated by $h^2 = 2(r_{MZ} - r_{DZ})$ where r is a specific inherited body measurement or metabolic variable.

Immunocytes   Cells involved in the immune system such as B- and T-cells.

Insulinoma   A cancer that replicates β-islet cells and secretes excess insulin.

Lymphokines   Cytokines released by lymphocytes.

Meta-analysis   The pooling of several studies relating a variant gene to disease.

Microarray (or array)   Thousands of copies of a segment of DNA or RNA on a square of glass (a chip) that serve as probes to determine whether a particular complementary sequence occurs in a sample DNA. The presence of specific SNPs can be determined in this manner.

Molecular mimicry   The same amino acid sequence appears on another protein. If antibodies or T-cells are produced that recognize this sequence then both proteins will be treated as foreign bodies and will be destroyed.

Monozygotic   Derived from one fertilized egg.

Pairwise concordance   Number of twins who both have T1D divided by the sum of the number of twins in which only one has T1D and the number of twins who have T1D. The probability that both twins have T1D in a population is at least one twin has T1D. Pairwise concordance $= C/(C + D)$ where C = number of concordant pairs and D = number of discordant pairs.

Probandwise concordance   The proportion of twins in which only one has T1D. Probandwise concordance $= 2C/(2C + D)$. See pairwise concordance symbols.

Proteosomes   A complex of proteases that degrade proteins to peptides.

Resistance   Protection

Single nucleotide polymorphism (SNP)   A single base substituting for another base in the DNA of a specific organism. Sometimes called a point mutation. The acronym SNP is pronounced snip.

T-cell receptor (TCR)   Site on helper T-cell that binds the HLA-antigen and secretes cytokines.

Trans-regulatory unit   Produces a protein that regulates the cis-regulatory unit by binding to the promoter region of the gene.

REFERENCES

1. Smith C. Concordance in twins: methods and interpretation. Am J Hum Genet 1974;26:454.

2. Hyttinen V, Kaprio J, Kinnunen L, et al. Genetic liability of Type 1 diabetes and the onset age among 22,650 young Finnish twin pairs. Diabetes 2003;52:1052. DOI: 10.2337/diabetes.52.4.1052.

3. Akaike H. A new look at the statistical model identification. System identification and time-series analysis. IEEE Trans Autom Control 1974;19:716.

4. Redondo MJ, Fain PR, Eisenbarth GS. Genetics of Type 1A diabetes. Recent Prog Horm Res 2001;2001:69.

5. Barnett AH, Eff C, Leslie DG, et al. Diabetes in identical twins: a study of 200 pairs. Diabetologia 1981;20:87.

6. Poulsen P, Kyvik KO, Beck-Nielsen H. Heritability of Type II (non-insulin-dependent) diabetes mellitus and abnormal glucose tolerance-a population-based twin study. Diabetologia 1999;42:139.

7. Harvald B, Hauge M. Hereditary factors elucidated by twin studies. In: Neel JV, Shaw MW, Schull WJ (eds.) *Genetics and Epidemiology of Chronic Diseases*. Public Health Service; 1965. Publication 1163. Washington DC: Public Health Service 1965, p. 65.

8. Gottlieb MS, Root HF. Diabetes in twins. Diabetes 1968;17:693.

9. Tattersall RB, Pyke DA. Diabetes in identical twins. Lancet 1972;2:1120.

10. Ganda OP, Soeldner SS. Genetic, acquired, and related factors in the etiology of diabetes mellitus. Arch Intern Med 1977;137:461 Symposium on Diabetes Mellitus.

11. Kahn CB, Soeldner JS, Gleason RE, et al. Clinical and chemical diabetes in offspring of diabetic couples. N Engl J Med 1969;281:343.

12. Rjasanowski I, Kolting I, Kerner W. Frequency of diabetes transmission from two Type 1 diabetic parents to their children. Diabetes Care 2003;26:2219. DOI: 10.2337/diacare.26.7.2219-a.

13. Tattersall RB, Fajans SS. Prevalence of diabetes and glucose intolerance in 199 offspring of thirty-seven conjugal diabetic parents. Diabetes 1975;24:452. DOI: 10.2337/diabetes.24.5.452.

14. Meigs JB, Cupples LA, Wilson PWF. Parental transmission of Type 2 diabetes: The Framingham Offspring study. Diabetes 2000;49:2201.

15. Mrena S, Virtanen SM, Laippala P, et al. Models for predicting Type 1 diabetes in siblings of affected children. Diabetes Care 2006;29:662. DOI: 10.2337/diacare. 29.03.06.dc05-0774.

16. Nepom GT. A unified hypothesis for the complex genetics of HLA associations with IDDM. Diabetes 1990;39:1153.

17. Sheehy MJ. HLA and insulin-dependent diabetes: a protective perspective. Diabetes 1992;41:123.

18. Mehers KL, Gillespie KM. The genetic basis for type 1 diabetes. Br Med Bull 2008;88:115.

19. Al-Mutairi HF, Mohsen AM, Al-Mazidi ZM, et al. Genetics of Type 1 diabetes mellitus. Kuwait Med J 2007;39:107.

20. Lambert AP, Gillespie KM, Thomson G, et al. Absolute risk of childhood-onset type 1 diabetes defined by human leukocyte antigen class II genotype: a population-based study in the United Kingdom. J Clin Endocrinol Metab 2004;89:4037.

21. Barker JM, Barriga KJ, Yu L, et al. Prediction of autoantibody positivity and progression tp type 1 diabetes: Diabetes Autoimmunity Studenty in the Young (DAISY). J Clin Endocrinol Metab 2004;89:3896.

22. Redondo MJ, Kawasaki E, Mulgrew CL, et al. DR- and DQ-associated protection from type 1A diabetes: Comparison of DRB1*1401 and DQA1*0102-DQB1*0602*. J Clin Endocrinal Metab 2000;85:3793.

23. Noble JA, Valdes AM, Varney MD, . Effect of HLA Class I and Class II alleles on progression from autoantibody positivity to overt type 1 diabetes in children with risk-associated class II genotypes. Diabetes 2010;59:2972.

24. Lipponen K, Gombos Z, Kiviniemi M, et al. Effect of HLA class I and class II alleles on progression from autoantibody positivity to overt type 1 diabetes in children with risk-associated class II genotypes. Diabetes 2010;59:3253.

25. Strawbridge RJ, Dupuis J, Prokopenko I, et al. Genome-wide association identifies nine common variants associated with fasting proinsulin levels and provides insights into the pathophysiology of type 2 diabetes. Diabetes 2011;60:2624.

26. Ugleholdt R, Poulsen M-LH, Holst PJ, et al. Prohormone convertase 1/3 is essential for processing of the glucose-dependent insulinotropic polypeptide precursor. J Biol Chem 2004;281:11050.

27. Zhu X, Orci L, Carroll R, et al. Severe block in processing of proinsulin to insulin accompanied by elevation of des-64,65 proinsulin intermediates in islets of mice lacking prohormone convertase 1/3. Proc Natl Acad Sci U S A 2002;99:10299.

28. Voight BF, Scott LJ, Steinthorsdottir V, et al. Twelve type 2 diabetes susceptibility loci identified through large-scale association analysis. Nat Genet 2010;42:579.

29. Saxena R, Elbers CC, Guo Y, et al. Large-scale gene-centric meta-analysis across 39 studies identifies type 2 diabetes loci. Am J Hum Gene 2012;90:410.

30. Florez JC, Jablonski KA, Bayley N, et al. TCF7L2 polymorphisms and progression to diabetes in the Diabetes Prevention Program. N Engl J Med 2006;355:241.

31. Stancakova A, Kuulasmaa T, Paananen J, et al. Association of 18 confirmed susceptibility loci for type 2 diabetes with indices of insulin release, proinsulin conversion, and insulin senstivity in 5,327 nondiabetic Finnish men. Diabetes 2009;58:2129.

32. Staiger H, Machicao F, Fritsche A, et al. Pathomechanisms of type 2 diabetes genes. Endocrinol Rev 2009;30:557.

33. Grarup N, Sparse T, Hansen T. Physiologic characterization of type 2 diabetes-related loci. Curr Diab Rep 2010;10:485.

34. McGlannan F. Diabetes: epidemiology suggests a viral connection. J Learn Diab 1975;8:27.

35. Gunderson F. Is diabetes of infectious origin? J Infect Dis 1927;41:197.

36. Craighead JE. Does insulin dependent diabetes mellitus have a viral etiology? Hum Pathol 1979;10:267.

37. Maugh TH II,. Research news: diabetes: epidemiology suggests a viral connection. Science 1975;188:147.

38. Gamble DR, Taylor KW. Seasonal incidence of diabetes mellitus. Br Med J 1969;13:631.

39. Jun H-S, Yoon J-W. Avances in diabetes through animal-related research: a new look at viruses in type 1 diabetes. Instit Lab Anim Res (ILAR) 2004;45:349.

40. Yeung W-CG, Rawlinson WD, Craig ME. Enterovirus infection and type 1 diabetes mellitus: systematic review and meta-analysis of observational molecular studies. BMJ 2011;342:d35.

41. Oikarinen M, Taurianen S, Oikarinen S. Type 1 diabetes is associated with enterovirus infection in gut mucosa. Diabetes 2012;61:687.

42. von Herrath M. Can we learn from the viruses how to prevent type1 Diabetes: The role of viral infections in the pathogenesis of type 1 diabetes and the development of novel combination therapies. Diabetes 2009;58:2.

43. Ginsberg-Fellner F, Witt ME, Yagihaski S, et al. Congenital rubella syndrome as a model for type 1 (insulin-dependent) diabetes mellitus: Increased prevalence of islet cell surface antibodies. Diabetologia 1984;27:87.

44. Karounos DG, Wolinsky JS, Thomas JW. Monoclonal antibody to rubella virus capsid protein recognizes a beta-cell antigen. J Immunol 1993;150:3080.

45. Hyoty H, Leinikki P, Reunanen A, et al. Mumps infections in the etiology of type 1 (insulin-dependent) diabetes. Diabetes Res 1988;9:111.

46. Alba A, Planas R, Verdaguer J, et al. Viral infections and autoimmune diabetes. Immunologia 2005;24:33.

47. Pak CY, McArthur RG, Eun HM, et al. Association of cytomegalovirus infection with autoimmune type 1 diabetes. Lancet 1988;332:1.

48. Hao W, Serreze DV, McCulloch DK, et al. Insulin autoantibodies from human IDDM cross-react with retroviral antigen. J Autoimmun 1993;6:787.

49. Honeyman M, Coulson BS, Stone NL, et al. Association between rotavirus infection and pancreatic islet autoimmunity in children at risk of developing type 1 diabetes. Diabetes 2000;49:1319.

50. Parkkonen P, Hyoty H, Ilonen J, et al. Antibody reactivity to an Epstein-Barr virus BERF4-encoded epitope occurring also in Asp-57 region of HLA-DQ8 beta chain. Clin Exp Immunol 1994;95:287.

51. Elliott RB, Martin JM. Dietary protein a trigger of insulin-dependent diabetes in the BB rat? Diabetologia 1984;26:297.

52. Virtanen SM, Rasanen L, Ylonen K, . Early introduction of dairy products associated with increased risk of IDDM in Finnish children. The Childhood in Diabetes in finland Study Group. Diabetes 1993;42:1786.

53. Karjalainen J, Martin JM, Knip M, . A bovine albumin peptide as a possible trigger of insulin-dependent diabetes mellitus. N Engl J Med 1992;327:302.

54. Wahlberg J, Vaarala O, Ludvigsson J. Dietary risk factors for the emergence of type 1 diabetes-related autoantibodies in 2 1/2 year old Swedish children. Br J Nutr 2006;95:603.

55. Norris JM, Beaty B, Klingensmith G, et al. Lack of association between early exposure to cows milk protein and beta cell autoimmunity: Diabetes Autoimmunity Study in the Young. JAMA 1996;276:609.

56. Couper JJ, Steele C, Beresford S, et al. Lack of association between duration of breast-feeding or introduction of cow's milk and development of islet immunity. Diabetes 1999;48:2145.

57. The Eurodiab Substudy 2 Study Group. Vitamin D supplement in early childhood and risk for type I (insulin-dependent) diabetes mellitus. Diabetologia 1999;42:51.

58. Hypponen E, Laara E, Reunanen A, et al. Intake of vitamin D and risk of type 1 diabetes: a birth-cohort study. Lancet 2001;358:1500.

59. Dina C, Meyre D, Gallina S, et al. Variation in FTO contributes to childhood obesity and severe adult obesity. Nat Genet 2007;39:724.

60. Frayling TM, Timpson NJ, Weedon MN, et al. A common variant in the FTO gene is associated with body mass index and predisposes to childhood and adult obesity. Science 2007;316:889. DOI: 10.1126/science.1141634.

61. Scott LJ, Mohike KL, Bonneycastle LL, et al. A genome-wide association study of type 2 diabetes in Finns detects multiple susceptibility variants. Science 2007;316:1341.

62. Tshritter O, Preissl H, Yokoyama Y, et al. Variation in the FTO gene locus is associated with cerebrocortical insulin resistance in humans. Diabetologia 2007;50:2602.

63. Han Z, Niu T, Chang J, . Crystal structure of the FTO protein reveals basis for its substrate specificity. Nature 2010;464:1205.

64. Small KS, Hedman AK, Grundberg E, et al. Identification of an imprinted master trans regulator at the KLF14 locus related to multiple metabolic phenotypes. Nat Genet 2011;43:561.

CHAPTER 10

TREATMENT

PART 1: MEDICINAL TREATMENT

Insulin is not a cure for diabetes; it is a treatment. It enables the diabetic to burn sufficient carbohydrates, so that proteins and fats may be added to the diet in sufficient quantities to provide energy for the economic burdens of life.

—Sir Frederick Grant Banting, from his Nobel Acceptance Lecture, September 15, 1925. Online (1)

INSULIN (EARLY TREATMENT)

In Chapter 2 we followed the history of insulin from Banting and Best's crude preparation to Frederick Sanger's determination of the entire amino-acid sequence of bovine (beef) insulin. Previous to Sanger's characterization of insulin the hormone had been carefully purified and crystallized. Purified bovine insulin was used for many years as the principal drug for both T1D and T2D diabetics. However, although produced in large quantities bovine insulin differs from human insulin by three amino acids. At position 8 in the A chain human insulin has threonine and in contrast bovine insulin has alanine. Position 10 in the A chain human insulin has isoleucine, whereas bovine insulin has valine. Finally position 30 in the B chain human insulin has threonine and bovine insulin has alanine. (You may refer to Fig. 4.1 to see this difference.) These substitutions cause an

Understanding Diabetes: A Biochemical Perspective, First Edition. Richard F. Dods.
© 2013 John Wiley & Sons, Inc. Published 2013 by John Wiley & Sons, Inc.

allergic response in many of its users. Eli Lilly and Company started marketing bovine insulin in 1923.

In 1934 David A. Scott at the University of Toronto reported that "In searching the literature, it was found that pancreas contains appreciable quantities of zinc, cobalt and nickel." In his investigations he learned that the beaker containing insulin in phosphate buffer and acetone to which was added zinc chloride produced rhombohedral insulin crystals (2). The next year Scott and Fisher (3) reported that insulin crystals synthesized in this manner did indeed contain zinc as "chemically combined constituents and not as impurities." Scott and Fisher licensed their discovery to a little known producer of insulin named Novo Nordisk. Novo Nordisk had been founded by Hans Christian Hagedorn (1888–1971) under a charter from the Danish government to produce insulin as a nonprofit foundation in 1926. Hagedorn and his colleague August Krough (1874–1949) had obtained the rights to insulin from Banting and Best.

Insulin had prolonged lives of children from 2 years after diagnosis of T1D to more than 30 years and had reduced quite considerably the cases of ketoacidosis that were encountered. Ironically, as insulin was purified to a greater extent the duration of the effect of insulin decreased. In 1936, Hagedorn and associates (4) discovered that the effects of insulin were lengthened by the addition of protamine[1] to insulin. A clinical trial (5) of zinc protamine insulin controlled the diabetes of their patients as a single dose per 24 h. The difference in the actions of zinc insulin alone and the protamine added insulin seemed to be due to the rates of absorption from the subcutaneous tissues where they are injected. "Soluble insulin begins to lower blood sugar from fifteen to thirty minutes after the time of injection. Zinc protamine insulin shows little, if any, effect on the blood sugar in the first three to six hours after injection." In addition, and most importantly, the action of insulin lasted less than 8–10 h, irrespective of the size of the dose. Zinc protamine insulin acts for 24 h when injected in moderate doses.

It took until 1946 for Nordisk to obtain the protamine porcine insulin in crystalline form. Porcine insulin differs from human insulin in position 30 of the B chain where alanine is present instead of threonine. Nordisk named the protamine porcine injectable insulin neutral protamine Hagedorn (NPH) (also known as *isophane insulin*) insulin. Another insulin that entered the market in the early 1950s was Lente. Lente was composed of porcine and bovine insulins with added zinc and was a slow-acting insulin.

During the 1950s and 1960s, additional insulin products such as semilente and ultalente reached the market. Their basic difference was the quantity of zinc they contained, which determined the time of onset, peak time, and duration of effect.

Parallel to these discoveries a scientist was working on a strategy for the synthesis of peptides that would revolutionize the field and earn him the Nobel Prize. R. Bruce Merrifield (1921–2006) reported a solid-phase method for the

---

[1]Protamine is a small-sized arginine and lysine-rich protein. The numerous arginine and lysine residues make the protein positively charged. It was initially obtained from the semen of river trout.

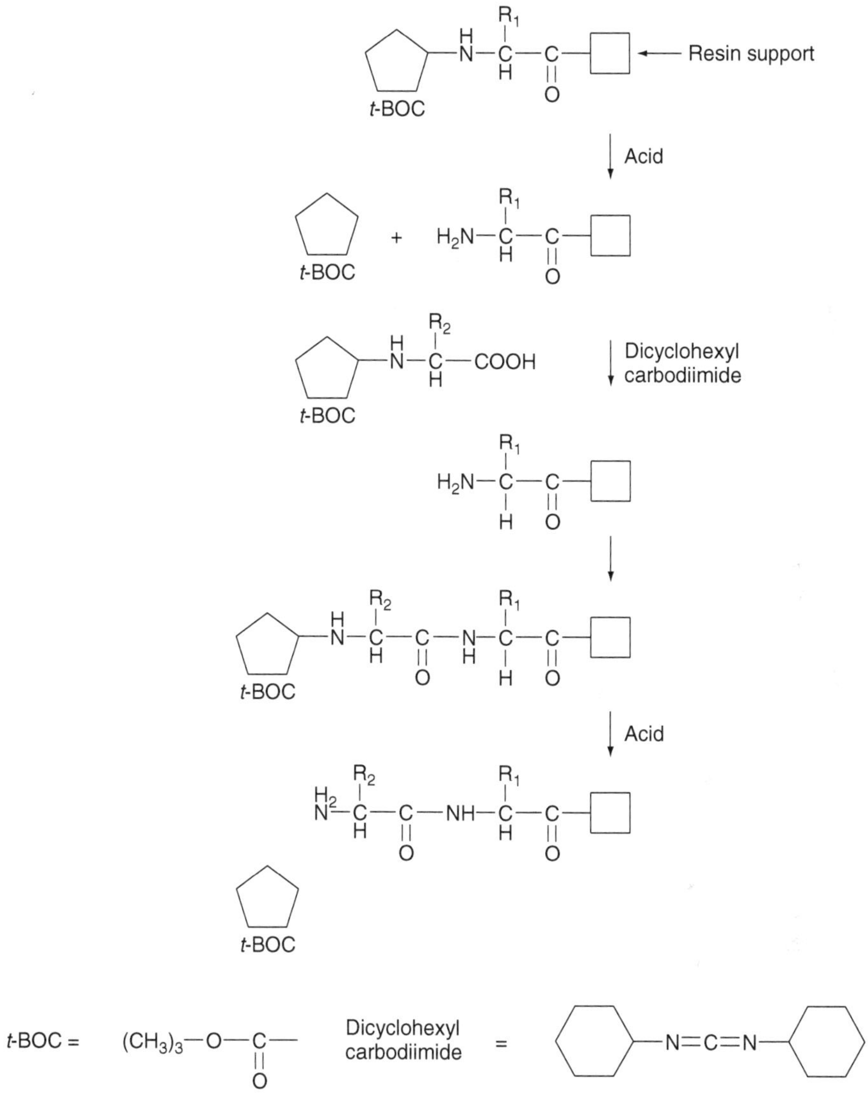

Figure 10.1 Merrified solid-phase approach to the synthesis of peptides.

synthesis of peptides in 1963 (6). As shown in Figure 10.1, Part 1, amino acids are added stepwise to a growing peptide chain. The chain is bound to an insoluble resin, which is separated by centrifugation from the liquid phase. The incoming amino-acid group is coupled to the amino acid of the growing chain by dicyl-cohexylcarbodiimide (DCC). The incoming amino group is protected by a t-Boc ($tert$-butyloxycarbonyl) group that is easily removed on acidification. The reaction for each step is greater than 98%, and therefore, the yield of a short peptide, bradykinin, a nine amino-acid peptide (1060 Da), is 85%. It was synthesized in 27 h. The A chain of insulin was synthesized in 8 days and the B chain was made in 11 days using this method. But the disulfide bridges connecting the A and B chains were not synthesized. In 1966 (7), a synthetic B chain was connected to a natural A chain of bovine insulin by disulfide bonds. The molecule that was created exhibited insulin activity. In 1993, human insulin was produced with the disulfide bridges being produced by utilizing methyltricholorosilane-diphenyl sulfoxide (8).

Unfortunately, these methods could not be scaled up to produce human insulin in bulk and inexpensively.

IT IS NOT YOUR FATHER'S INSULIN ANY MORE
MODERN-DAY HUMAN INSULIN

In 1972, investigators (9) at the School of Medicine, University of California, San Diego, found a method by which porcine insulin could be altered to produce human insulin. As mentioned previously, porcine insulin differs from human insulin by one amino acid at the end (position 30) of the B chain. They replaced chemically and enzymatically the last eight amino acids of the porcine B chain and replaced it with an octapeptide, which had been produced synthetically identical to the last eight amino acids of the human B chain. Thus, human insulin could be produced from porcine insulin inexpensively and in bulk quantities. Novo Nordisk brought this insulin into the market in 1982.

In 1975, Ciba-Geigy of Basel, Switzerland, synthesized human insulin. The Ciba-Geigy synthetic human insulin was used in a study of six patients and was found to be as efficacious as animal-derived insulin (10).

In 1978, a little-known biotechnology company, Genentech, established in 1976, made a revolutionary discovery. It would change forever the way human insulin was produced. They used the procedure that is outlined in Figure 10.2, Part 1, to produce human insulin in massive amounts and at low cost. As you can see from the figure Genentech used recombinant DNA technology to produce human insulin. In 1980, collaboration between Guy's Hospital in London and Lilly Research Centre, also located in the United Kingdom, compared recombinant DNA human insulin to porcine insulin in healthy men (11). The study suggested that the two insulins had similar effects but that the depression of blood glucose for human insulin was a little more potent than the porcine insulin at low doses. The report concluded that the recombinant DNA human insulin was "safe and effective." However, it was soon learned that a disadvantage of human insulin is that it does not alert diabetics to the subtle signs and symptoms that they are slipping into hypoglycemia, which animal insulin does. In 1982 Eli Lilly obtained FDA approval for what were called *Humulin R* and *Humulin N*. The reader should consult a good Biochemistry textbook to learn more about recombinant DNA technology.

1. The human Insulin gene is isolated.

2. The gene is copied multiple times using the polymerase chain reaction (PCR).

3. Circular pieces of DNA (plasmids) are removed from *E. coli* cells.

4. The plasmid is opened and the gene for insulin is inserted producing a recombinant DNA molecule.

5. The recombinant DNA containing the human insulin gene is replaced in the *E.coli*.

6. The *E.coil* are subjected to conditions where they multiply rapidly thus duplicating the insulin gene manifold.

7. The human insulin produced by the *E.coil* is isolated and purified.

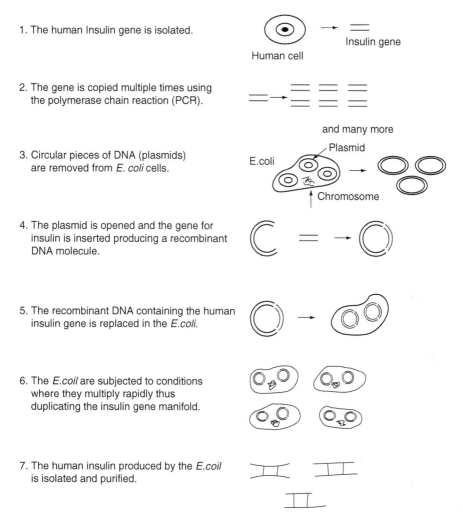

Figure 10.2 Human insulin from recombinant DNA technology. (*See insert for color representation of the figure.*)

Genetically Engineered Insulin Derivatives

Rapid advances in DNA technology resulted in development of new kinds of insulin. As you can see from Figure 10.3, Part 1, after a meal, non-diabetic insulin levels increase rapidly up to 30–45 min and then decrease more slowly to return to basal levels at 2–3 h.

The properties of the insulin preparations on the market do not follow this pattern. But scientists with new DNA technologies at hand set about the search for the insulin that followed the time sequences of natural insulin. Figure 10.4, Part 1, shows the time pattern of some of the insulins.

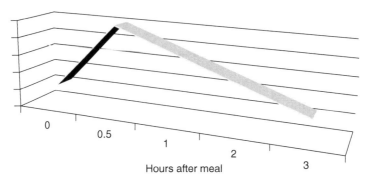

Figure 10.3 Blood Insulin concentrations in healthy persons after a meal. (*See insert for color representation of the figure.*)

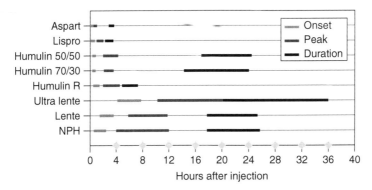

Onset - Length of time before insulin begins to reduce blood glucose
(as a range) (individual variability)

Peak - Length of time when insulin is most effective (as a range) (individual variability)

Duration - Length of time that insulin acts (as a range) (individual variability)

• Humulin 50/50 - consists of 50%, NPH + 50% recombinant DNA human insulin

• Humulin 70/30 - consists of 70%, NPH + 30% recombinant DNA human insulin

Figure 10.4 Onset, peak, and duration of various types of injectable insulin. (*See insert for color representation of the figure.*)

Lispro was the first genetically engineered insulin. Lispro reversed the sequence in the B chain of proline (position 28) with lysine (position 29). The tendency of insulin molecules to form polymers (self-association) was eliminated and therefore Lispro was more easily absorbed resulting in reduced onset times, higher peaks, and shorter duration times. As you can see from Figures 10.3 and 10.4, Part 1, new insulins came closer in their pharmacokinetics to natural insulin. Insulin Aspart, which had aspartic acid replaced by proline, also reduced self-association of the insulin. It too provided a shorter-acting insulin derivative.

Insulin with greater duration of activity was achieved by lengthening the C-terminal end of B chain by two glycine molecules. This produced insulin glargine. The pharmacokinetics of this genetically engineered insulin was not only similar to natural human insulin but also showed a glucose decrease that lasted 24 h. Lower incidences of hypoglycemia were reported from those using glargine and only one dose was needed per day (12).

Insulin glulisine is a rapid acting insulin derivative. Marketed in 2004, it has asparagine substituted for by lysine in position 3 of the B chain and lysine substituted for by glutamic acid in position of 29 of the B chain. It reaches the blood earlier than natural insulin and therefore it is recommended for the patient to inject it 15 min prior to a meal.

In 2006 insulin determir was marketed. Insulin determir is a long-acting human insulin. An alternate approach to prolonging insulin action is to covalently bind the insulin molecule to nonesterified fatty acids. The insulin then complexes with albumin. The complex decreases absorption of the insulin. Insulin determir has myristic acid bound to lysine in position 29 of the B chain. Studies showed insulin determir to have the same incidence of hypoglycemia of other insulins being marketed but with less weight gain.

Other studies of genetically engineered insulin are presently in the animal stages or human clinical trial stages. The ideal insulin would be one with early onset, long duration, no hypoglycemic side effect, and no weight gain.

Summary Box 10.1

- Early injectable insulin was from cows.
- Bovine insulin differed from human insulin by three amino acids and produced for some diabetics an immune response.
- In 1934 insulin was found to contain zinc as a key component.
- Insulins were developed that contained zinc, and protamine.
- Insulin that was derived from pigs came into use in the 1940s. It was only one amino acid different from human insulin.
- In 1963 Bruce Merrifield discovered a solid-phase method for the synthesis of insulin.
- In 1972 researchers learned how to replace the amino acid that made porcine insulin differ from human insulin with the correct amino acid.
- Genentech in the early 1980s discovered how to use recombinant DNA technology to manufacture human insulin in huge quantities and at low cost.
- Since then there have been numerous insulins produced with differences in their amino acid sequences that lead to differences in the onset of action and duration of effect.

Other Modes of Delivering Insulin: Tablets or Capsules, Inhalable Insulin and Nasal Spray Insulin

Production of solid oral or inhalable insulins has been very difficult. As of today only the subcutaneous or intravenous form of insulin is available. For production of solid oral insulin, the insulin must not be released until it passes into the small intestine. There are presently coatings, which can surround the capsule or tablet containing the insulin and protect its journey through the stomach to the small intestine. But the passage of the intact insulin through the intestinal wall into the blood stream is more challenging. Enhancers to permit the insulin molecule to pass into the blood stream are available but no single approach has proved viable. But the work goes on and eventually a tablet or capsule insulin will be produced.

Another route to deliver insulin into the blood stream is inhalation as a powder through the lungs. In fact an inhalable spray has been marketed and became available as Exubera. A study using inhaled insulin and insulin glargine in T2D diabetics has been reported (13). The T2D subjects had poorly controlled glucose levels even though they were on insulin therapy. They received treatment with inhaled insulin powder delivered by an inhaler, particle size between 1 and 5 μm in diameter, reaching deeply into the lungs where alveoli absorbed the insulin into the blood stream. In addition, they were injected with insulin glargine at bedtime. The comparison group was subjects who were treated with injectable biaspart insulin. HbA_{1c} was followed for both groups. After 52 weeks HbA_{1c} was similar in both groups. Those using inhaled insulin had fewer instances of hypoglycemia and gained less weight than the group that used only injections of insulin. However the group that inhaled insulin had a greater number of upper respiratory infections and experienced coughing fits for about 10 min after inhalation. The inhaler was cumbersome being 6-inches long when collapsed and almost a foot long during use and furthermore, it was costly. In 2007, Exubera was taken off the market for lack of sales. At this time some other companies are developing more useful inhaled forms of insulin.

A third method of insulin delivery is the nasal spray referred to in Chapter 8, Reference 39. But it has so far only been used to provide insulin to the brain in experiments for Alzheimer disease.

Closed-Loop Insulin Delivery (Artificial Pancreas)

During recent years we have been closing in on an artificial pancreas for T1D patients. Simply such a delivery system for insulin would consist of a glucose sensor: a transmitter that sends blood glucose readings to a modulator, which, using control algorithms, determines how much insulin should be delivered from an insulin pump into the blood stream. As all of these components must be located on the diabetic, they must be small and wireless. Modern day technologies make this setup feasible. The glucose sensor is attached to the abdominal skin. It sends a wireless signal to the modulator, which in turn sends a signal to the pump. The modulator and the pump are in the subject's pocket with a small diameter tube capped by a thin needle stretching from the pump to the abdominal subcutaneous skin layer. This arrangement of insulin delivery components is called a *closed-loop insulin delivery system*. The glucose sensor can be implanted or not even directly connected to the blood circulation using ultrasound or dielectric technology to read blood glucose levels (called *transcutaneous glucose reading*). The heart of the closed-loop delivery system or artificial pancreas is the modulator. The system that takes the data from the glucose sensor and translates them to the insulin output is called an *algorithm*. There are numerous algorithms developed for this system; but none that is absolutely accurate. Two principal strategies are used. Proportional integral derivative (PID) control adjusts insulin delivery by using a targeted level of glucose. PID reacts to current glucose levels. In contrast to this, there is model predictive control (MPC), which uses a mathematical formula to predict the quantity of insulin needed at times such as after meals and physical activity (14).

A study (15) using 166 adults and 78 children on an artificial pancreas contrasted with 163 adults and 78 children on injection therapy for 1 year showed that HbA_{1c} levels had decreased to 7.5% for those on the pump and 8.1% for those on a schedule of daily insulin injections. These values were reduced from an average of 8.3% for both groups at the start of the study.

A bihormonal closed-loop artificial pancreas is also being studied. This pump using both insulin and glucagon (remember ying–yang) adds glucagon as a strategy to avoid hypoglycemia. A study of this apparatus utilizing a total of 11 T1D subjects demonstrated the safe use of this pump. Six subjects had average blood levels of 140 mg/dl and showed no signs of hypoglycemia but five subjects did exhibit hypoglycemia that necessitated treatment. For the insulin-glucagon pump there appears to be more studies that need to be conducted (16).

Islet Transplantation and Stem Cell Therapy

Techniques that need another 5–10 years or more before they become remedies for T1D are islet transplantation and stem cell therapies. For T1D diabetics, islet transplantation offers an exciting approach to obtaining normal glucose levels without the dependency on insulin injections. The technical problems related to islet isolation and transplantation have been essentially resolved. But the problem of rejection by the immune system remains unresolved. A recent study (17) reported subjecting eight T1D patients to hepatic intraportal islet transplants. The transplants consisted of β-islets from deceased donors that were transferred to the hepatic portal vein. Powerful immunosuppressants were used to prevent rejection of the islets. HbA_{1c} levels were monitored in all patients and were found to decrease after transplantation. One of the immunosuppressants, efalizumab, was withdrawn from the market and therefore had to be replaced with other immunosuppressants. The use of islet transplantation is an important step in combatting T1D but the long-term consequences of immunosuppression need to be determined.

Embryonic stem cell formation of pancreatic cells seems to be far off into the distant future. Our understanding about how cells proliferate, differentiate, and pattern into an organ is far from complete. An excellent review of what is presently known is presented in (18). Perhaps in the next edition of this textbook we shall have more to say about this topic.

Summary Box 10.2

- Detriments to producing a tablet or capsule insulin delivery system include the requirement that the insulin be released in the small intestine, and production of insulin that is capable of passing from the intestines to the blood stream.
- Inhalable insulin has been produced and found effective. But the instrument used to propel it into the lungs is somewhat cumbersome.
- Nasal sprays have found use in experimental settings. They bring insulin to the brain.
- Artificial pancreas consists of a glucose sensor; a transmitter sends that reading from the sensor to the modulator and a pump that delivers the insulin to the blood stream with a message from the modulator. The modulator has a program that determines when and how much insulin should be pumped into the blood.
- An artificial pancreas that adds glucagon and insulin to the blood stream is in development.
- Islet transplantation and stem cell therapy are futuristic remedies for T1D.

ANTIDIABETIC ORAL DRUGS

Until 1956 the only drug available for T2D was insulin. Insulin is used today for T2D upon initial diagnoses to bring glucose levels under control before going to oral drugs. As discussed earlier, T2D is a disease where there is resistance to insulin. As resistance to insulin increases the regimen of insulin must increase. The first drugs marketed for T2D were the sulfonylureas (tolbutamide) and biguanides (metformin and phenformin). Because our emphasis is not pharmacological, we shall only briefly outlined the many antidiabetic oral drugs that were and are currently the market. We will present them according to the chemical class of the drug.

Sulfonylureas

Sulfonylureas such as tolbutamide were discovered by German pharmaceutical scientists in their search for new sulfa antibiotic drugs during the early 1940s. World War II intervened and next the Berlin Wall sequestered the sulfonylureas in East Germany. It was not until the 1950s before the sulfonylureas went into clinical trials and the early 1960s before they were marketed in the United States. The most popular of these drugs was tolbutamide. Tolbutamide is a potassium channel blocker and thus stimulates insulin secretion. All sulfonylureas bind to receptors in the β-cell, which then activate Epac2. As you may recall in Chapter 4 and Figure 4.11 Epac2, when activated, causes the VDCC to open and permit calcium ions to enter the cells leading to the activation of GLUT4 and the docking of LDCV to the plasma membrane. Both the sulfonylurea receptor and Epac2 are necessary for ATP-dependent potassium channels to close and activate the exocytosis mechanism by which insulin leaves the LDCV and passes through the plasma membrane into the intracellular space (not shown in Fig. 4.11) (19).

The onset of action of tolbutamide is 1 h and its duration of action is 6–24 h. There are several other sulfonylureas that have also been developed and marketed. They include tolazamide, chlorpropamide, glyburide, glipizide, and glimepiride.

Biguanides

The biguanides act primarily to decrease the release of glucose from the liver. Biguanides derive from an herb, *Galega officinalis*, which dates back to ancient Egypt (20).

The basic structure of the biguanides is

$$(R)_2-N-\overset{\overset{\displaystyle N}{\|}}{C}-NH-\overset{\overset{\displaystyle NH}{\|}}{C}-NH_2$$

where R=CH$_3$ in metformin

Metformin (dimethylbiguanide) was not the first substance with hypoglycemic properties isolated from *G. officinalis*. Guanidine was too toxic for medicinal use, Synthalin A and B[2] was discontinued in the early 1930s, and phenformin and buformin were used until the frequency of lactic acidosis prevented their further use. Discovered in the 1920s metformin was overshadowed by advances in insulin therapy. Metformin made a comeback in the 1940s. Metformin has become the drug of choice to treat T2D patients. Metformin is especially used to treat overweight and obese subjects because it does not promote weight gain. The drug is unusually devoid of side effects the most common side effect being GI upsets. There is also evidence that it prevents diabetic cardiovascular disease. The only disadvantages of the drug are that its onset of action is 2 h and its duration only 6 h. However, metformin is marketed in an extended release tablet form, which has a duration of action up to 24 h although the onset of action for the first time user is 1–2 days.

---

[2]Both Synthalins had a long chain alkyl group placed between two guanidine moieties. Guanidine is

$$NH_2-\overset{\overset{\displaystyle NH}{\|}}{C}-NH_2$$

Currently metformin is the most widely prescribed drug for the treatment of T2D. Metformin is a double barreled drug. Not only does it lower glucose levels but patients who use it have 25–40% less cancer occurrences than those using insulin or sulfonylurea drugs (21).

Chapter 4 describes AMPK in detail. In muscle and adipose cells metformin inhibits Complex 1 of the electron transport system, thus leading to decreased synthesis of ATP. If you recall from Chapter 4, decreased ATP leads to an increase in the AMP/ATP ratio. LKB1 (also called *STK11*) kinase activates AMPK by phosphorylation and thereby stimulates GLUT4 glucose uptake and transport. In the liver metformin induces small heterodimer partner (SHP) expression. SHP is a factor that represses transcription of several enzymes. Specifically in this instance, SHP inhibits the pathway for the expression of the gluconeogenesis enzymes, pyruvate carboxykinase and glucose 6-phosphatase (22) (Refer to Chapter 3). With the glycolytic pathway enhanced and gluconeogenesis inhibited, the quantity of blood glucose is remarkably decreased.

In 2002, an interesting study appeared in which metformin was investigated as to its potential to prevent or delay the onset of T2D (23). In this study there were 3234 nondiabetic subjects who had values of 95–125 mg/dl (5.3–6.9 mmol/l) for fasting blood glucose and 140–199 mg/dl (7.8–11.0 mmol/l) for 2 h OGTT. The investigators randomly assigned the subjects to a placebo group, metformin twice a day, or an exercise and diet protocol with goals of 150 minutes of exercise and at least a 7% weight loss. After an average of 2.8 years, the incidence of overt diabetes as measured as a fasting blood glucose of ≥ 126 mg/dl (7.0 mmol/l) or 200 mg/dl (11.1 mmol/l) for a 2-h OGTT was 58% lower in the exercise–diet group, and 31% lower in the metformin group than in the placebo group. This study suggests that lifestyle changes prevented or delayed the onset of diabetes to the greatest extent. Next to this was the use of metformin alone.

Metformin is also prescribed with a second medication. The second medication uses a different response mechanism than metformin. Included in the dual medications have been sulfonylureas and thiazolidinediones (TZD) (discussed next in this chapter).

Summary Box 10.3

- Tolbutamide was the first medication taken orally for T2D. Until the discovery that tolbutamide had hypoglycemic properties insulin injections were used for T2D.
- Tolbutamide is a potassium channel blocker.
- Biguanides decrease glucose release from the liver.
- Metformin (dimethylbiguanide) acts by decreasing ATP and via AMP-activated protein kinase GLUT4 glucose uptake. In the liver metformin induces SHP expression and thereby inhibits gluconeogenesis. Glycolysis enhancement and gluconeogenesis inhibition reduce blood glucose levels.

Thiazolidinediones

TZD (frequently called *glitazones*) have a checkered history as glucose-reducing drugs. The three TZDs that have been marketed are rosiglitazone, pioglitazone, and troglitazone. The mechanism of TZDs involves the activation of peroxisome proliferator-activated receptor gamma (PPARγ[3]) which was discussed in Chapter 5. In the nucleus, activated PPARγ sends signals that increase the expression of GLUT4, reduce blood free fatty acids, and increases low and high density lipoprotein cholesterol. It is the latter discovery that was partly responsible for rosiglitazone being placed under restrictions in the United States and withdrawn from the market in Europe. Furthermore a study of 227,571 Medicare Part D (prescription drug plan) participants, mean age of 74.4 years who were prescribed rosiglitazone or pioglitazone suggested (24) that rosiglitazone was associated with 25% greater risk of heart failure and 27% greater risk of cardiovascular accident (stroke) than those who used pioglitazone. Pioglitazone has been taken off the market in France and Germany due to a study that found that the drug raised the risk of bladder cancer. Troglitazone was removed from the market because of reports of liver damage.

A recent study (25) suggests the TZDs may have an opportunity to again become potent T2D drugs. In this study investigators found that in mice, when obesity is induced by high fat meals, protein kinase cyclin-dependent kinase (CDK5) in adipose cells is expressed. CDK5 phosphorylates PPARγ. When phosphorylated PPARγ reduces insulin-sensitizing adiponectin. Adiponectin is a 244 amino-acid protein hormone that, when secreted from fat cells, regulates glucose metabolism by increasing the uptake of glucose by GLUT4. Phosphorylation is blocked by rosiglitazone. Inhibition of CDK5 is associated with antidiabetic effects of rosiglitazone (listed earlier). The conclusion of the investigators is that CDK5 phosphorylation of PPARγ may be involved in insulin-resistance of T2D.

INCRETIN-BASED INHIBITORS

For this section, you may want to refer to Chapter 4. The pharmaceutical companies, realizing the importance of the discovery of incretins, immediately synthesized several substances that had the potential of increasing the half-life of incretins or mimicking their effect on the incretin receptor site. So far there are two incretin-based therapies: first, competitors (agonists) to the receptors for GLP-1R and the second inhibitors to dipeptidyl-peptidase-4 (DPP-4). At present, the GLP-1R agonists are exenatide, liraglutide, albiglutide, and taspoglutide. The DPP-4 inhibitors are sitagliptin, vildagliptin, and saxagliptin. The GLP-1R agonists mimic GLP-1. The DPP-4 inhibitors cause accumulations of GLP and GIP. As you would expect, the effects of these drugs are to increase and extend the physiological actions common to GLP-1 and GIP.

---

[3]In Chapter 5, we discussed the α type of PPAR. The γ type is expressed in a multitude of tissues and is similar to α.

GLP-1R agonists are injectable and DPP-4 inhibitors are taken orally. Presently there are several pharmaceutical companies that are developing oral GLP-1R medications.

GLP-1 receptors are found in many different tissues and organs including brain, central and peripheral nervous system, stomach, gastrointestinal tract, heart and endothelium tissue, liver, pancreas, adipose tissue, and skeletal muscle. As you can see from Table 10.1, Part 1, the effects of the GLP-1R agonists and DPP-4 inhibitors are diverse and depend on the organ or tissue that is affected. In summary the principal factors that are accentuated by these medications are blood glucose lowering, a more normal pattern of blood glucose levels after a meal, body weight loss, and elimination of hypoglycemia. Not yet proven in humans are β-cell protection and perhaps β-cell formation (26, 27).

Next we will briefly discuss the pharmacology of exenatide, liraglutide, sitagliptin, vildagliptin, and saxagliptin.

TABLE 10.1   Biochemical Effects of GLP-1R Agonists and DPP-4 Inhibitors

Organ or Tissue	Biochemical Effect
Pancreas	• Inhibits glucagon secretion • Promotes insulin synthesis and secretion • Promotes glucose transport into cell • Promotes extrusion of LDCVs from the β-cell • Beta cell mass is increased (in animal studies)
Gastrointestinal tract	• Slows gastric emptying by effecting motility thus preventing exaggerated blood glucose levels • Increases feelings of fullness • Decreases hunger sensation • Hence loss of body weight
Hypothalamus	• Reduces appetite • Reduces food intake
Heart	• Lowers systolic and diastolic blood pressure • Modulates heart beat
Liver	• Increases glycogen synthesis • Decreases hepatic gluconeogenesis via inhibition of glucagon secretion
Kidney	• Increases sodium excretion. This aids in the decrease in blood pressure
Arterial wall (endothelial cells)	• Improves endothelial function • Reduces atherosclerosis lesions (atheromas) by decreasing monocyte and macrophage accumulation at the arterial wall
Overall effects	• Reduces weight • Decreases HbA$_{1c}$ by 0.8–1.1% • Does not have a history of hypoglycemia except when used with other diabetes drugs such as sulfonylureas

Exenatide

Exenatide was the original GLP-1R agonist approved for the treatment of T2D. It is often coupled with metformin or a sulfonylurea. Originally isolated from Gila monster saliva, it has 53% of its amino-acid sequence identical to that of GLP-1. It is administered subcutaneously and peaks in the blood at about 2 h. Patients receiving exenatide had a significant weight loss of 3.3–6.6 lb (1.5–3.0 kg). Hypoglycemia was exhibited only when exenatide was coupled with a sulfonylurea. Since its approval, pancreatitis has been reported in a small group of users and as exenatide is eliminated by the kidneys, its use is not recommended for patients with renal insufficiency. A long-acting exenatide has been developed providing for once-per-week injections.

Liraglutide

Liraglutide differs from GLP-1 by having a side-chain extending from lysine in position 26 consisting of glutamic acid linked to palmitic acid. In addition, arginine replaces lysine in position 34. Refer to Figure 4.10 for the structure of GLP-1. Liraglutide is administered subcutaneously and peaks later than exenatide and has a longer duration of action. It provides a greater weight loss than does exenatide. With respect to renal impairment liraglutide does not show any negative effects. The frequency of hypoglycemic episodes was similar to placebos as long as sulfonylurea was not used concomitantly.

Albiglutide and Taspoglutide (Long-Acting Release)

Albiglutide and taspoglutide are long-acting GLP-1R agonists. Albiglutide is formed from the fusion of two GLP-1 repeats of the amino-acid sequence histidine to the penultimate amino acid, arginine. Both amino-acid sequences are attached to albumin thereby forming a dimer. Substitution of glycine for alanine at position 8 (the amino acid that occurs after histidine, which is considered to be position 7) renders the protein resistant to hydrolysis by DDP-4. Refer to Figure 4.10. Taspoglutide has two amino-acid substitutions, at positions 8 and 35. Aminoisobutyrate replaces alanine at position 8 and arginine at position 35 making the peptide resistant to hydrolysis by DDP-4. Albiglutide has a half-life of 5–8 days, which permits weekly injections and taspoglutide was found to maximize at 14 days post injection and was still effective up to 28 days.

Sitagliptin, Vildagliptin, and Saxagliptin

The structures of these three DPP-4 inhibitors appear in Figure 10.5, Part 1. I think you must admit that these are complex structures. Sitagliptin was the first DPP-4 inhibitor approved for the treatment of T2D. Sitagliptin, vildagliptin, and saxagliptin lower HbA_{1c} in a manner comparable to the GLP-1R agonists. They can be used alongside metformin or a sulfonylurea. Hypoglycemic episodes were

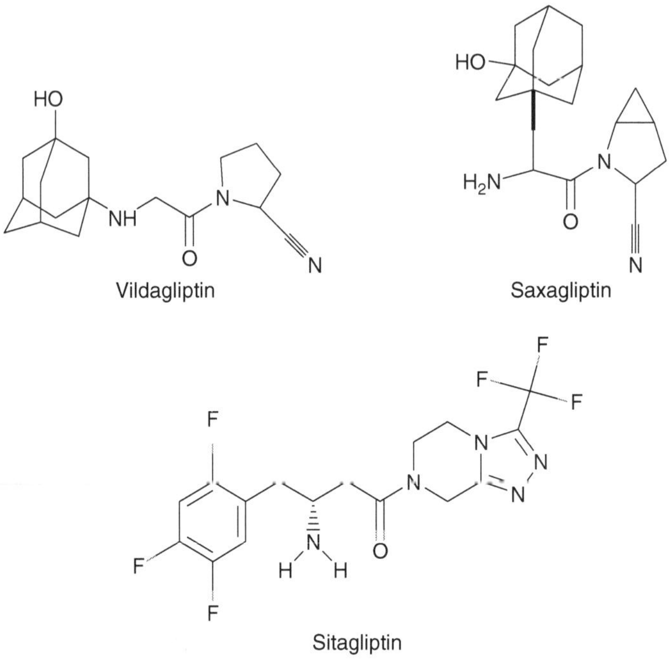

Figure 10.5 Structures of the DPP-4 inhibitors.

identical to the use of a placebo. By inhibiting DPP-4, sitagliptin, vildagliptin, and saxagliptin produce an increase in GLP-1 levels. The three drugs did not have any effects on weight. The side effects of the drugs are benign: headache, skin rashes, dizziness, etc. The drugs also did not negatively affect patients with renal impairment. They are administered orally.

Summary Box 10.4

- TZD activates peroxisome proliferator-activated receptor gamma and increases GLUT4 expression.
- Rosiglitazone was associated with an increase in heart failure and was taken off the market. Pioglitazone was taken off the market due to a study that associated the drug with bladder cancer. Troglitazone was taken off the market due to liver damage.
- There are two classes of incretin-based inhibitors: competitors to the receptor for GLP-1R and inhibitors to DPP-4.
- GLP-1 agonists are exenatide, liraglutide, albiglutide and taspoglutide.
- DPP-4 inhibitors are sitagliptin, vildagliptin and saxagliptin.
- The effects of GLP-1R and DPP-4 inhibitors are diverse.

- GLP-1R mimics GLP-1 and binds to the GLP-1 receptor.
- DPP-4 inhibitors inhibit DPP-4 preventing the destruction of GLP and resulting in its accumulation.
- The GLP-1 agonists differ according to their peak action time, duration of action, and the degree to which they cause weight loss. Another positive effect is that there are few cases of hypoglycemia unless the medication is used with a sulfonylurea.

AMYLIN DERIVATIVES (PRAMLINTIDE)

Review amylin discussion in Chapter 4 and its structure shown in Figure 4.12. Also compare the structures of amylin and pramlintide. As you may remember from Chapter 4, amylin is secreted from β-islet cells in proportion to insulin. Thus individuals with T1D secrete neither insulin nor amylin. T2D patients have a relative scarcity of both insulin and amylin. Using insulin replacement therapy the individual either has very low amylin levels or none at all. This would not matter if amylin did not have biochemical properties affecting glucose levels. Amylin affects glucose in the following ways:

- In skeletal muscle amylin reduces insulin-activated glycogenesis.
- Amylin decreases postprandial glucagon secretion and thereby lowers blood glucose by reducing liver glycogenolysis. T1D and T2D patients have an increase in postprandial glucagon secretion.
- Amylin slows gastric emptying, thus decreasing the flow of glucose into the blood after a meal.
- It also reduces food intake by inducing suppression of appetite.

As amylin is very insoluble and has to a tendency to self-aggregate, it has not found use in replacement therapy. But its derivative, pramlintide, does not have these characteristics. Pramlintide is administered by subcutaneous injection and is used in conjunction with insulin. Its effects are similar to amylin.

- Reduces surges in postprandial blood glucose within the first 30–60 min by slowing gastric emptying.
- When used with insulin injections pramlintide caused reductions in HbA_{1c} of 0.5–1.0% from base values of individuals on a regimen of insulin and a placebo in both T1D and T2D. Other studies have shown similar effects on HbA_{1c} (28).
- When used with insulin injections pramlintide caused reductions in weight of 1.6 kg (3.5 lb) in T1D and 2.4 kg (5.3 lb) in T2D from base values of individuals on a regimen of insulin and a placebo. The average BMI for T1D patients was greater than 27 kg/m^2 and for T2D patients was greater

than 35 kg/m^2 at the outset of the study. The study was conducted for 26 weeks (29).

• Unfortunately pramlintide increases the risk of insulin-induced hypoglycemia. Therefore patients who are already using insulin need to decrease their dosage by about 50% and monitor themselves for hypoglycemia. Those who are not presently being administered insulin and are being placed on a regimen of insulin and pramlintide need to start at low dosages and be titrated slowly to higher levels. Pramlintide apparently causes higher frequency of hypoglycemia when used in concert with insulin.

GLUCOKINASE ACTIVATORS (GKA): POTENTIAL ANTI DIABETIC COMPOUNDS

Refer to Chapter 3 and Table 3.3 for our initial discussion of glucokinases (GKs). I left the greater part of our discussion about GK to this section of this book. GK (sometimes called *hexokinase IV*) found primarily in the liver and β-cell plays a key role in the regulation of glucose homeostasis. As you may recall from Chapter 3 it is the principal enzyme for five metabolic pathways: glycolysis, the hexose monophosphate shunt, the uronic acid pathway, the hexosamine biosynthesis pathway, and glycogenesis. Both glycolysis and glycogenesis are defective in T2D. GK, discovered independently by three laboratories in 1964 was found to be present in beta islet tissue in 1968 (30). In 1990 a regulatory (inhibitor) protein was discovered (31) and the mechanism by which it inhibited GK was determined (32). Thus the site for which activation of the enzyme was defined. Small organic molecules that reversed the inhibition and thereby stimulated GK could be screened. As you may recall from Chapter 6 mutations in GK were discovered in 1993 to be the cause of MODY-2.

GK controls the β-cell threshold for glucose-stimulated insulin release (GSIR). As described in Chapter 4 and Figure 4.11 GK releases insulin by working in cooperation with the ATP sensitive K$^+$ (K$_{ATP}$) channels and VDCCs. GK determines the ATP/ADP ratio by its control of the glycolytic pathway which in turn leads into the tricarboxylic acid cycle and then the electron transport system and oxidative phosphorylation. When the ATP/ADP ratio increases to a threshold level the K channel is closed depolarizing the cell. Next when the membrane potential threshold is reached the VDCC channel opens and causes the release of insulin. You must admit this is a beautiful physiological mechanism (32). After this brief review of the mechanism by which GK regulates insulin secretion, we shall discuss the development of glucokinase activators (GKAs).

Clearly in order to develop a drug that would affect GK and increase insulin secretion you need to find an activator of the enzyme. Pramlintide comes close but is itself a hormone not an activator of a hormone or enzyme. When you are dealing with activation you have to modify and tweak the degree of activation. This is where X-ray diffraction plays a role. GK has a pocket that serves as an allosteric site.[4] Recall the regulatory protein site for GK. Molecules (effectors) for the GK allosteric site can be synthesized once the site is sequenced. By judicious manipulation of the effector's structure the activity of the enzyme can be tuned to the investigators needs. The first strategy that most pharmaceutical companies use is to screen their library of small compounds. Hoffman-LaRoche Inc. screened their library of 120,000 compounds and found a compound that increased the enzymatic activity of GK. This led to the synthesis of a compound called *RO-28-1675* which was a potent GKA (33).

In animal studies GKAs consistently lowered blood glucose levels and stimulated insulin secretion with no effect on weight. Human trials using a GKA called *piragliatin* (Roche Pharmaceuticals) indicated that blood glucose levels were reduced with increased insulin secretion. ARRY 403 confirmed these findings. Since RO-28-1675, piragliatin and ARRY 403 a number of additional GKAs have been developed and are presently going through phase II and III evaluations.

Summary Box 10.5

- Amylin is secreted in equal proportions to insulin from β-islet cells and causes a reduction in glycogenesis in skeletal muscle, a decrease in postprandial glucagon secretion, slows gastric emptying, and suppresses appetite.

- Amylin is very insoluble and tends to self-aggregate thus making it a poor choice for hormone replacement. But its derivative pramlintide has the biochemical effects of amylin and is amenable to subcutaneous injection.

- GK (hexokinase IV) is found predominantly in liver and β-cells and mediates the phosphorylation of glucose to glucose 6-phosphate.

- GK stimulates glucose-stimulated insulin release.

- GKAs have been found which lower blood glucose levels with increased insulin secretion.

α-GLUCOSIDASE INHIBITORS

α-Glucosidase inhibitors are drugs that prevent the digestion of carbohydrates. In the intestinal tract there are enzymes that hydrolyze complex polysaccharides to monosaccharides (Refer to Chapter 3). There are two inhibitors used for the treatment of T2D; acarbose and miglitol. Both drugs inhibit the α-glucosidase found in the brush border of the small intestine. Acarbose also inhibits α-amylase. You may remember α-amylase and its action from Chapter 3. Both acarbose and

---

[4]Allosteric binding sites are sequences of amino acids in an enzyme that bind relatively small compounds which change the conformation of the enzyme thereby making the active site more or less receptive to its natural substrate.

miglitol are taken orally and their onset of action is immediate. A limiting factor for both is gastrointestinal side effects; essentially bloating, clamping and diarrhea. They are not recommended during pregnancy and breast feeding. Acarbose and miglitol do not cause hypoglycemia.

OTHER NEW STRATEGIES THAT ARE IN THE CLINICAL TRIALS PHASE[5]

SGLT2 Inhibitors

Among the other routes that pharmaceutical companies are investigating are sodium-glucose transporter-2 (SGLT2) inhibitors. You may recall from Chapters 3 and 4 that SGLT2 are alternate transporters of glucose to the GLUTs. The drug dapagliflozin is an example of this class of inhibitor. It is used alone or with metformin. Dapagliflozin and insulin were administered to T2D subjects who were not reaching acceptable blood glucose levels with insulin alone. Statistically significant decreases in HbA_{1c} were seen. Dapagliflozin led to weight loss. Incidence of hypoglycemia was low and when it occurred was probably due to the insulin.

11β-Hydroxysteroid Dehydrogenase Type 1 Inhibitors

Cortisone is reduced to cortisol by 11β-hydroxysteroid dehydrogenase type 1 (11β-HSD_1). The cortisol formed binds to the glucocorticoid receptor and leads to glucose synthesis and adipose cell formation in the liver. It has long been suspected that cortisol plays a role in T2D. An inhibitor to 11β-HSD_1 AMG 221, in animal studies blocked the enzyme and resulted in reduced glucose levels. In obese humans a trial study conducted over 24 h showed strong inhibition of 11β-HSD_1 by orally administered AMG 221. Blood glucose was not followed. In June 2008 the drug entered phase I trials.

Summary Box 10.6

- Two inhibitors of α-glucosidase are acarbose and miglitol. They inhibit the conversion of polysaccharides to monosaccharides.
- α-Glucosidase inhibitors are oral medications. They do not cause hypoglycemia but do cause gastrointestinal upset.

---

[5]The protocol for clinical phase studies as outlined by the US National Library of Medicine, National Institutes of Health. *Trial Phase I*: Test is conducted in a small group of people to evaluate safety, safe dosage and identify side effects. *Trial Phase II*: A larger number of subjects are used to see if the drug is effective and further evaluate safety. *Trial Phase III*: Drug is given to a large number of subjects to confirm effectiveness, side effects, compare to commonly used drugs, and to collect data to allow drug to be used safely. *Trial Phase IV*: After the drug is marketed information is gathered on its effect in various populations and to determine side effects associated with long-term use.

- Dapagliflozin is an inhibitor of SGLT2.
- 11β-HSD_1 reduces cortisone to cortisol.
- An inhibitor of 11β-HSD_1, AMG 221 causes reduced blood glucose levels. As of March, 2012, the drug is in phase 1 trials.

REFERENCES

1. Banting Sir FG. Official Web Site of the Nobel Prize. Online. Available at www.nobelprize/nobel_prizes/medicine/laurates/1923/banting-lectures.html. Accessed 2012 Feb 10.

2. Scott DA. CCXI. Crystalline insulin. Biochem J 1934;28:1592.

3. Scott DA, Fisher AM. CXXXI. Crystalline insulin. Biochem J 1935;29:1048.

4. Hagedorn HC, Jensen BN, Krarup NB, et al. Protamine insulinate. JAMA 1936;106:177. DOI: 10.1001/jama.1936.02770030007002.

5. Lawrence RD, Archer N. Zinc protamine insulin: a clinical trial of the new preparation. Brit Med J 1937;1:487.

6. Merrifield RB. Solid phase peptide synthesis 1. The synthesis of a tetrapeptide. JACS 1963;85:2149. DOI: 10.1021/ja00897a025.

7. Katsoyannis PG, Tometsko A, Ginos JZ, et al. Insulin peptides. XI. The synthesis of the B chain of human insulin and its combination with the natural A chain of bovine insulin to generate insulin acitivity. JACS 1966;88:164. DOI: 10.1021/ja00953a032.

8. Akaji K, Fujino K, Tatsumi T, et al. Total synthesis of human insulin by regioselective disulfide formation using the silyl chloride-sulfoxide method. JACS 1993;115:11384.

9. Ruttenberg MA. Human insulin: Facile synthesis by modification of porcine insulin. Science 1972;18:623. DOI: 10.1126/science.177.4049.623.

10. Teusche A. The biological effect of purely synthetic human insulin in patients with diabetes mellitus (translated). Schweiz Med Wochenschr 1979;19:743. Available at www.ncbi.nlm.gov/pubmed/451498.

11. Keen H, Pickup JC, Bilous RW, et al. Human insulin produced by recombinant DNA technology: safety and hypoglyaemic potency in healthy men. Lancet 1980;316:398. DOI: 10.1016/S0140-6736(80)90443-2.

12. Vajov Z, Duckworth WC. Genetically engineered insulin analogs: Diabetes in the new millenium. Pharmacological Reviews 2000;52:1.

13. Rosenstock J, Lorber DL, Gnudi L, et al. Prandial inhaled insulin plus insulin glargine versus twice daily biaspart insulin for type 2 diabetes: a multicentre randomised trial. Lancet 2010;375:2244. DOI: 1016/S0140-6736(10)60632-0.

14. Elleri D, Dunger DB, Hovorka R. Closed-loop insulin delivery for treatment of type 1 diabetes. BMC Med 2011;9:1741.

15. Bergenstal RM, Tamborlane WV, Ahmann A, et al. Effectiveness of sensor-augmented insulin-pump therapy in type 1 diabetes. NEngJMed 2010;363:311. DOI: 10.1056/NEJMoa1002853.

16. El-Khatib FH, Russell SJ, Nathan DM, et al. A bihormonal closed loop artificial pancreas for type 1 diabetes. SciTranslMed 2010;2:27ra27. DOI: 10.1126/scitranslmed.3000619.

17. Posselt AM, Bellin MD, Tavakol M, et al. Islet transplantation in type 1 diabetics using an immunosupressive protocol based on the anti-LFA-1 antibody efalizumab. Am J Trans 2010;10:1870.

18. Mfopou JK, Chen B, Sui L, et al. Recent advances and prospects in the differentiation of pancreatic cells from human embryonic stem cells. Diabetes 2010;59:2094. DOI: 10.2337/db10-0439.

19. Hinke SA. Epac2: A molecular target for sulfonylurea-induced insulin release. Sci Signal 2009;2:pe54. DOI: 10.1126/scisignal.285pe54.

20. Bailey CJ, Day C. Metformin: its botanical background. Pract Diab Int 2004;21:115.

21. Dowling JO, Goodwin PJ, Stambolic V. Understanding the benefit of metformin use in cancer treatment. BMC Med 2011;9:33.

22. Kim YD, Park K-G, Lee Y-S, et al. Metformin inhibits hepatic gluconeogenesis through AMP-activated protein kinase-dependent regulation of the orphan nuclear receptor SHP. Diabetes 2008;57:306.

23. Diabetes Prevention Program Research Group. Reduction in the incidence of type 2 diabetes with lifestyle intervention or metformin. N Eng J Med 2002;346:393.

24. Graham DJ, Ouellet-Hellstrom R, MaCurdy TE, et al. Risk of acute myocardial infarction, stroke, heart failure, and death in elderly Medicare patients treated with rosiglitazone or pioglitazone. JAMA 2010;304:411. DOI: 10.1001/jama.2010.920.

25. Choi JH, Banks AS, Estall J, et al. Anti-diabetic drugs inhibit obesity-linked phosphorylation of PPARgamma by Cdk5. Nature 2010;466:451. DOI: 10.1038/nature09291.

26. Cernia S, Raz I. Therapy in the early stage: Incretins. Diabetes Care 2011;34:S264. DOI: 10.2337/dc11-s223.

27. Koliaki C, Doupis J. Incretin-based therapy: a powerful and promising weapon in the treatment of type 2 diabetes mellitus. Diabetes Ther 2011;2:101. DOI: 10.1007/s13300-011-0002-3.

28. Ryan CJ, Jobe LJ, Martin R. Pramlintide in the treatment of type 1 and type 2 diabetes mellitus. Clin Ther 2005;27:1500.

29. Buse JB, Weyer C, Maggs DG. Amylin replacement with pramlintide in type1 and type 2 diabetes: A physiological approach to overcome barriers with insulin therapy. Clin Diab 2002;20:137. DOI: 10.2337/diaclin.20.3.137.

30. Matschinsky FM, Magnuson MA, Zelent D, et al. Perspectives in diabetes; The network of glucokinase homeostatasis and the potential of glucokinase activators for diabetes therapy. Diabetes 2006;55:1.

31. Vandercammen A, Van Schaftingen E. The mechanism by which rat liver glucokinase is inhibited by the regulatory protein. Eur J Biochem 1990;191:483.

32. Matschinsky FM, Zelent B, Doliba N, et al. Glucokinase activators for diabetes therapy. Diabetes Care 2011;34 Supplement 2:S236.

33. Grimsby J, Sarabu R, Corbett WL, et al. Allosteric activators of glucokinase: Potential role in diabetes therapy. Science 2003;301:370.

PART 2: PREVENTION, DELAY AND MANAGEMENT

As the call, so is the echo.

From "Random House Dictionary of Popular Proverbs and Sayings." Compiled by Gregory Y. Titelman. Published by Random House, New York in 1996. The Russian equivalent of "What goes around, comes around."

PREVENTION AND DELAY

In Part 2 of this chapter, we return to the basic conclusion made in Chapter 1 that overweight and obesity is the leading cause of the worldwide epidemic of diabetes. It seems fitting to end this book with that which started it.

Several studies such as the Diabetes Prevention Program (DPP) and Look AHEAD (Action for Health in Diabetes) demonstrate moderate weight loss through diet and exercise reduces risk of T2D to a significant degree. The DPP study was described in this chapter, Part 1, Medicinal Treatment. This study strongly supported the conclusion that the incidence of diabetes in individuals at high risk is delayed and possibly prevented by lifestyle changes such a low fat low calorie diet and 150 min/week of moderate exercise such as walking. Lifestyle changes were even more effective than drug therapy (administration of metformin).

An ongoing study (1) of 5145 overweight or obese individuals demonstrated that intensive lifestyle intervention that caused sustained weight loss resulted in improved fitness, glucose control, and improved cardiovascular risk factors. The study called *Look AHEAD* is still ongoing and in its fifth year. It intends to look at the development of cardiovascular disease in subjects who have reduced the risk factors for the disease. A summary of the results of the first 4 years of the study and a description of the study appears next:
Subjects

- 55–76 years of age were recruited through 16 centers in the nation. Subjects include T2D patients.
- BMI 25 or more (≥ 27 for patient receiving insulin injections).
- HbA_{1c} less than 11.0%.
- Systolic pressure less than 160 mm Hg.
- Diastolic pressure less than 100 mm Hg.
- Triglycerides less than 600 mg/dl.

Intervention

- Subjects were randomly assigned to the intensive lifestyle intervention (ILI) group or the control group, diabetes support and education (DSE) group.

- ILI group were assigned 1200–1800 kcal/day with less than 30% of the calories from fat (<10% from saturated fat) and 15% or more from protein.
- ILI exercise goal was 175 min/week of moderate exercise such as walking.
- DSE subjects were invited to three sessions each year focusing on diet, physical activity, and social support.

Results

- ILI subjects
  - had a greater percentage of weight loss, 6.15%, than DSE subjects, 0.88%.
  - showed greater improvement in treadmill fitness than DSE subjects.
  - HbA_{1c} was lower 0.36% than DSE, 0.09%.
  - Systolic pressure was lower 5.33% than DSE 2.97%. Diastolic pressure was lower 2.92% than DSE 2.48%.
  - Triglycerides were lower 25.56 mg/dl versus DSE 19.75 mg/dl.
  - HDL cholesterol increase was 3.67 mg/dl versus DSE 1.97 mg/dl.[1]
  - Low density lipoprotein cholesterol decrease was 11.27 mg/dl versus DSE 12.84 mg/dl
- Percentages indicate the average percentage change that occurred from the initial value before the trial began.

What may be startling is the greater decrease in low-density lipoprotein cholesterol for the DSE group than the ILI group. This effect is explained by the greater use of lipid-lowering drugs in the DSE group. This study appears to affirm that dieting and physical activity increase glycemic control and improve risk factors for cardiovascular disease. As this is an ongoing study with the punch line yet to be determined the Look AHEAD trial should be closely followed by those interested in diabetes.

Numerous other well-conducted and controlled studies clearly show that T2D can be at least delayed if not prevented by moderate exercise and low calorie, low fat, and low salt diets.

Exercise

For patients with prediabetes impaired glucose tolerance (IGT), impaired fasting glucose (IFG) or an HbA_{1c} of 5.7–6.4% The ADA (2) recommends an exercise program targeting weight loss of 7% of body weight. Physical activity should be increased to at least 150 min/week of moderate activity such as walking.

Individuals with diabetes should perform at least 150 min/week of moderate aerobic physical activity or 50–70% of maximum heart rate. They also should perform resistance training three times per week. Aerobic and resistance

---

[1]HDL cholesterol is the "good" lipoprotein and increases in HDL benefit the individual.

exercise in combination is more effective than one or the other alone in promoting glycemic control. Aerobic exercise is walking, running, swimming, bicycling, and sports such as basketball and tennis. Resistance exercise is defined as performing against an opposite force. The body or limb can be moving against the opposite force in which case it is called *isotonic* or the body or limb is trying to hold still against the force in which case the exercise said to be isometric. An example of the former is when the participant is pushing against a wall. An example of the latter is clasping your hands above your head and having one arm push the other down.

The ADA and the American College of Sports Medicine (ACSM) in a joint statement (3) are more specific about their exercise recommendations. They stipulate that healthy adults and T2D patients engage in moderate aerobic exercise conducted at least 3 days/week, a minimum total of 150 min/week. The aerobic exercise should be spaced out in episodes of 10 min. There should not be more than two consecutive days between the exercises. They further state that healthy adults and T2D patients need to be involved in resistance exercise training at least two times a week if not three times in addition to aerobic exercise. The intensity of the exercise should be moderate and involve at the least 10–15 repetitions involving the basic muscle groups (upper, lower, and core). These exercises should be near fatiguing and should progress to more challenging exercises overtime. Weights such as dumbbells and barbells are illustrated as useful.

A few of the papers that support exercise as effective in promoting insulin sensitivity and improving glucose tolerance appear below.

Evidence

An early publication published in 1994 (4) described lowering of body fat, decreases in fasting insulin levels and decreases in insulin during the oral glucose test (OGTT) that occurred in a strength training program of 16 weeks duration in 11 normal men, ages 50–63 years. Fasting glucose levels and glucose levels during the OGTT remained constant. The conclusion of the study was that strength training increases insulin sensitivity and reduces blood insulin levels in middle-aged and older men.

Newer studies support these conclusions. A study in Helsinki (5) randomly assigned 172 men and 350 women with IGT, ages of 48–62 years with a mean BMI of 31 to two groups, the intervention group and the control group and found that T2D is preventable by changes in lifestyle. Each person in the intervention group received counseling on reducing weight and exercise. Included in the instructions was advice on how to reach the goals of the study. Specifically, each person in the intervention group was to reduce their weight by 5% or more, limit their intake of fat to less than 30% of calories consumed and saturated fat to less than 10% of calories consumed, increase fiber to least 15 g/1000 kcal, and to participate in moderate exercise for 30 min/day. Also recommended was frequent ingestion of whole-grain products, vegetables, fruits, low fat milk and meat products, margarines, and vegetables oils rich in monounsaturated fatty acids. Each person in the control group was given a general oral and a two-page leaflet on good diet and exercise. No individual programs were given to them. Reduction in weight, intake of fat, and saturated fat, fiber, and exercise was followed with an OGTT conducted annually. The mean follow-up was 3.2 years. The weight loss at the end of 1 year for the intervention group was 5.25 greater than the control group. By 2 years it was 4.38. After 4 years the incidence of diabetes was 6–15% in the intervention group and 17–29% in the control group. The conclusion of the study was that the incidence of diabetes was directly proportional to changes in lifestyle.

Another study (6) which appeared in 2009 reviewed the "most current and reliable literature" regarding the effects of resistance exercise training (RET) in the treatment of diabetes and obesity. It concluded that RET promotes insulin sensitivity and glucose tolerance in a wide range of the studies it surveyed. But in addition the study concluded that the improved glucose uptake was not only due to the increase in muscle associated with RET but also a consequence of "qualitative changes in resistance-trained muscle." No specifics to the latter statement were offered.

Yet another study (7) looked at structured exercise training (aerobic and/or resistance) and physical activity advice with or without dietary changes and their effect on HbA_{1c}. The investigators in this study made a systematic review and meta-analysis of 47 randomized clinical trials to evaluate structured exercise training (SET) with physical therapy advice (PTA) with or without dietary restrictions. SET was defined in this study as "patients ... engaged in planned, individualized, and supervised exercise..." PTA was defined as "... patients were partially or not engaged in supervised exercise training..." but were told or read instructions to exercise regularly with or without explicit instructions. Included in the study were randomized clinical trials that compared any category of SET and/or PTA that included T2D patients and evaluated HbA_{1c}. The conclusions of the study were that SET consisting of aerobic and/or resistance training is associated reductions of HbA_{1c} in T2D patients. SET of greater than 150 min/week reduced HbA_{1c} more greatly than that of 150 min/week or less. PTA was associated with a reduced HbA_{1c} only when coupled with dietary advice.

Now are you convinced about the association between T2D and exercise? I am going out for a long walk after writing this section.

Summary Box 10.1

- The DPP and Look AHEAD studies support the conclusion that T2D is delayed and possibly avoided by individuals who exercise and eat nutritious, low fat, and low calorie meals.
- Intensive lifestyle intervention was shown to cause significant weight loss, lower triglyceride and HbA_{1c} levels, improved systolic and diastolic pressures, and higher HDL levels than a purely educational support program.

- The ADA suggests for prediabetics or those with an HbA_{1c} of 5.7–6.4% an exercise program of at least 150 min per week of moderate activity such as walking. For diabetics they recommend an additional program of resistance exercise three times per week.
- In conjunction with the ADA, the ACSM suggests more specific recommendations.
- Several studies demonstrate that aerobic exercise concomitant with resistance exercise training is most effective in curbing diabetes.

Diet

A study reported in 2011 (8) followed a total of 120,877 healthy, nonobese subjects from 1986 to 2006. The participants came from three studies: 50,422 women from the Nurses' Health Study (NHS), 47,898 women from the Nurses' Health Study II (NHSII) and 22,557 men from the Health Professionals Follow-up Study (HPFS). The participants were followed up during the periods; 1986–2006, 1991–2003, and 1986–2006. The subjects were studied in three separate groups (NHS, NHSII, and HPFS) and evaluated at 4-year intervals. A 4-year weight gain was significantly associated with consumption of potato chips, potatoes, sugar-sweetened beverages, unprocessed red meats and processed meats as opposed to individuals who consumed increased quantities of vegetables, whole grains, fruits, nuts and yogurt. The average weight gain for the three groups was 3.35 lb or 2.4% of body weight (for each 4-year period). Those who were involved in physical activity had 1.76 fewer pounds gained for each 4-year period which is extrapolated to 16.8 lb for 20 years. Weight gain from the consumption of refined grains was equal to that of sweets and desserts. Greater weight gain occurred in subjects who slept less than 6 h and with those who slept more than 8 h. Time watching television was directly proportional to weight gain. Overall weight was directly proportional to alcohol use but there obviously were differences associated with type of beverage, volume and whether more or less usage occurred. Those who quit smoking initially showed weight gain but little change thereafter.

Biochemistry of the Beneficial Effects of Exercise

Glucose utilization is enhanced in individuals who exercise regularly. As we have discussed, exercise reduces weight, strengthens and tones muscles, and concomitant with a nutritious diet delays and possibly prevents diabetes. But in addition to weight loss what also accompanies it is the effect it has on health. A very interesting hypothesis involves a phenomenon called *autophagy* (9). Autophagy involves the normal process by which cells recycle. When a cell dies it develops an internal membrane that encapsulates its contents. It forms a membrane indentation called a *phagophore* that engulfs the cytoplasm and traps proteins, fats, mitochondria, and so on. When the membrane encircles back on itself and becomes spherical it is called an *autophagosome*. Next it attaches to a lysosome. The lysozymes found in the lysosome hydrolyze the proteins to amino acids, nucleic acids to nucleotides, lipids to fatty acids, and so on. Autophagy is a system to eliminate aged or damaged cells but yet maintain their component parts for reuse. The origins of the autophagosome membrane are still a mystery and are being researched.

Now where does exercise and diabetes fit into this picture? A research group has reported that exercise stimulates autophagy (10, 11). Furthermore it has been shown that in animal models (essentially our favorite rodent, mice), short-term exercise activates muscle AMPK. Using mutant mice in which autophagy has been genetically removed exercise does not induce AMPK. You may remember from Chapter 4 that AMPK maintains energy balance for the whole body. Among many biochemical effects AMPK stimulates GLUT4 and therefore glucose uptake.

It must be emphasized that the above hypothesis is in its infant stage and a lot of collaborative evidence must be forthcoming before it becomes the explanation for the benefits of exercise.

GASTRIC BYPASS SURGERY (A CURE FOR T2D?)

In 1987 Walter Pories, a surgeon with East Carolina University School of Medicine in Greenville, North Carolina published the first report of the efficacy of gastric bypass surgery in reversing T2D to normal (12). He performed a type of gastric bypass surgery known as the Greenville Bypass on 397 morbidly obese patients. The bypass surgery was effective in controlling weight gain and maintained "satisfactory" weight for at least 6 years; the length of the study. Preoperative weight averaged 290 lb and after 6 years 205 lb. One hundred and forty-one of the patients initially had "abnormal glucose metabolism." T2D was present in 88 (22%) of patients. Fifty three patients were "glucose impaired." Within 4 months of the gastric bypass operation all but two had normal blood glucose levels without medication or special diet.

Pories studied 42 T2D patients more intensively and found that after surgery their fasting blood glucose, fasting blood insulin, and HbA_{1c} were normal. His conclusion was that the reversal of T2D to normal was not only due to weight loss but due to some other factor, possibly bypass of the antrum and duodenum.

Thousands of gastric bypass surgeries have been conducted in the United States and globally. Gastric bypass surgery has come in various forms. The common one is to connect the stomach to the small intestine several feet lower than normal. This restricts the stomach immensely. Another approach is gastric banding in which most of the stomach is restricted by a band that seals it off. The principal side-effect is hypoglycemia which can occur at any time after the bypass surgery.

There have been many hypotheses as to how the gastric bypass surgery causes remission of T2D. Weight loss itself should cause improvement in T2D patients. But this explanation of the effect of gastric bypass surgery on insulin secretion does not hold true because the T2D disappears soon after the operation before weight loss occurs. In addition gastric bypass experiments performed on lean animals bred to have T2D caused the animals to reverse their T2D to normal. Weight loss was clearly not a factor in this research. A second hypothesis involves ghrelin, a hormone produced in the stomach that promotes appetite. Another hypothesis is that GLP-1 hypersecretion occurring after gastric bypass surgery enhances glucose metabolism (13, 14).

The latest hypothesis (15) involves amino acids, particularly branched-chain amino acids. In a study in which severely obese, sedentary, T2D subjects lost equal amounts of weight either after gastric bypass surgery or dietary intervention, blood concentrations of total amino acids, branched-chain amino acids, oxidized derivatives of branched-chain amino acids, and acylcarnitines were measured. Gastric bypass surgery resulted in a decrease in total amino acids and branched-chain amino acids. Oxidized derivatives of branched-amino acids also were decreased. These constituents remained constant in dietary intervention that resulted in the same approximate weight loss as for gastric bypass surgery. Proinsulin levels, C-peptide (a measure of insulin levels) response to an OGTT, and insulin sensitivity were inversely related to the branched-chain amino acids and their metabolites and acylcarnitines, whereas insulin resistance was directly related to branched-chain amino acids and metabolites.

The data from this research suggest that the decrease in blood amino acids after gastric bypass surgery contributes to improvement in glycemic control. This improvement is not caused by weight loss.

The ADA (2) recommends consideration of gastric bypass surgery for those adults with BMI greater than 35 and T2D especially if their diabetes is hard to control with medicinal and lifestyle changes. The ADA states that for those with BMI less than 35 "there is currently insufficient evidence to generally recommend surgery..." Gastric bypass does have its dangers. Serious side effects such as infections and hernias are possible. The death rate from gastric bypass although low is still significant. Hypoglycemia is an ongoing problem for those who have had gastric bypass surgery. Essentially you pay your money and you take your chances.

Summary Box 10.2

- A studied weight gain in health care professional over 4-year periods demonstrated that those whose diets were rich in starch, sugars, and meat gained more weight than those whose diet was predominantly vegetables, fruits, whole grains, nuts, and yogurt.
- Weight gain from refined grains was equal to those who eat sweets and desserts. Time watching TV was directly proportional to weight gain. Little change in weight occurred after an initial weight gain after quitting smoking. Overall weight gain was associated with alcohol use. Those involved in

exercise showed a lower weight gain. Greater weight gains occurred in subjects who slept less than 6 h and more than 8 h.
- Autophagy occurs in the normal process of cell death. It recycles cell components for reuse. Physical activity stimulates autophagy.
- AMPK is induced by autophagy and thereby stimulates GLUT4 for glucose uptake.
- T2D is returned to normal after gastric bypass surgery.
- The cause is not directly related to weight loss as it occurs almost immediately after surgery before weight loss can occur.
- There are a number of explanations including a recent one that involves a reduction in branched-chain amino acids and their metabolites.
- The ADA recommends gastric bypass surgery for adults with BMIs greater than 35 and T2D that it not responding well to conventional treatment.

PROJECT

Canvas your friends and relatives. For those who have prediabetes or T2D, determine the type of treatment they are receiving especially the medication they are being prescribed. What benefits are provided by the medicine? What are the side effects?

GLOSSARY

Allosteric binding sites   Are sequences of amino acids in an enzyme that bind relatively small compounds which change the conformation of the enzyme thereby making the active site more or less receptive to its natural substrate.

High density lipoprotein cholesterol   Is the "good" lipoprotein and increases in HDL benefit the individual.

PPARγ   A type of PPAR. The γ type is expressed in a multitude of tissues and is similar to PPARα.

Protamine   Is a small-sized arginine and lysine-rich protein. The numerous arginine and lysine residues make the protein positively charged.

Synthelin   Has a long chain alkyl group placed between two guanidine moieties.

REFERENCES

1. The Look AHEAD Research Group. Long-term effects of a lifestyle intervention on weight and cardiovascular risk factors in individuals with type 2 diabetes mellitus. Arch Intern Med 2010;170:1566.

2. American Diabetes Association. Standards of medical care in diabetes-2011. Diabetes Care 2011;34(Suppl 1):S11. DOI: 10.2337/dc11-5011.

3. Colberg SR, Sigal RJ, Fernhall B, et al. Exercise and type 2 diabetes: The American College of Sports Medicine and the American diabetes Association: Joint position statement. Diabetes Care 2010;33:e147. DOI: 10.2337/dc10-9990.

4. Miller JP, Pratley RE, Goldberg AP, et al. Strength training increases insulin action in healthy 50- to 65-yr-old men. J Appl Phys 1994;77:1122.

5. Tuomilehto J, Lindstrom J, Eriksson J, et al. Prevention of tyupe 2 diabetes mellitus by changes in lifestyle among subjects with impaired glucose tolerance. N Eng J Med 2001;344:1343.

6. Tresierras MA, Balady GJ. Resistance training in the treatment of diabretes and obesity: mechanisms and outcomes. J Cardiopulml Rehab Prev 2009;29:67.

7. Umpierre D, Ribeiro PAB, Kramer CK, et al. Physical activity advice only or structured exercise training and association with HbA1c levels in type 2 diabetes: a systematic review and meta-analysis. JAMA 2011;305:1790.

8. Mozaffarian D, Hao T, Rimm EB, et al. Changes in diet and lifestyle and long-term weight gain in women and men. N Eng J Med 2011;364:2392.

9. Leslie M. How do hungry cells start eating themselves? Science 2011;334:1049.

10. Garber K. Autophagy: explaining exercise. Cellular "self-eating" may account for some benefits of exercise. Science 2012;20:281.

11. He C, Bassik MC, Moresi V, et al. Exercise-induced BCL2-regulated autophagy is required for muscle glucose homeostasis. Nature 2012;481:511. DOI: 10.1038/nature10758.

12. Pories WJ, Card JF, Flickinger EG, et al. The control of diabetes mellitus (NIIDM) in the morbidityobese with Greenville gastric bypass. Ann Surg 1987;206:316.

13. Rubino F, Forgione A, Cummings DE, et al. The mechanism of diabetes control after gastrointestinal bypass surgery reveals a role of the proximal small intestine in the pathophysiology of type 2 diabetes. Ann Surg 2006;244:741. DOI: 10.1097/01.sla.0000224726.61448.1b.

14. Mason EE. Ileal [correction of ilial] transportation and enteroglucagon/GLP-1 in obesity (and diabetic?). Obes Surg 1999;9:223.

15. Laferrere B, Reilly D, Arias S, et al. Differential metabolic impact of gastric bypass surgery versus dietary intervention in obese diabetic subjects despite identical weight loss. Sci Transl Med 2011;3:80re2. DOI: 10.1126/scitranslmed.3002043.

POSTSCRIPT

THE FUTURE

As I stated in the Preface of this book "I think we are on the brink of curing diabetes." But it will not be a single cure; it will take on many facets. It will involve stem cell research, insulin pumps that recognize blood glucose levels and rapidly respond to them, new ways of delivering insulin to the body, newer medications with less troublesome side effects, more widespread and effective weight control programs, medications that lessen appetite and control weight, and discoveries we cannot as yet imagine.

Richard F. Dods, Ph.D., D.ABCC

APPENDIX A

General Assembly

Sixty-first session

Agenda item 113

Resolution adopted by the General Assembly

[*without reference to a Main committee (A/61/L.39/Rev.1 and Add.1)*]

61/225. World Diabetes Day

The General Assembly,

Recalling the 2005 World Summit Outcome[1] and the United Nations Millennium Declaration,[2] as well as the outcomes of the major United Nations conferences and summits in the economic, social and related fields, in particular the health-related development goals set out therein, and its resolutions 58/3 of 27 October 2003, 60/35 of 30 November 2005 and 60/265 of 30 June 2006,

Recognizing that strengthening public-health and health-care delivery systems is critical to achieving internationally agreed development goals including the Millennium Development Goals,

Recognizing also that diabetes is a chronic, debilitating and costly disease associated with severe complications, which poses severe risks for families, Member States and the entire world and serious challenges to the achievement of internationally agreed development goals including the Millennium Development Goals,

Recalling World Health Assembly resolutions WHA42.36 of 19 May 1989 on the prevention and control of diabetes mellitus[3] and WHA57.17 of 22 May 2004 on a global 4 strategy on diet, physical activity and health,[4]

Welcoming the fact that the International Diabetes Federation has been observing 14 November as World Diabetes Day at a global level since 1991, with co-sponsorship of the World Health Organization,

Recognizing the urgent need to pursue multilateral efforts to promote and improve human health, and provide access to treatment and health-care education,

1. *Decides* to designate 14 November, the current World Diabetes Day, as a United Nations Day, to be observed every year beginning in 2007;

2. *Invites* all Member States, relevant organizations of the United Nations system and other international organizations, as well as civil society including non-governmental organizations and the private sector, to observe World Diabetes Day in an appropriate manner, in order to raise public awareness of diabetes and related complications as well as on its prevention and care, including through education and the mass media;

3. *Encourages* Member States to develop national policies for the prevention, treatment and care of diabetes in line with the sustainable development of their health-care systems, taking into account the internationally agreed development goals including the Millennium Development Goals;

4. *Requests* the Secretary-General to bring the present resolution to the attention of all Member States and organizations of the United Nations system.

83rd plenary meeting
20 December 2006

The White House
Office of the Press Secretary

For Immediate Release
October 29, 2010
Presidential Proclamation--National Diabetes Month

BY THE PRESIDENT OF THE UNITED STATES OF AMERICA
A PROCLAMATION

Today, nearly 24 million Americans have diabetes, and thousands more are diagnosed each day. During National Diabetes Month, we recommit to educating Americans about the risk factors and warning signs of diabetes, and we honor all those living with or lost to this disease.

Diabetes can lead to severe health problems and complications such as heart disease, stroke, vision loss, kidney disease, nerve damage, and amputation. Type 1 diabetes, which can occur at any age but is most often diagnosed in young people, is managed by a lifetime of regular medication or insulin treatment. Type 2 diabetes is far more common, and the number of people developing or at elevated risk for the disease is growing at an alarming rate, including among our Nation's children. Risk is highest among individuals over the age of 45, particularly those who are overweight, inactive, or have a family history of the disease, as well as among certain racial and minority groups. While less prevalent, gestational diabetes in expectant mothers may lead to a more complicated or dangerous delivery, and can contribute to their child's obesity later in life. With more Americans becoming affected by diabetes and its consequences every day, our Nation must work together to better prevent, manage, and treat this disease in all its variations.

Obesity is one of the most significant risk factors for Type 2 diabetes. National Diabetes Month gives Americans an opportunity to redouble their efforts to reduce their chances of developing Type 2 diabetes by engaging in regular physical activity, maintaining a healthy weight, and making nutritious food choices. For people already living with diabetes, these lifestyle changes can help with the management of this disease, and delay or prevent complications.

We must also do more to reverse the climbing rates of childhood obesity so all America's children can grow into healthy, happy, and active adults. Through her "Let's Move!" initiative, First Lady Michelle Obama is helping to lead an Administration-wide effort to solve the epidemic of childhood obesity within a generation. "Let's Move!" promotes nutritious foods and physical activities that lead to life-long healthy habits. I encourage all parents, educators, and concerned Americans to visit www.LetsMove.gov for more information and resources on making healthy choices for our children.

The new health insurance reform law, the Affordable Care Act, adds a number of tools for reversing the increase in diabetes and caring for those facing this disease. Insurance companies are no longer able to deny health coverage or exclude benefits for children due to a pre-existing condition, including diabetes. This vital protection will apply to all Americans by 2014. Also, all new health plans and Medicare must now provide diabetes screenings free of charge to patients, and Medicare covers the full cost of medical nutritional therapy to help seniors manage diabetes. This landmark new law also requires most chain restaurants to clearly post nutritional information on their menus, ensuring that Americans have consistent facts about food choices and can make more informed, healthier selections.

In recognition of National Diabetes Month, I commend those bravely fighting this disease; the families and friends who support them; and the health care providers, researchers, and advocates working to reduce this disease's impact on our Nation. Together, we can take the small steps that lead to big rewards -- a healthier future for our citizens and our Nation.

NOW, THEREFORE, I, BARACK OBAMA, President of the United States of America, by virtue of the authority vested in me by the Constitution and the laws of the United States, do hereby proclaim November 2010 as National Diabetes Month. I call upon all Americans, school systems, government agencies, nonprofit organizations, health care providers, and research institutions to join in activities that raise diabetes awareness and help prevent, treat, and manage the disease.

IN WITNESS WHEREOF, I have hereunto set my hand this twenty-ninth day of October, in the year of our Lord two thousand ten, and of the Independence of the United States of America the two hundred and thirty-fifth.

BARACK OBAMA

APPENDIX B

PROBLEMS

CHAPTER 1

- Define diabetes mellitus.
- Describe the prevalence of diabetes in the American adult population.
- Describe the prevalence of diabetes for the different races that compose the American population.
- Describe the predicted increase in diabetes cases by the year 2050.
- List the reasons for this increase in diabetics by 2050.
- Describe the evidence that diabetes is becoming more prevalent in American children and adolescents.
- Describe the prevalence of diabetes in all American racial groups.
- Discuss the cost of the diabetes epidemic.
- Describe the diabetes epidemic worldwide.
- List countries in order of number of diabetes cases.
- List countries in order of diabetes prevalence.
- Explain the difference in cost of medical care for diabetics among countries in the world.
- Define overweight and obese in terms of the BMI index.
- List the nine states in the United States that had prevalence of obesity ≥ 30 finding in 2009.

Understanding Diabetes: A Biochemical Perspective, First Edition. Richard F. Dods.
© 2013 John Wiley & Sons, Inc. Published 2013 by John Wiley & Sons, Inc.

- Summarize the statistics in Table 2.
- Describe the results of the BRFSS survey of 2006–2008.
- List from most prevalent to least the order of obesity for American ethnic groups.
- Explain the degree that overweight and obesity has increased worldwide.
- Describe the extent of worldwide overweight and obesity in children.
- Describe the early reports that obesity is linked to diabetes.
- Describe the earliest publication that used measurements to associate obesity with diabetes.

CHAPTER 2

- Discuss the significance of the Ebers Papyrus.
- Discuss the implications of a region on Neandertal DNA that corresponds to a modern human gene associated with type 2 diabetes.
- Give details of how Hippocrates, Aretaeus, and Demetrius contributed to the early understanding of diabetes.
- Explain why the author believes that Galen may have seen diabetes insipidus and not diabetes mellitus.
- Describe how the "Sushruta Samhita" written by Sushruta relates to diabetes.
- Describe how "The Canon of Medicine" written by Ibn Sina relates to diabetes.
- Outline the passages in the "The Yellow Emperor's Classic of Internal Medicine" that describe diabetes.
- Describe how and why ancient Japanese medicine is similar to that of ancient Chinese medicine.
- Describe Paracelsus' contribution to diabetes.
- Describe how the name mellitus was tagged onto diabetes.
- Discuss how Brunner (although he did not know it) "scooped" Mehring and Minkowski.
- Explain how Dobson determined erroneously that diabetes was a dietary disease.
- Describe the treatment of diabetes by Rollo and Cruickshane.
- List the medicines prescribed by Crawley to treat Allen Holford's diabetes.
- Describe what Chevreul concluded about urinary sugar.
- Give an explanation for the reasons Bernard claimed that the liver was the center for diabetes.
- Discuss the reasons for the inclusion of Langerhans in the history of diabetes.

- Describe how Mehring and Minkowski linked the pancreas to diabetes.
- Describe the early approaches to the qualitative determination of glucose in urine.
- Describe Moore's test for glucose.
- Describe Trommer's test for glucose.
- Explain how Barreswil's test differed from that of Fehling's.
- Explain how Pavy's test differed from that of Fehling's.
- Discuss what modifications of Pavy's test were made by Benedict.
- Explain what modification was made so that blood glucose could be quantified.
- Explain the changes in the Folin–Wu procedure made by Haden and subsequently by Nelson that permitted the quantitation of blood sugar.
- Name the person who first used the term *insulin* in describing the factor that regulates glucose.
- Discuss the procedure that Banting, Best, and MacLeod used to isolate the factor produced by the pancreas that regulates glucose.
- Give details about Leonard Thompson, the first recipient of the crude extract from the pancreas that regulates glucose.
- Discuss how Abel's crystallization of insulin led to Sanger's determination of the amino acid sequence of insulin.
- Briefly describe the methodology that Sanger used to determine the structure of insulin.
- Describe how insulin causes physiological effects.

CHAPTER 3

- Describe the structure of a carbohydrate.
- Describe the structures of glucose, starch, amylose, and amylopectin.
- List the steps in the digestion of carbohydrates.
- Define disaccharides.
- Identify and characterize three disaccharidases.
- Explain the mechanism by which glucose and galactose are absorbed into the blood stream.
- Describe the manner in which fructose is absorbed.
- Define aerobic and anaerobic glycolysis.
- Identify the pathways that shut down in the absence of oxygen.
- Explain the function of the shunting of pyruvate to lactate.
- Give details about the mechanism by which glucose is produced in the absence of dietary sources of glucose.

- Explain how ATP serves as a source of energy for endergonic reactions.
- List the metabolic pathways that glucose enters.
- Explain reasons why glucose cannot simply diffuse across the plasma membrane of a cell.
- Describe how glucose is transported across the cell membrane.
- Upon entering the cell, what occurs next to the glucose molecule that makes it amenable to metabolism.
- Name the five metabolic pathways that glucose can now proceed to.
- Give an explanation of how metabolic pathways reverse.
- List the steps that constitute glycogenesis.
- Give details about the conversion of glycogen synthase a to b via protein kinase.
- Describe the conversion of protein kinases via cyclic AMP.
- Describe how cyclic AMP is produced.
- Identify how glycogen synthase is regulated.
- Illustrate using a drawing how transferase causes branching of glycogen.
- List the steps in glycogenolysis.
- Explain how the debranching enzyme acts in a multifunction manner.
- Describe how glycogen phosphorylase acts.
- Describe how phosphorylase kinase is activated.
- Explain the function of $\alpha(1 \rightarrow 4)$-glucosidase.
- Identify how glycogenesis and glycogenolysis are synchronized.
- Give an explanation for the fact that PFK is termed the *principal enzyme* of the glycolytic pathway.
- Explain how aldolase produces two three-carbon compounds from the six-carbon fructose 1,6-bisphosphate.
- Describe the effects of glyceraldehyde 3-phosphate dehydrogenase on glyceraldehyde 3-phosphate.
- Explain the role of the Rapoport–Luebering shunt in erythrocytes.
- Determine the net number moles of ATP per mole of glucose produced by glycolysis and indicate the steps in which ATP is produced and used.
- List the four pathways that pyruvate enters.
- Explain the statement: lactate is similar to glycogen as a storage molecule for glucose.
- Discuss how the tricarboxylic acid cycle uses reducing units to produce ATP.
- Explain how pyruvate dehydrogenase operates as a multicomponent complex to convert pyruvate to produce acetyl CoA.
- List the steps and enzymes involved in the conversion of pyruvate to form acetyl CoA.

- The tricarboxylic acid next goes through eight steps from acetyl CoA to oxaloacetate. Identify the steps that produce reducing units and those that produce carbon dioxide.
- Explain at which steps fatty acids and amino acids enter the pathway.
- Describe how ATP is formed directly in the cycle.
- Give details on the moles of ATP, NADH, $FADH_2$, and CO_2 produced directly by glycolysis and the tricarboxylic acid cycle.
- Give details about cytochromes, CoQ, iron–sulfur proteins, and flavin mononucleotide.
- Describe the components of Complex I–IV.
- Give details about the Q cycle.
- Describe how oxidative phosphorylation produces ATP.
- Explain how reoxidation of NADH and $FADH_2$ produce ATP.
- Explain how the proton gradient (chemiosmotic theory) provides the energy for the synthesis of ATP from ADP and inorganic phosphate.
- Outline how the glycerol 3-phosphate and malate-aspartate shuttles produce different moles of ATP.
- Determine the total moles of ATP produced in aerobic metabolism.
- List the principal products of the phosphogluconate oxidative cycle.
- Describe the function of transaldolase.
- Describe the function of transketolase.
- Describe the fate of glyceraldehyde 3-phosphate.
- Give details about the function of the uronic acid pathway.
- List the principal products of the hexosamine biosynthesis pathway.
- Define the initial step in this pathway.
- Describe the utilization of the amino acid glutamine in this pathway.
- Outline the steps in gluconeogenesis and indicate those enzymes that are shared with the glycolytic pathway and those that are unique to gluconeogenesis.
- Describe how pyruvate produced in the cytoplasm crosses the mitochondrial membrane.
- Explain why fructose 1,6-bisphosphatase and phosphofructose kinase are called *traffic directors* for gluconeogenesis and glycolysis.
- Clarify the role ADP and ATP play in gluconeogenesis.
- Give details on the role that epinephrine, cortisol, glucagon, and insulin play in the regulation of glycolysis and gluconeogenesis.

CHAPTER 4

- Give details on the biosynthesis of insulin.
- Describe what happens to proinsulin after it reaches at the Golgi apparatus.

- Explain the function of LDCVs.
- Give details on the function of SNARES.
- Define the terms priming, docking, and fusion.
- Define exocytosis and endocytosis.
- Explain the current hypothesis for the movement of insulin from the β-cell.
- Discuss the "kiss and run" alternative hypothesis.
- Explain how the secretion of GLUT4 complies with the mechanism for the secretion of insulin.
- Outline the Akt pathway from the binding of insulin with the insulin receptor to the activation of Akt.
- Outline the negative feedback loop that exists within the Akt pathway.
- Outline the two insulin signaling pathways for GLUT transport.
- Outline the insulin stimulation for glycogenesis from the activation of Akt.
- Outline the insulin stimulation of gluconeogenesis from the activation of Akt.
- Outline the insulin stimulation of protein synthesis from the activation of Akt.
- Outline the insulin stimulation of lipogenesis from the activation of Akt.
- Outline the insulin stimulation of lipolysis from the activation of Akt.
- Describe how the components of insulin signaling find each other.
- Give details on the names, origins, and functions of the incretin hormones.
- Give details on how incretins, once they have bound to receptor sites, stimulate insulin synthesis and secretion and promote glucose transport into cells.
- Discuss the relation of amylin to insulin.
- Define the Chinese philosophical concept of yin yang with respect to glucagon and insulin.
- List the steps that occur after glucagon binds to its receptor.
- List the steps in the synthesis of epinephrine.
- Give details about the effect of epinephrine on insulin secretion, glycogenolysis, glycolysis, lipolysis, and glucagon and ACTH secretion.
- Describe the overall effect of epinephrine on blood circulation.
- Explain what is meant by "fight or flight hormones."
- List the steps for the biosynthesis of epinephrine from tyrosine.
- Describe the functions of epinephrine.
- Give details on how epinephrine increases blood glucose levels.
- List the cell types that have receptor sites for somatotropin.
- Outline the steps for somatotropin signaling.
- List the physiological effects of somatotropin.
- Describe the synthesis of somatostatin from preprosomatostatin.

- List the effects of somatostatin.
- List the somatostatin receptors and the processes they regulate.
- Illustrate the hypothalamus–pituitary–adrenal physiological pathway.
- List the effects cortisol has on glucose metabolism.
- Explain the origins of ACTH.
- Illustrate the regulation of ACTH.
- Explain the effect ACTH has on glucose metabolism when it binds with its receptor.
- Illustrate the hypothalamus–pituitary–thyroid physiological control system.
- Name the two thyroid hormones.
- Give details on the synthesis of the thyroid hormones from the protein, thyroglobulin.
- Outline the metabolic pathways affected by the thyroid hormones.
- Describe how the thyroid hormone affect glucose metabolism.
- List the functions of insulin-like growth factor.
- Name a functional similarity with insulin and a functional difference with insulin.
- Name the hormone secreted by cells in the small intestine when they are in contact with bile acids.
- List the functions of this hormone.
- Describe the effect of this hormone on glucose metabolism.
- Describe the structure of AMPK.
- Describe its effects on fatty acids, GLUT4 and gluconeogenesis.
- Describe how AMPK is activated 100-fold.
- Describe the structure and function of the Bateman domains.
- Give an explanation of how AMPK serves as a sensor for biological stress.
- Explain how AMPK affects glucose levels.

CHAPTER 5

- Explain how Wrenshall confirmed the observations of Aretaeus, Sushruta, and Ibn Sina that there were two types of diabetes.
- Describe how Bornstein and Lawrence confirmed Wrenshall's observations.
- Describe the evolution of the terms *growth onset* and *maturity onset* to describe the two different types of diabetes to the present nomenclature of T1D and T2D.
- Explain the cause of T1D.
- Explain how increased levels of glucose appear in the blood of type 1 diabetic individuals.

- Discuss the implications of low insulin levels and high insulin levels.
- Give details on how pancreatic β-cell mass changes through a human's life.
- State how pancreatic β-cell mass in T1D differs from that of obese and T2D individuals.
- Outline the events that lead to pancreatic β-cell mass being decreased in T1D and T2D individuals.
- Discuss the hypothesis that IRS-2 regulates pancreatic β-cell mass.
- Discuss the connection between obesity and T2D that involves PKCζ.
- Identify the principal defects in T1D.
- Identify the principal defects in T2D.
- Outline the differences among normal subjects, T1D, and T2D subjects in regard to glycogen synthesis and breakdown.
- Define glycogen cycling.
- Briefly outline the four procedures in measuring glycogen cycling.
- Discuss what has been learned about glycogen cycling in T1D from these methods.
- Give details about the production of glucose, lactate, and alanine in T2D patients.
- Explain the mechanisms that lead to the production of excess blood glucose in T2D patients.
- List the differences among lean T2D, obese T2D, and normal subjects in relation to gluconeogenesis and glycogenolysis.
- Discuss the postprandial release of glucose in T1D and T2D.
- Discuss gastric emptying in T1D subjects contrasted with normal subjects.
- Explain how insulin infusion experiments demonstrate that the defect in hexokinase activity is inherited.
- Explain the experimental techniques used to study glycolysis, glycogenesis, and glucose oxidation.
- Explain how conducting experiments that vary glucose and insulin levels allow determination of glycolysis and glucose oxidation in T2D subjects relative to normal subjects.
- Outline the conclusions that derive from these experiments.
- Discuss the properties of the UCP family of compounds and relate them to T2D.
- Discuss the ATP/ADP ratio and its relation to T2D.
- Explain how the sirtuin group of proteins may cause insulin resistance.
- Explain the current hypothesis of how increased FFA oxidation leads to insulin resistance.
- Discuss the evidence that the HBP is in communication with the insulin secretion mechanism.
- Outline the steps by which the HBP augments gluconeogenesis.

- Describe the hyperinsulinemic-euglycemic clamp as used in animals and as used in human subjects.
- Describe the use of vastus lateralis muscle biopsy.

CHAPTER 6

- Define the commonalities in all the classes of diabetes.
- Describe T1D and its causes.
- Explain how LADA is similar to T1D and how it is different from T1D.
- Explain why LADA was originally diagnosed as T2D.
- Describe the principal symptoms of T2D and how they differ from the principal symptoms of T1D.
- Discuss how hybrid diabetes occurs and what are its principal characteristics.
- Give details about idiopathic diabetes.
- With regard to the genetic defects of β-islet function:
  - Describe the eight MODY defects.
  - List the most prevalent.
- List the eight diseases of the exocrine pancreas that cause diabetic symptoms
- List the endocrinopathies that cause diabetic symptoms.
- List some of the drugs, hormones, and toxins that cause diabetes.
- List the viral infections that may initiate diabetes.
- List the autoimmune disorders that cause diabetes.
- List genetic syndromes that cause diabetes.
- Define the two classes of prediabetes.
- Give an explanation as to why the data for 2003 is ominous.
- Define gestational diabetes (GDM).
- List complications that occur to the gestational diabetes mother and her fetus.
- Define potential abnormality of glucose tolerance and potential abnormality of glucose tolerance. Discuss how they differ from each other.
- List the cluster of risk factors found in metabolic syndrome.
- Discuss the ominous data that derives from the NHANESIII study.
- Discuss the problem that ensues from placing an individual into one of the classes described earlier.

CHAPTER 7

- Illustrate the bell-shaped frequency distribution curves (Gaussian distribution curves) for a laboratory test that accurately distinguishes a healthy population from an abnormal population.

- Using the curves you designed above show a region in the curves where there is a 100% probability that the test result is abnormal or normal.
- Define the prevalence of a disease.
- Illustrate the prevalence of disease.
- Define the normal range.
- Define sensitivity and specificity of a laboratory test.
- Illustrate how sensitivity and specificity relate to Gaussian curves for normal and abnormal populations for a specific laboratory test.
- Define the predictive value of a normal result and that of an abnormal result.
- Write the equations that represent sensitivity, specificity, predictive value of a normal result, predictive value of an abnormal result, prevalence, and incidence.
- Discuss how adjustments to the normal range affect the above parameters.
- Discuss how the normal range is manipulated to diagnose a life-threatening disease that is treatable.
- Define the truthfulness (efficiency) of a laboratory test.
- Describe a non-Gaussian laboratory test.
- Explain how the reproducibility of a test affects sensitivity and specificity.
- Explain how a laboratory test can increase but still be in the normal range.
- Define parallel and series multiparameter testing.
- Describe early methods for the deproteinization of proteins from biological fluids.
- Describe the ferricyanide technique.
- Discuss the advantages of enzymic methods such as the GOD/POD and hexokinase methods.
- Explain how the oxidase/peroxidase method was made into a visible method.
- Discuss the "Achilles heel" of the oxidase/POD method.
- Define coupling as described in the chromogen, and enzymic approaches to the measurement of glucose in biological fluids.
- Explain how the hexokinase method is measured.
- Define glycated hemoglobin.
- Illustrate a Schiff base.
- Discuss the fact that glucose freely enters the erythrocyte and through a nonenzymic mechanism binds to the amino groups on the globin portion of hemoglobin.
- Define plasma and serum.
- Name and explain the actions of chemicals used to prevent coagulation.
- Explain why it is important to refrigerate whole blood.
- Define the definitive method or gold standard for a specific assay.

- Define the reference method for a specific assay.
- Discuss how breakthroughs in technology relate to the discovery of assays in clinical medicine.
- Discuss what needs in the clinical laboratory the autoanalyzer, Technicon SMA, and Technicon SMAC solve.
- Discuss the features provided by modern instruments.
- List the symptoms that suggest diabetes testing.
- List the risk factors that asymptomatic adults should possess in order to be a candidate for diabetes testing.
- List the risk factors that asymptomatic children should possess in order to be a candidate for diabetes testing.
- Discuss the efficacy of urinary glucose for the diagnosis of diabetes.
- Describe the dietary restrictions for fasting blood glucose.
- List the upper limit of normal for fasting blood glucose and the cut point that suggests diabetes.
- List the standard conditions for the oral glucose test (OGTT).
- List the upper range of normal for the OGTT and the cut point that suggests diabetes.
- Define impaired fasting glucose.
- Describe how HbA_{1c} changed from an assay to monitor diabetes to the test to diagnose diabetes.
- Discuss the advantages of using HbA_{1c} instead of fasting blood glucose or OGTT for the diagnosis of diabetes.
- Describe the best approach to determine the upper limit of a clinical laboratory test.
- Summarize the cut points for diabetes diagnoses for FBG, 2-h PG, and HbA_{1c}.
- Define the Pedersen hypothesis.
- Describe the effect of increasing FBG, 1-h PG, and 2-h PG with abnormalities in neonates.
- Discuss how cord blood C-peptide correlates with fetal growth.
- Outline the new recommendations by the ADA for GDM.
- Describe the reservations that Dr. Ryan has with the new recommendations.
- Discuss the four principal autoimmune antibodies and their role in T1D and LADA.
- Briefly describe the methods used to measure autoimmune antibodies.
- Describe the association of autoantibodies with T1D.
- Discuss the evidence that T2D is also an autoimmune disease.

CHAPTER 8

- Describe how diabetic retinopathy manifests itself as a complication of diabetes.
- Describe the current trend for diabetic retinopathy.
- List the symptoms of neuropathy.
- Define gastric emptying and gastroparesis.
- Describe the sexual complications caused by diabetes.
- Describe how bladder functions can be affected by diabetes.
- Describe the diabetes complication of nephrology.
- Define DKD.
- Define end-stage renal disease.
- Describe how the glomerular filtration rate is measured.
- Describe the creatinine clearance test.
- Define microalbuminuria.
- Describe the studies on the prevalence of DKD and the prevalence of diabetes.
- List the conditions that cardiovascular disease encompasses.
- Define HR.
- Define RR.
- Define hypertension.
- Define coronary heart disease.
- Define the two types of CVA.
- Describe the four layers that compose the walls of the blood vessel.
- List the three principal functions of the endothelium.
- Define shear stress.
- Give details about the production of NO from the endothelium and its function.
- Describe an atheroma.
- Identify how ESS defines the movement of blood in the blood vessels.
- Describe the effect of ESS on endothelial cells.
- Define the term ketone bodies.
- Describe how ketone bodies are formed.
- List the components that comprise HHNC syndrome.
- Discuss the situations that lead to hypoglycemia in diabetics.
- Describe the frequency of infection in diabetics.
- Discuss the connection between diabetes and AD.
- Discuss the hypotheses that associate AD with diabetes.
- Discuss the connection between diabetes, obesity, and cancer.

- List the three hypotheses for the pathophysiology of the complications for diabetes.
- Describe how glycation of proteins can cause some of the complications such as retinopathy and nephropathy.
- Describe the sorbitol pathway.
- Describe how the sorbitol pathway can cause complications.
- Explain how the sorbitol pathway results in glycation of proteins.
- Describe a Reactive Oxygen Species (ROS).
- Describe how ROS can be involved in diabetes.

CHAPTER 9

- Clarify the statement "Diabetes develops from a complex interplay of environment and genetics."
- Define the terms concordance, discordance (or disconcordance), pairwise concordance, and probandwise concordance.
- Describe the pairwise and probandwise concordances for monozygotic (MZ) and dizygotic (DZ) twins.
- Explain the limitations of pairwise and probandwise concordance studies.
- Explain the effect including IGT has on pairwise and probandwise concordance studies of diabetics.
- Clarify the meaning of heritability.
- Describe the frequency of IGT in the offspring of parents who both have T2D.
- Discuss the environmental factors that contribute to the development of T2D.
- Summarize studies that examine the transmission of T1D and T2D from parents to their offspring.
- Summarize studies that examine the presence of T1D in siblings.
- Discuss the HLA chromosome region, its classes, and the gene products it produces.
- Describe the functions of the T cell (Tc), T helper cell (Th), and TCR.
- Explain what is meant by susceptibility and resistant (or protective) HLAs.
- Provide two explanations for the mechanism of susceptibility and protection against T1D by HLAs.
- Describe the HLA that has the strongest association with T1D.
- Describe how the change in the amino acids located at positions 57 and 52 of the DQ α and β chains respectively affects susceptibility to T1D.
- Explain how the appearance of certain autoantibodies forecasts T1D.

- Describe how genome-wide association studies (GWAS) are used to search for genes that are associated with T2D.
- Describe how meta-analysis aids in the search for genes associated with T2D.
- Define the term single nucleotide polymorphism (SNP).
- Explain how the microarray is used to determine SNPs.
- Briefly describe four studies that used meta-analysis of GWAS and microarrays of SNPs to relate loci associated with susceptibility to T2D.
- Describe the evidence for the association of T1D and viruses.
- List and describe five proposed mechanisms for an autoimmune response to viral infections.
- List and describe the evidence for enteroviruses, rubella virus, mumps virus, CMVs, retrovirus, reovirus, rotavirus, and EBV as a cause of T1D.
- Explain the hypothesis that feeding newborns cow's milk as opposed to breast feeding can give rise to T1D.
- Describe the evidence for the association of early exposure to cow's milk and the onset of T1D.
- Describe the evidence that challenges the above hypothesis.
- Describe the association of vitamin D_3 and T1D.
- Describe the factors that currently must be taken into account if studies on cow's milk and vitamin D_3 and the onset of T1D are to be convincing.
- Explain the reasons to suspect that the FTO and the KLF14 are linked to the onset of obesity.
- Explain the early evidence on the function of FTO gene.
- Explain how the KLF14 gene is hypothesized to act in regulating pathways that lead to obesity.

CHAPTER 10

- Define the differences between human insulin and bovine insulin.
- Describe how zinc was determined to be a component of insulin.
- Trace the history of insulin until 1972.
- Describe the advent of recombinant DNA to produce insulin with the human sequence of amino acids.
- Describe how genetically engineered insulin advanced the therapy of T1D and T2D diabetics.
- Describe the obstacles to delivering insulin as a tablet or capsule.

- Give details on delivering insulin as a powder through the lungs.
- Discuss the delivery of insulin through a nasal spray.
- Identify the component parts of an artificial pancreas.
- Describe the obstacles that have to be avoided in obtaining the ideal artificial pancreas.
- Discuss the pitfalls of islet transplantation and stem cell therapy.
- Identify the first medications used for T2D.
- Describe the mechanism by which the sulfonylureas and biguanides control blood glucose levels.
- Discuss the efficacy of metformin in T2D and why it is the most widely prescribed drug for T2D.
- Describe the mechanism by which metformin controls blood glucose levels.
- Describe the mechanism by which TZD act.
- Discuss the reasons that three of the TZD were taken off the market.
- Review the biochemical actions of incretins.
- Name the two types of drugs for diabetes that are derived from the biochemical effects of incretins.
- Describe the actions of exenatide, liraglutide, albiglutide, taspoglutide, sitagliptin, vildagliptin, and saxagliptin.
- Review the biochemistry of amylin presented in Chapter 4.
- Describe the effects of amylin on blood glucose.
- Describe the effects of pramlintide.
- Utilizing the information presented in Chapter 3 describe the role of GK in glucose homeostasis including the basis for MODY-2 (Chapter 6).
- From Chapter 4 describe the basis for GK control of glucose-stimulated insulin release.
- Describe how the GKAs, piragliatin, RO-28-1675, and ARRY 403 were discovered.
- Describe the actions of α-glucosidase inhibitors.
- Describe the actions of SGLT2 inhibitors.
- Describe the actions of $11\beta\text{-HSD}_1$.
- Describe the studies that support the conclusion that diabetes can be delayed and possibly avoided by diet and physical activity.
- Give details of the ADA recommendations for exercise, both aerobic and resistance.
- Discuss the evidence that supports these recommendations.

- Describe the findings of the study that compared individuals who consumed potato chips, potatoes, sugar-sweetened beverages, unprocessed red meats, and processed meats to those who consumed primarily vegetables, whole grains, fruits, nuts, and yogurt.
- Describe autophagy.
- Describe how exercise affects autophagy.
- Describe the effect of gastric bypass surgery on T2D.
- Discuss the hypotheses that have been offered for the effect of gastric bypass surgery on T2D.

INDEX

A page reference followed by the letter "n" denotes that the page referred to includes the item as a footnote. The number following "n" indicates the number of the footnote.

A page reference in **boldfaced** type refers to the page where the term is defined.

A page reference followed by the letter "f" denotes the page where the term occurs in a figure.

A page reference followed by the letter "t" denotes the page where the term occurs in a table.

Understanding Diabetes: A Biochemical Perspective, First Edition. Richard F. Dods.
© 2013 John Wiley & Sons, Inc. Published 2013 by John Wiley & Sons, Inc.